Living in Britain

No 31

Results from the 2002 General Household Survey

Leicha Rickards

Kate Fox

Caroline Roberts

Lucy Fletcher

Eileen Goddard

London: TSO

Contact points

For enquiries about this publication,
contact Lesley Sanders:
Tel: 020 7533 5444
E-mail: lesley.sanders@ons.gsi.gov.uk

To order this publication, call TSO on
0870 600 5522. See also inside back
cover.

For general enquiries, contact the
National Statistics Customer Contact
Centre on 0845 601 3034
(minicom: 01633 812399)
E-mail: info@statistics.gsi.gov.uk
Fax: 01633 652747
Letters: Room D115,
            Government Buildings,
            Cardiff Road,
            Newport NP10 8XG

You can also find National Statistics on the
internet – go to www.statistics.gov.uk.

This report is available on the National
Statistics website
www.statistics.gov.uk/lib

For further information about access to GHS
data deposited at The Data Archive
contact:
            The Data Archive
            University of Essex
            Wivenhoe Park
            Colchester
            Essex CO4 3SQ

Tel: 01206 872001
Fax: 01206 872003
E-mail: archive@essex.ac.uk
Website: www.data-archive.ac.uk

About the Office for National Statistics

The Office for National Statistics (ONS) is
the government agency responsible for
compiling, analysing and disseminating
many of the United Kingdom's economic,
social and demographic statistics, including
the retail prices index, trade figures and
labour market data, as well as the periodic
census of the population and health
statistics. The Director of ONS is also the
National Statistician and the Registrar
General for England and Wales, and the
agency administers the registration of
births, marriages and deaths there.

A National Statistics publication

National Statistics are produced to high
professional standards set out in the
National Statistics Code of Practice.
They undergo regular quality assurance
reviews to ensure that they meet
customer needs. They are produced
free from any political interference.

Contents

page

Acknowledgements

We would like to thank everybody who contributed to the Survey and the production of this report. We were supported by our specialist colleagues in ONS who were responsible for sampling, fieldwork, coding and editing.

Our thanks also go to colleagues who supported us with administrative duties and during publication. Particular thanks are due to the interviewers who worked on the 2002 survey, and to all those members of the public who gave their time and co-operation.

Notes to tables

1. **Harmonised outputs**: where appropriate, tables including marital status, living arrangements, ethnic groups, tenure, economic activity, accommodation type, length of residence and general health have adopted the harmonised output categories described on the National Statistics website. However, where long established time series are shown, harmonised outputs may not have been used.

2. **Classification variables:** variables such as age and income, are not presented in a standard form throughout the report partly because the groupings of interest depend on the subject matter of the chapter, and partly because many of the trend series were started when the results used in the report had to be extracted from tabulations prepared to meet different departmental requirements.

3. **Nonresponse and missing information:** the information from a household which co-operates in the survey may be incomplete, either because of a partial refusal (eg to income), or because information was collected by proxy and certain questions omitted because considered inappropriate for proxy interviews (eg marriage and income data), or because a particular item was missed because of lack of understanding or an error.

Households who did not co-operate at all are omitted from all the analyses; those who omitted whole sections (eg marriages) because they were partial refusals or interviewed by proxy are omitted from the analyses of that section. The 'no answers' arising from omission of particular items have been excluded from the base numbers shown in the tables and from the bases used in percentaging. The number of 'no answers' is generally less than 0.5% of the total and at the level of precision used on GHS the percentages for valid answers are not materially affected by the treatment of 'no answers'. Socio-economic classification and income variables are the most common variables which have too many missing answers to ignore.

4. **Base numbers:** Very small bases have been avoided wherever possible because of the relatively high sampling errors that attach to small numbers. In

general, percentage distributions are shown if the base is 50 or more. Where the base is 20-49, the percentages are shown in square brackets. For some analysis several years data have been combined to increase the sample size to enable appropriate analysis.

5. **Percentages:** A percentage may be quoted in the text for a single category that is identifiable in the tables only by summing two or more component percentages. In order to avoid rounding errors, the percentage has been recalculated for the single category and therefore may differ by one percentage point from the sum of the percentages derived from the tables.

The row or column percentages may add to 99% or 101% because of rounding.

6. **Conventions:** The following conventions have been used within tables:
.. data not available
- category not applicable
0 less than 0.5% or no observations
[] the numbers in square brackets are percentages on a base of 20-49. See note 4.

7. **Statistical significance:** Unless otherwise stated, changes and differences mentioned in the text have been found to be statistically significant at the 95% confidence level.

8. **Mean:** Throughout the report the arithmetic term 'mean' is used rather than 'average'. The mean is a measure of the central tendency for continuous variables, calculated as the sum of all scores in a distribution, divided by the total number of scores.

9. **Weighting:** All percentages and means presented in the tables in the substantive chapters are based on data weighted to compensate for differential nonresponse. Both the unweighted and weighted bases are given. The unweighted base represents the number of people/ households interviewed in the specified group. The weighted base gives a grossed up population estimate in thousands. Trend tables show unweighted and weighted figures for 1998 to give an indication of the effect of the weighting. For the weighted data (1998 and 2000 to

2002) the weighted base (000's) is the base for percentages. Unweighted data (up to 1998) are based on the unweighted sample.

Missing answers are excluded from the tables and in some cases this is reflected in the weighted bases, ie these numbers vary between tables. For this reason, the bases themselves are not recommended as a source for population estimates. Recommended data sources for population estimates for most socio-demographic groups are: ONS mid-year estimates, the Labour Force Survey, or Housing Statistics from the Office of the Deputy Prime Minister. See Appendix D for details of weighting.

Chapter 1 Introduction

The General Household Survey (GHS) is a multi-purpose continuous survey carried out by the Office for National Statistics (ONS). It collects information on a range of topics from people living in private households in Great Britain. The survey started in 1971 and has been carried out continuously since then, except for breaks to review it in 1997/1998 and to re-develop it in 1999/2000.

- An overview of the General Household Survey
- The 2002/2003 survey
- Content of the interview
- Sports and leisure
- Contraception
- Hearing
- Weighting and grossing
- Inclusion of extra households
- Disseminating the results
- Content of the report
- The availability of unpublished data

An overview of the General Household Survey

The survey presents a picture of households, families and people living in Great Britain. This information is used by government departments and other organisations, such as educational establishments, businesses and charities, to contribute to policy decisions and for planning and monitoring purposes.

The main aim of the GHS is to collect data on a range of core topics. These are:
* households, families and people;
* housing tenure and household accommodation;
* access to and ownership of consumer durables, including vehicles;
* employment;
* education;
* health and use of health services;
* smoking;
* drinking;
* family information, including marriage, cohabitation and fertility;
* income;
* demographic information about household members;
* migration.

For the past 32 years, the GHS has documented the major changes that have occurred in households and families. These include the decline in average household size and the growth in the proportion of the population who live alone, the increase in the proportion of families headed by a lone parent and in the percentage of people who are cohabiting. It has also recorded changes in housing, such as the growth of home-ownership, and the increasing proportion of homes with such household facilities and goods as central heating, washing machines, and, more recently, home computers and access to the Internet. The survey also monitors trends in the prevalence of smoking and drinking.

Fieldwork for the GHS is conducted on a financial year basis[1], with interviewing taking place continuously throughout the year. A sample of approximately 13,000 addresses is selected each year from the Postcode Address File. All adults aged 16 and over are interviewed in each household that responds. Demographic and health information is also collected about children in the household. For 2002/2003, the survey response rate was 69%, with an achieved sample size of 8,620 households and 20,149 people of all ages (see Appendix B).

The survey is sponsored by the Office for National Statistics, Department of Health, Office of the Deputy Prime Minister, Department of Transport, Department of Culture, Media and Sport, Department for Work and Pensions, Inland Revenue, Department for Education and Skills, Scottish Executive, Government Actuary's Department and the Health Development Agency, a public sector organisation.

The 2002/2003 survey

The 2002/2003 survey was the third year of fieldwork following the introduction of the recommendations of the GHS review in 1997. Following the review, the survey was re-launched from April 2000 with a different design. The relevant development work and the changes made are described in the 2000 survey report[2].

Content of the interview

The GHS now consists of two elements: the Continuous Survey and Trailers.

The Continuous Survey is to remain unchanged for the five-year period April 2000 to March 2005, apart from certain essential modifications that take into account, for example, changes in benefits and pensions. It consists of a household questionnaire, to be answered by the household reference person (see Appendix A) or spouse, and an individual questionnaire to be completed by all adults aged 16 and over who are resident in the household. The household questionnaire covers the following topics:
* demographic information about household members;
* household and family information;
* household accommodation;
* housing tenure;
* access to and ownership of consumer durables, including vehicles;
* migration.

The individual questionnaire includes sections on:
* employment;
* pensions;
* education;
* health and use of health services;
* smoking;
* drinking in the last seven days;
* family information, including marriage, cohabitation and fertility history;
* income.

The modular structure of the GHS allows for a number of trailers to be included each year to a plan agreed by its sponsors. The trailers included in the 2002/2003 survey were:

- sports and leisure;
- contraception;
- hearing.

Sports and leisure

Since 1987, the survey has periodically included questions about the extent to which people take part in different types of sports and leisure activities. These questions were asked most recently in 1996. The module asks all respondents about their participation in sports or physical activities and the type of facility used for each sport. It includes questions on club membership, competitive participation and tuition.

The information collected has enabled the GHS both to provide a measure of the effectiveness of campaigns encouraging people to take more exercise, and to help monitor the growth or decline of particular sports. In 2002 over half of men (51%) and over a third of women (36%) had participated in some sport or physical activity (excluding walking) during the four weeks before interview[3]. Further results will be presented in a separate report to be published in 2004.

Contraception

Questions on contraception have been included in the survey in 1983, 1986, 1989, 1991, 1993, 1995 and 1998. In 1983 they were asked of women aged 18 to 49 and of women aged 16 to 17 who were or who had been married. Since 1986 the questions have been asked of all women aged 16 to 49. The results are presented in Chapter 10.

Hearing

Questions on difficulty with hearing and type of aid worn have been included in the survey periodically since 1979, most recently in 1998. The results are presented in Chapter 11.

Weighting and grossing

In 2000 the relaunched survey introduced a dual weighting and grossing scheme (a full description of this major methodological change can be found in the 2000 report). First, weighting to compensate for non-response in the sample based on known under-coverage in the Census-linked study of non-response[4] (Foster, 1994). Second, the sample, which has been weighted for non-response, has been grossed up to match known population distributions (as used in the *Labour Force Survey*, LFS).

During the writing of the 2001 report, the results from the 2001 Census were published. They indicated that previous mid-year estimates of the total UK population were around 900,000 too high, with disparities being most apparent among men aged 25 to 39. These disparities were larger than expected, so it was decided that the 2001 weighting would need to be revisited post publication as the reporting process was too far advanced for them to be incorporated.

In February 2003, the ONS published revised mid-year population estimates for 1991 to 2000. This brought them into line with the post 2001 Census-based mid-2001 population estimates (hereafter referred to as the 2001-based intermediate population estimates). In September 2003 there was a relatively small upward revision to the mid-2001 population estimate of men (mainly in those aged 25 to 34 of around 190,000). There have been, and will be, further revisions to some local authority population estimates.

ONS examined the effects on GHS estimates of re-weighting the data using the 2001-based intermediate population estimates. The effects were sufficiently small that it was decided that it was unnessessary to issue revised weighting post-publication. However, GHS 2001 data with weights derived from the 2001-based intermediate population estimates are available from the UK Data Archive (www.data-archive.co.uk). The substantive chapters of the report present data for 2002/2003 in weighted form only. For further details of the weighting see Appendix D.

Weighted and unweighted bases are shown in the tables in this report. Weighted bases should primarily be considered as bases for the percentages shown rather than estimates of population size. Details of data presentation in report tables can be found in 'Notes to tables'.

Inclusion of extra households

Each year on the GHS there is a small proportion of responding households where one or more members do not complete the individual questionnaire. For these households there is a complete set of data at household level and a full individual interview for at least one adult in the household. Before 2001, these households were excluded from the analysis, which resulted in the loss of some data each year. Since 2001, these households have been included in the analysis. This increases both the household sample size and the individual sample size.

The non-response weighting was calculated based on a responding sample that excluded these households. ONS methodologists judged that the effect of including these households was negligible since the numbers involved are small (around 240 households). The middle response rate quoted for 2002/2003 excluded these households to provide direct comparison with previous years. They are included in the overall response rate - see Appendix B for full details.

Disseminating the results

The GHS 2000 Living in Britain report was the first major ONS report to be published as a web-designed publication and we continue to use this successful format. A hard copy of the report is also available.

Content of the report

The report is based on the data collected by the GHS in 2002/2003 and provides information across a wide range of topics. It also includes a number of tables presenting data on trends and changes measured by the GHS since it began.

Technical information is provided in the appendices. These include:
- a glossary of definitions and terms used throughout the report and notes on how these have changed over time (A);
- information about the sample design and response (B);
- sampling errors (C);
- weighting and grossing (D);
- the household and individual questionnaires used in 2002, excluding self-completion forms and prompt cards (E);

- a list of the main topics covered by the survey since 1971 (F).

The availability of unpublished data

ONS can make unpublished GHS data available to researchers for a charge, if resources are available and provided the confidentiality of informants is preserved. Individuals are responsible for any work they produce based on the GHS data, but ONS should be given the opportunity to comment in advance on any report or paper using GHS data, whether prepared for publication or for a lecture, conference or seminar.

Subject to similar conditions, copies of GHS datasets are also available for specific research projects, through the Data Archive at the University of Essex[5].

Notes and references

1   Prior to 1988 fieldwork was conducted on a calendar year basis.

2   Walker A et al. *Living in Britain Results from the 2000 General Household Survey.* The Stationery Office (London 2002). Also available on the web: www.statistics.gov.uk/lib

3   These participation rates are based on weighted data. Unweighted data show similar rates (50% of men and 37% of women had participated in some sport or physical activity, excluding walking, during the four weeks before interview in 2002).

4   Foster, K (1994) *The General Household Survey report of the 1991 census-linked study of survey non-respondents.* OPCS (*unpublished paper*).

5   For further information, see:

www.data-archive.co.uk or contact:
Data Archive
University of Essex
Wivenhoe Park
Colchester Essex CO4 3SQ
Tel: 01206 872 001
Fax: 01206 872 003
e-mail: archive@essex.ac.uk

Chapter 2 A summary of changes over time

The GHS, produced regularly since 1971, remains a key source of information about changes in the demographic, social and economic characteristics of households, families and people in Great Britain. This chapter presents a summary of the main changes that the survey has measured between 1971 and 2002.

- Households
- Families with dependent children
- Population
- Housing tenure
- Cars
- Consumer durables
- Marriage and cohabitation
- Pensions
- Self-reported illness
- Use of health services
- Smoking
- Drinking
- Contraception
- Hearing

Among the key changes that the GHS has recorded during this time are:

- a decline in household size and changes in household composition;
- a growth in the proportion of lone-parent families;
- an increase in the proportion of people living alone;
- an increase in the proportion of people who are cohabiting;
- an increase in home ownership and a decline in the proportion of households living in social housing;
- an increase in the household availability of consumer durables;
- an increase in the prevalence of self-reported longstanding illness or disability;
- a decline in the prevalence of smoking; and
- changes in the proportion of respondents belonging to occupational pension schemes.

This chapter presents a summary of the main changes that the survey has measured between 1971 and 2002. More detailed analyses of life in Britain in 2002 are given in the subsequent chapters. Changes and additions to the wording of survey questions mean that the time period for which information is available varies between topics. The introduction of weighting for non-response has had a small effect on some of the trend data. Details can be found in Appendix D. Changes to the classification of socio-economic status and ethnicity have also affected the presentation of time-series data. Details can be found in Chapter 3 (Households, families and people).

Households

Between 1971 and 1991 there was a decline in the average size of household in Great Britain, from 2.91 persons to 2.48. It continued to decline though at a slower rate throughout the next decade, falling to 2.32 by 1998. Since then it has remained fairly constant. In 2002 the average number of persons per household was 2.31.

Since 1971 there have also been changes in the composition of households. In particular, these have included an increase in the proportion of one-person households, and of households headed by a lone parent. Between 1971 and 1998, the overall proportion of one-person households almost doubled from 17% to 31%, and the proportion of households consisting of one person aged 16 to 59 tripled from 5% to 15%. Over the last five years there have been no statistically significant

changes in the overall proportion of adults living in one-person households, and among people aged 65 and over the proportion living alone has remained relatively stable since the mid-1980s.

The proportion of households containing a married or cohabiting couple with dependent children declined from just under one third of all households (31%) in 1979 to just over one fifth (21%) in 2002. By comparison, the proportion of households with dependent children headed by a lone parent rose from 4% of all households in 1979 to 7% in 1993. It has remained constant since then.

Families with dependent children

As well as measuring changes in the composition of households, the GHS also provides information about the composition of families. The two measures have followed similar trends over time. There has been a decline in the proportion of families headed by a married or cohabiting couple and a corresponding increase in the proportion headed by a lone parent. In 2002 73% of families in Great Britain consisted of a married or cohabiting couple and their dependent children. This is a proportion that has declined steadily since 1971, when 92% of families were of this type.

The large growth in the proportion of lone-parent families (from 8% of families in 1971 to over a quarter of families [27%] in 2002) has mainly been among families headed by a lone mother. Lone-father families have accounted for 1% to 3% of families since 1971, whereas the percentage of lone-mother families has risen from 7% in 1971 to 24% in 2002. The percentage of families headed by mothers who have never married (i.e. single) has risen from 1% in 1971 to 12% in 2002. The percentage of families headed by mothers who were previously married, and are now divorced, widowed or separated, has risen from 6% to 12% during the same period.

Population

The percentage of people aged 75 and over[1] rose from 4% in 1971 to 7% in the mid-1990s. Since then this proportion has remained constant. By contrast, there has been a decrease in the percentage of children under the age of 16, from 25% in 1971 to 20%, since 1998.

Chapter 3

Housing tenure

Between 1971 and 2002 home ownership increased from 49% to 69%, with most of the increase occurring in the 1980s. The increase has levelled off since then. The 'right to buy' scheme introduced in the early 1980s may have contributed to the increase in home ownership, as it allowed local authority tenants to buy their own home. Corresponding to this, the percentage of households renting council homes increased from 31% in 1971 to 34% in 1981, but then gradually declined during the 1980s to 24% in 1991. The percentage continued to decrease and in 2002 14% of all households rented from the council. The decline during the 1990s may in part be a result of the transfer of housing stock from local authorities to housing associations (or Registered Social Landlords [RSLs]), which occurred at this time. The percentage of households renting from a housing association increased from 1% in 1971 to 3% in 1991, continuing throughout the 1990s to 7% in 2002. Another contributing factor to this rise may be the concentration of public funding for new housing in the RSL sector since the mid-1990s.

The percentage of private renter households decreased from 19% in 1971 to 10% in 1995. Since 1995 there have been no significant changes and it has remained relatively constant (between 10% and 11%).

Cars

In 1972, just over half (52%) of households had access to at least one car or van. This proportion had increased to 71% by 1995, and in 2002 almost three quarters (73%) of all households had access to a car. The proportion of households with one car has remained relatively constant during this time, and was 45% in 2002. However, the proportion of households with two or more cars or vans has increased substantially. In 1972, 8% of households had two cars and 1% had three or more cars. By 1995 22% had two cars and 4% had three or more. These figures have remained relatively constant since then (22% and 5% respectively in 2002).

Consumer durables

Since the early 1970s, the GHS has recorded a steady increase in the ownership of consumer durables. Ownership of a refrigerator rose from 73% of households in 1972 to 95% in 1985. Other household amenities that were available only to a minority of households in the early 1970s were also more widespread by 2002. For example, the percentage of households with central heating rose from 37% in 1972 to 93% in 2002. By the mid-1990s, most homes had access to a freezer, a washing machine, a telephone and a television. The proportion of households with access to more recently introduced items (such as the dishwasher, tumble drier and microwave) continues to rise.

Since their introduction to the survey, entertainment items have become much more widely available. Access to a television has always been highly prevalent (93% of households in 1972, rising to 99% in 2002). The proportion of households using satellite, cable or digital television was first measured by the GHS in 1996 when 18% of households received satellite, cable or digital television. This figure rose to 44% in 2002.

Ownership of home computers increased from 13% in 1985 to 34% in 1998. Over the last five years the percentage of households with home computers has increased further, to 54% of all households in 2002. Access to the Internet at home has also increased in recent years, from 33% of households in 2000 to 44% in 2002.

Under half (42%) of all households had a fixed telephone in 1972. In 2000 98% had a phone. Since then, the proportion of households with fixed telephones has remained almost constant. There has however been a significant increase in the availability of mobile phones. The proportion of households owning mobile telephones increased from nearly three fifths (58%) in 2000 to three quarters (75%) of households in 2002.   **Chapter 4**

Marriage and cohabitation

As questions on marriage and cohabitation were initially part of the GHS only for women aged 18 to 49, much of the longer-term trend data refers to this group. The proportion of women aged 18 to 49 who were married has declined continuously since 1979, from almost three quarters (74%) to less than one half (49%) in 2002. During this same period, the proportion of single women has more than doubled from 18% in 1979 to 38% in 2002.

The proportion of non-married women who were cohabiting at the time of interview has increased from

11% in 1979 to 29% in 2002. Among single women, the proportion cohabiting has almost quadrupled (from 8% in 1979 to 31% in 2002). The proportion of divorced women who were cohabiting at the time of interview rose from one in five (20%) in 1979 to around one in three (between 32% and 35%) since 1998. **Chapter 5**

Pensions

Since 1989 trends in participation in employer pension schemes have differed for both men and women, and for those working part time and full time. The proportion of men working full time who were members of their current employer's occupational pension scheme decreased from 64% in 1989 to 55% in 2002. To an extent this reflects changes in the proportion of employees offered a pension scheme by their employer.

The percentage of women working full time who were members of an occupational pension scheme showed the opposite pattern, rising from 55% in 1989 to 60% in 2002. Among women working part time, however, the proportion who were members of an occupational pension scheme has more than doubled from 15% in 1989 to 33% in 2002. This is partly due to a rise in the proportion of employers providing such a scheme.

Among the self-employed, between 1991 and 1998 the take-up of personal pension schemes by men working full time stayed fairly stable at around two thirds. In more recent years the proportion has dropped from 64% in 1998 to 52% in 2002. **Chapter 6**

Self-reported illness

Over the lifetime of the GHS, the prevalence of self-reported longstanding illness in adults and children has risen from 21% in 1972 to 35% in 2002. The prevalence increased steadily in the 1970s and early 1980s, continuing to increase gradually until 1996. Since then the proportion has fluctuated between 32% and 35%, with no clear pattern over time.

The prevalence of limiting longstanding illness in adults and children has risen from 15% in 1975 to 21% in 2002. It has shown a similar trend to the prevalence of longstanding illness, although the overall increase in prevalence has been less over time. The proportion of all persons reporting restricted activity due to illness or injury in the two weeks prior to interview doubled from

8% in 1972 to 16% in 1996. The proportion has since remained relatively stable (15% in 2002).

Self-reported illness is based on the respondent's own assessment of their own, or their children's health. Therefore an increase in reported prevalence may reflect changes in the expectations people have about their health, as well as changes in the actual prevalence of sickness.

Use of health services

The proportion of adults and children who saw an NHS GP in the 14 days before interview has increased from 12% in 1972 to 15% in 2002, reaching a peak in the mid 1990s. However, the average number of NHS GP consultations per person per year has remained relatively constant over time, at between four and five since 1972. The proportion of GP consultations that took place at home decreased from 22% in 1971 to 5% in 2002. This coincided with an increase in both surgery and telephone consultations. Surgery consultations increased from 73% in 1971 to 86% in 2002, while telephone consultations more than doubled from 4% in 1971 to 9% in 2002.

In 2002 14% of all respondents had attended an outpatient or casualty department in the three months before interview. This figure increased from 10% in 1972 to 16% in 1998 before declining. The proportion of people attending hospitals as day patients has doubled since this question was first asked in 1992 (from 4% in 1992 to 8% in 2002). There has been little change since 1982 in the proportion of adults and children who had inpatient stays during the 12 months prior to interview, ranging from 8% to 10%. The proportion has remained at 8% since 2000. **Chapter 7**

Smoking

The prevalence of cigarette smoking decreased substantially in the 1970s and early 1980s, from 45% of all men and women aged 16 and over in 1974 to 35% in 1982. Since then the proportion smoking decreased more gradually until the early 1990s, since when it has levelled out at around 26% to 28%. In 2002 26% of people aged 16 and over were cigarette smokers.

In 1974 51% of men smoked cigarettes compared with 41% of women. The difference between the proportions of men and women smoking has gradually reduced, although it has not disappeared completely. For example, 38% of men and 33% of women were smokers in 1982, compared with 27% of men and 25% of women in 2002.

Although there has consistently been a greater proportion of men smoking than women, this is not the case in every age group. In recent years there has been a significant drop in the proportion of men aged 16 to 19 smoking cigarettes (from 30% in 2000 to 22% in 2002). Women aged 16 to 19 were significantly more likely to smoke cigarettes than men in this age group, with 29% smoking in 2002. Since 1974, the greatest percentage decrease in the proportion smoking has been among people aged 60 and over, where the prevalence has more than halved from 34% to 15% in 2002. However, this reflects the fact that people in this age group are more likely to have been regular smokers in the past who have given up.

There has been an increase in the proportion of men and women aged 16 and over who have never smoked a cigarette. Among men, the proportion who have never smoked rose from 25% in 1974 to 46% in 2002. The increase in the proportion of women who have never smoked has been smaller, from 49% in 1974 to 54% in 2002.

The GHS has consistently shown that cigarette smoking is more prevalent among people in manual occupational groups than those in non-manual groups. In the 1970s and 1980s, the proportion of cigarette smokers in non-manual occupations fell more sharply than that for manual occupations. Since the 1990s, however, proportions in both groups have remained relatively constant. In 2002, 20% of those classified as non-manual workers[2] smoked cigarettes, compared with 31% of those classified in the manual group. The introduction of the NS-SEC classification means that any comparisons over this period should be treated with caution (see Chapter 8).

Although the prevalence of cigarette smoking changed little during the 1990s, the GHS has shown a continuing fall in the reported number of cigarettes smoked. The fall in consumption has occurred mainly among younger smokers, while the number of cigarettes smoked by those aged 50 and over has changed very little since the mid-1970s. Most of the decline in consumption in the 1990s is due to a reduction in the proportion of

heavy smokers. The proportion of respondents smoking on average 20 or more cigarettes a day fell from 14% of men in 1990 to 11% in 1998, and from 9% to 7% of women over the same period. It has since remained virtually unchanged among both men and women. The proportion of respondents who were light smokers also changed little throughout the 1990s.

Filter cigarettes continue to be the most widely smoked type of cigarette. However, there was an increase in the 1990s in the proportion of people smoking hand-rolled cigarettes. Among men the proportion increased from 18% in 1990 to 33% in 2002. Among women it increased from 2% to 13% during the same period. This increase may be partly due to the rising price of packaged cigarettes and the reduction of tar yield in packaged cigarettes (hand-rolled may be made with a higher tar yield).

A decline in the prevalence of pipe and cigar smoking among people aged 16 and over has been evident since the survey began, with most of the decrease occurring in the 1970s and 1980s. The proportion of men smoking pipes fell from 12% in 1974 to 6% in 1986. In 2002 it was 1%. The proportion of men smoking at least one cigar a month more than halved from 34% in 1974 to 16% in 1978. By 2002 it had reduced to 5%. Only 3% of women smoked cigars in 1974, and since 1978 the proportion of women who smoke cigars has scarcely been measurable by the GHS.   **Chapter 8**

Drinking

The GHS uses two measures of alcohol consumption:
- maximum daily amount drunk last week; and
- average weekly alcohol consumption.

When drinking questions were first asked on the GHS in 1978, weekly-based measures of alcohol consumption were reported. Questions relating to maximum daily amount have been asked since 1998, reflecting the move in 1995 from weekly-based to daily-based guidelines from the Department of Health. Longer-term trend data are therefore currently only available for weekly-based measures.

Maximum daily amount drunk last week

There were no statistically significant changes overall between 1998 and 2002 in the proportion of men drinking more than their recommended number of daily units on any day of the week prior to interview. However, in the youngest age group of men (aged 16 to 24) there is evidence of a slight downward trend in the percentage drinking more than four units on at least one day in the week prior to interview. Between 1998 and 2002, the proportion fell from 52% to 49%, while the proportion drinking more than eight units fell from 39% to 35%. In women, however, there was a significant increase in the proportion drinking more than six units on at least one day in the previous week (from 8% in 1998 to 10% in 2000). It has remained at that level since. Most of the increase occurred in the younger age groups, for example, among women aged 16 to 24 the proportion drinking more than 6 units rose from 24% in 1998 to 28% in 2002.

Weekly alcohol consumption level

During its history the GHS has recorded a slight increase in overall weekly alcohol consumption among men and a much more marked one among women. Weekly alcohol consumption among women rose from 10% drinking over the recommended weekly benchmark of 14 units a week in 1988 to 17% in 2002. The increase was greatest among 16 to 24 year old women, almost doubling between 1992 and 2002 from 17% to 33%. Average weekly consumption for women in this age group increased from 7.3 to 14.1 units during the same 20-year period. **Chapter 9**

Contraception

The 2002 GHS included questions on the use of contraception for the first time since 1998. The earliest trend data are available from 1986. Since then, the most common methods for avoiding pregnancy used by women aged 16 to 49 have been the contraceptive pill, surgical sterilisation (both male and female) and the male condom. In 2002, 72% of women aged 16 to 49 used at least one form of contraception, a figure that has remained relatively constant since it was first measured.

There has been a small increase in the proportion of women using the contraceptive pill, from 23% in 1986 to 26% in 2002. This increase was only statistically significant among women aged 16 to 17, 30 to 34 and 45 to 49 however. Trend patterns in the use of surgical sterilisation vary according to the woman's age. For example, among women aged 45 to 49 and their partners, use of sterilisation has increased significantly from 35% in 1986 to 44% in 2002. There has been a

steady rise in the proportion of women whose partners use condoms, from 13% in 1986 to 19% in 2002. Again, this trend was not observed across all age groups. Among women aged 45 to 49, condom use has declined since 1986. It has remained fairly constant among women aged 35 to 44 and increased among women under the age of 35. **Chapter 10**

Hearing

Between 1979 and 2002 there was an increase in the proportion of adults reporting hearing difficulties, from 13% to 16%. Most of this change occurred between 1979 and 1992, mainly among the older age groups. It has remained relatively stable since.

The percentage of people reporting that they wore a hearing aid doubled from 2% in 1979 to 4% in 2002. Hearing difficulties are strongly associated with age and sex. For example, the likelihood of a man aged 75 and over wearing a hearing aid almost doubled between 1979 and 2002 (from 12% to 23%). During the same period, the proportion of women aged 75 and over wearing a hearing aid rose from 12% to 17%. **Chapter 11**

Notes and references

1 The GHS interviews all people aged 16 and over in private households. This means that the population figures presented here do not included those living in institutions and residential homes.

2 The National Statistics socio-economic classification (NS-SEC) is assessed on the basis of the occupation of the household reference person.

Chapter 3 Households, families and people

Since 1971, the GHS has provided data about households, families and people. This chapter looks at how the composition of households and families has changed over the last 30 years and describes the socio-economic groups and ethnic origin of the people within these households.

- Trends in household size and composition

- One-person households

- Households with two or more adults without dependent children

- Households and families with dependent children

    Dependent children

    Age of youngest child

- Stepfamilies

- Household income

- Population

- Socio-economic classification

- Ethnic group

    Ethnic group and age

    Ethnic group and region

    Ethnic group and household size

Tables in this chapter look at three different levels of data: households, and the families and people who comprise households.

Trends in household size and composition

Over the period 1971 to 2002 there have been a number of changes to the size and composition of households.
* Between 1971 and 1981 mean household size declined steadily from 2.91 to 2.70 persons. The decline continued throughout the next decade, falling to 2.48 by 1991. By 1998 household size had fallen to 2.32[1], but since then it has remained fairly constant. In 2002 mean household size was down to 2.31, representing no statistically significant change since 2001.
* Household composition has also changed little since 1998. As in 2001, just over three in ten households contained one person only (31%) in 2002, while over a third contained two people (35%). The remaining 34% consisted of three or more people. By contrast, a little over half of households in 1971 (51%) consisted of three or more people.
* As in 2001, just over a quarter of households (27%) included children under the age of 16, compared with nearly two fifths (39%) in 1971. One or two adults aged 60 or over accounted for 30% of households.

Tables 3.1–3.2, Figure 3A

Figure 3A   **Mean household size: 1971–2002**

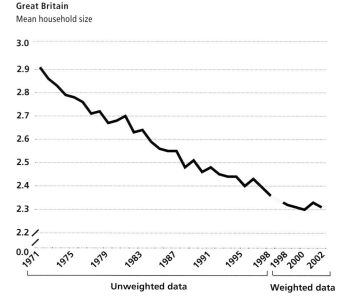

One-person households

The overall decline in the average household size has resulted from a large increase in the proportion of one-person households, which almost doubled from 17% in 1971 to 31% in 1998. The biggest change has been in the proportion of households containing one adult aged 16 to 59, which tripled from 5% in 1971 to 15% in 1998. Since 1998, however, there has been no statistically significant change in this proportion, which has ranged between 15% and 16%.   **Tables 3.1–3.2, Figure 3A**

Among all adults aged 16 and over, 17% lived alone in 2002. This proportion has remained fairly constant over the past five years, halting the upward trend observed since 1973 when 9% of adults aged 16 and over lived alone.

Changes over time in the proportion of people living alone varied according to age.
* In 2002 just under half of people aged 75 and over[2] lived alone (48%), compared with just over one in ten people aged between 25 and 44 (12%).
* The proportion of people aged 65 and over living alone has remained relatively stable since the mid-1980s.
* Among people aged 25 to 44, the proportion living alone increased from 2% in 1971 to 12% in 1998. It has remained stable since then.   **Table 3.3**

In 2002 there was no difference in the proportion of men and women aged 16 to 24 living alone (5%). However, there were statistically significant differences in the 25 to 44 age group between the relative proportions of men (16%) and women (8%) living alone. The situation was reversed for those aged 65 and over, partly reflecting the longer lifespan of women compared with men. Among women aged 65 to 74, 34% lived alone compared with 18% of men. Women aged 75 and over were more than twice as likely as men of the same age to be living alone (60% of women, 29% of men).   **Table 3.4**

Households with two or more adults without dependent children

There has been little change since the late 1970s in the proportion of households containing two or more adults without dependent children[3].

- Over a third (37%) of households in 2002 comprised two or more adults without dependent children. The same proportion was found in 1979.
- At 28%, the proportion of households consisting of a married or cohabiting couple with no children of any age has remained stable since the late 1970s.
- The proportion of households containing two or more unrelated adults was down from 3% in 2001 to 2% in 2002. Since 1979 it has ranged between 2% and 4%.
- The proportion of households consisting of a married or cohabiting couple with non-dependent children only has decreased slightly since the late 1970s, but has remained at 6% since the mid 1990s.
- In 2002 just 2% of households comprised lone parents with non-dependent children only, consistent with the slow decline from 4% in 1979.

Table 3.5, Figure 3B

Households and families with dependent children

In 2002, 18% of all households consisted of a married couple and their dependent children, while 3% consisted of a cohabiting couple and their dependent children. These figures have remained fairly constant over the past five years. Before 1998, however, there was a decline in the proportion of such households, which together represented nearly one third of all households (31%) in 1979. This trend is partly explained by the rise in one-person households, alongside a decline in the proportion of married couples with dependent children (from 23% in 1996 to 18% in 2002). The proportion of cohabiting couples with dependent children has remained unchanged at 3% during this period.

The proportion of households consisting of lone parents with dependent children has also remained steady at around 7% since the mid-1990s.   **Table 3.5, Figure 3B**

Table 3.6 looks at the different types of families with dependent children. In 2002, 73% of families with dependent children in Great Britain were headed by a married or cohabiting couple. This proportion has declined steadily since the GHS was first conducted in 1971, when 92% of families were of this type. During this period there has been a large increase in the proportion of lone-parent families, from less than one in ten in 1971 (8%) to the current level of over one in four (27%). This increase is explained by the substantial rise in lone mothers. In 1971 they headed 7% of families with dependent children, but accounted for 24% of these families in 2002. The proportion of lone fathers has ranged between 1% and 3% over the past decade.

The rise in the proportion of families with dependent children headed by lone mothers is explained by an increase in both the proportions of previously married lone mothers (widowed, divorced or separated) and of lone mothers who have never married (single).

- In 1971 lone mothers who had previously been married headed 6% of families with dependent children, while lone mothers who had never married headed only 1%.
- The proportion of families with dependent children headed by previously married lone mothers had increased to 11% by 1991, and has fluctuated around this level since.
- Lone mothers who had previously been married headed 12% of families with dependent children in

Figure 3B   Households by type of household: 1979 and 2002

Great Britain

Households

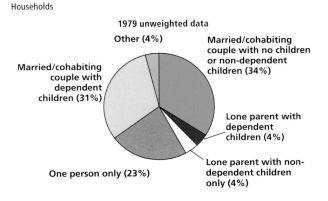

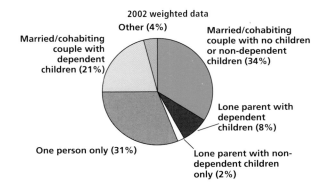

2002. The proportion of families with dependent children headed by single lone mothers has continued to rise since 1971 (1%) to its 2002 level of 12%.

Table 3.6, Figure 3C

Dependent children

Table 3.7 examines the trends since 1972 in the proportions of dependent children living in different types of family. The proportion of dependent children living in a married or cohabiting couple family has decreased steadily from 92% in 1972 to 75% in 2002. As a consequence, the proportion living in lone-parent families has risen from 8% to 25% during the same period.

Of all dependent children in families, 27% were the only dependent child in their family in 2002, compared with just 18% in 1972. This proportion has risen since 2001 when 24% of dependent children were the only dependent child in their family.

Table 3.7

The average number of dependent children in families declined from 2.0 in 1971 to 1.7 in 2002 (not statistically significantly different from 2001). Over the past two decades the average number of dependent children in families has remained relatively constant at around 1.8. The average number of children living in families headed by a married couple is greater than the average

number of children living in families headed by either a cohabiting couple or a lone parent. In 2002 the average number of children in families headed by a married couple was 1.8, while that for both cohabiting and lone parents was 1.6.

Table 3.8

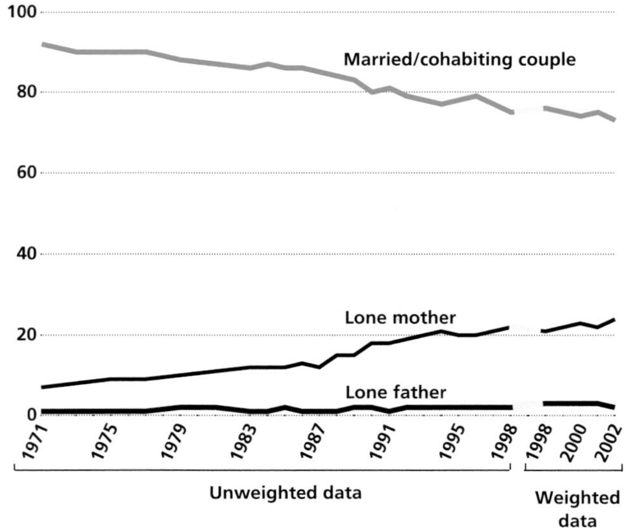

Age of youngest child

Data from 2001 and 2002 were combined to provide a sample large enough to analyse the relationship between the age of youngest child and family type. Among families with dependent children, married or cohabiting couples were more likely to have children under the age of five than families headed by a lone parent (40% compared with 33%). The dependent children of lone fathers were more likely to be older than those of lone mothers or married or cohabiting couples.

- The youngest child was under five in 35% of families headed by a lone mother. This compared with only 13% of families headed by a lone father.
- In 59% of families headed by a lone father, the youngest dependent child was aged ten or over, compared with 34% of both lone-mother and two-parent families.

Table 3.9

Stepfamilies

The GHS asks people aged 16 to 59 whether they have any stepchildren living with them. As has been shown in previous years, the majority of stepfamilies comprise couples with children from the woman's previous relationship. In 2002:

- 87% consisted of a couple with at least one child from the woman's previous relationship;
- 11% were a couple with at least one child from the man's previous relationship;
- 3% were families with children from previous relationships of both partners.

Table 3.10

Household income

There were statistically significant differences in gross weekly household income[4] between different types of families with dependent children.

- Families consisting of married couples with dependent children had the highest gross weekly incomes. Of these, 71% had household incomes of over £500, compared with 50% of cohabiting couples with dependent children and 16% of lone parents.
- Among lone mothers, 15% had household incomes over £500, compared with 28% of lone fathers.

Figure 3C **Families with dependent children by family type: 1971–2002**

Great Britain
Percentage of families with dependent children

Lone parents had lower gross weekly household incomes than married or cohabiting couples with dependent children.

- Over a third (35%) of lone-parent families had a gross weekly household income of £150 or less, compared with only 5% of married couples and 8% of cohabiting couples with dependent children.
- Lone mothers (36%) were more likely than lone fathers (27%) to have a gross weekly household income of £150 or less.

Lone mothers who had never been married were particularly likely to have low gross weekly household incomes. Just under half of single mothers (49%) had incomes of £150 or less compared with 19% of divorced lone mothers and 30% of separated lone mothers.

Table 3.11

Population

Since the GHS began over 30 years ago, it has reflected changes in the age of the British population. In 1971 the proportion of people aged 65 and over was 12%. In 2002, it was 16%. This increase in the proportion of older people in the population corresponds with a more substantial decrease in the proportion of children under the age of 16, from 25% in 1971 to 20% in 2002. Figure 3D illustrates the changes in the proportions of the population aged under 16 and 65 and over between 1971 and 2002.

Figure 3D **Children and older people – the proportion of the population aged under 16 or 65 and over in 1971 and 2002**

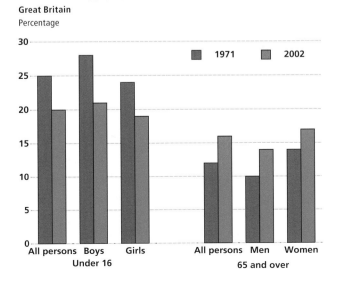

Table 3.13 shows the male/female distribution across different age groups. For most age groups the distribution is relatively even. However, there are more women than men aged 65 and over. While this is partly explained by the greater life span of women, it also reflects the large proportion of men in this cohort who died during World War Two. Women represent over three fifths (62%) of people aged 75 and over.

Tables 3.12–3.13, Figure 3D

Socio-economic classification

Based on the National Statistics Socio-economic Classification (NS-SEC)[5], Table 3.14 shows the distribution on the classification for 2002. Overall, 10% of people aged 16 and over in 2002 were classified in the higher managerial and professional group. At 15%, men were three times more likely than women (5%) to be in this group.

In the lower managerial and professional group, comprising 22% of adults, no such sex differences were found. The intermediate occupation group (which includes clerical workers, secretaries, call centre workers, nursery nurses and nursing auxiliaries) accounted for 13% of the adult population. Men were less likely than women to be classified in this group (just 6% of men compared with 19% of women). These proportions were similar to those recorded in 2001.

Of people aged 16 and over, 8% were in the small employers and own account workers group, and 10% in the lower supervisory and technical group (including occupations such as train drivers, plumbers, fitters, inspectors and printers). Compared with women, men were more likely to be in the small employers and own account workers group, and more likely to be in the lower supervisory and technical group.

In 2002:
- 11% of men were in the small employers and own account workers group, compared with 5% of women;
- 15% of men were classified in lower supervisory and technical occupations, compared with 6% of women.

A greater proportion of women than men was in the semi-routine occupational group – just under a quarter (23%), compared with 12% of men. The routine occupation group (which included machine operators, packers, cleaners, labourers, sales assistants, HGV drivers and bar staff) accounted for 15% of adults. There were no statistically significant differences found in the proportion of men and women in this group.

As in 2001, the proportion of people in the higher managerial and professional group peaked between the ages of 25 and 54. Only 4% of those aged 16 to 24 were in this group, rising to 13% of people aged 25 to 34 and 12% of people aged 35 to 54. This pattern was similar for both men and women. A similar distribution was found for the lower managerial and professional group. The proportion of people employed in semi-routine and routine occupations was highest among the youngest and oldest groups. Among 16 to 24 year-olds, 44% were in these groups, compared with only 26% of people aged 25 to 34. The proportions were the same for both men and women. **Table 3.14**

Ethnic group

The GHS sample has always been too small to analyse single-year data on ethnicity, and prior to 2001 data were added across three years to produce large enough sample sizes for analysis. The introduction of the National Statistics classification of ethnicity in 2001 meant that it was not possible to do this for that year. This report provides tables based on data for both 2001 and 2002 combined.

Table 3.15 shows the overall distribution of the 2001 and 2002 samples in terms of the National Statistics' ethnic classification. Among all persons:

- 88% of people described themselves as White British and 3% as Other White, making a total of 91% who described themselves as White;
- 4% of the sample described themselves as Asian (2% Indian and 2% Pakistani or Bangladeshi and 1% other Asian[6]);
- 2% described their ethnic origin as Black Caribbean or Black African comprising 1% in each group. Only a very small number described themselves as being of Other Black ethnic backgrounds;
- 1% described their ethnic background as Mixed and 1% as being of other ethnic backgrounds. **Table 3.15**

Ethnic group and age
Table 3.16 shows the age distribution among different ethnic groups. The White British population showed an older age structure than other ethnic groups, with the highest proportion of people aged 45 and over. Those describing their ethnic identity as Black Caribbean had the oldest age structure of non-White groups.

- 41% of the White British population were aged 45 and over, compared with 33% of those describing themselves as Black Caribbean, 15% of those describing themselves as Black African, 26% of those describing themselves as Indian and 15% of those of Pakistani or Bangladeshi background.
- Those describing themselves as mixed race showed the youngest age distribution, with just 8% aged 45 and over, 42% aged between 16 and 44, and the remaining half under the age of 16 (50%).

Tables 3.16–3.17

Ethnic group and region
Compared with those describing themselves as White British, people from other ethnic backgrounds were more likely to live in London than other regions of Britain.

- Over half of those describing themselves as Black Caribbean (56%) and four fifths of those describing themselves as either Black African or of other Black ethnic groups (80% in each) lived in London, compared with only 9% of the White British population.
- Those from other non-British White backgrounds were more likely to be located in London (39% compared with 9% of the White British population).
- 36% of those describing themselves as Indian and 25% of those describing themselves as Pakistani or Bangladeshi lived in London.

People from Indian, Pakistani and Bangladeshi, and other Asian backgrounds were more widely dispersed across the regions of England than other ethnic groups (other than the White British population). Of the Indian population, 19% were in the West Midlands, 14% in the North West and 12% in the East Midlands. Of those describing themselves as Pakistani and Bangladeshi, 21% were in the West Midlands, 15% in Yorkshire and the Humber, and 12% in the North West. **Table 3.18**

Ethnic group and household size

The average size of household in the combined years 2001 and 2002 differed depending on the ethnic group of the household reference person (HRP). For those describing their ethnic group as Indian, Pakistani or Bangladeshi , the average household size was higher than that for other ethnic groups (3.4 persons for Indian HRPs and 4.28 for Pakistani or Bangladeshi HRPs, compared with the total average of 2.32 persons).   For those describing their ethnic background as Black African, the average household size was 2.56 persons and households headed by Black Caribbean HRPs averaged 2.24 persons. By comparison, households in which the HRP described their ethnic group as White (British or Other) averaged 2.30 persons.   **Table 3.19**

Notes and references

1   The introduction of weighting for nonresponse from 1998 has had an important effect on estimates of the proportion of one-person households. This effect suggests that the increase seen over time has been slightly underestimated in the past. Consequently, estimates of mean household size were found to be smaller partly as a result of the weighting procedure.

2   As the GHS interviewers all people aged 16 and over in private households, the figures presented for those living alone aged 75 and over do not take account of the proportions of people in this age group who live in institutional and residential homes.

3   Dependent children are persons aged under 16 or single persons aged 16 but under 19, in full-time education, in the family unit and living as part of the household.

4   Usual gross weekly household income is the sum of usual gross weekly income for all adults in the household. Those interviewed by proxy are also included.

5   The new National Statistics Socio-economic Classification (NS-SEC) was introduced across all official statistics and surveys in 2001. It replaced Social Class based on occupation and Socio-economic Groups. For more information see Walker A et al. *Living in Britain*: *Results from the 2001 General Household Survey: Appendix E*. The Stationery Office (London 2003), also available on the web: www.statistics.gov.uk/lib.

6   In order to avoid rounding errors, the percentage has been recalculated for the single category and therefore may differ by one percentage point from the sum of the percentages derived from the tables.

Table 3.1 Trends in household size: 1971 to 2002

(a) Households and (b) Persons

Great Britain

Number of persons in household (all ages)	Unweighted								Weighted			
	1971	1975	1981	1985	1991	1995	1996	1998	1998	2000	2001	2002
					Percentage of households of each size							
(a) Households	%	%	%	%	%	%	%	%	%	%	%	%
1	17	20	22	24	26	28	27	29	31	32	31	31
2	31	32	31	33	34	35	34	36	34	34	34	35
3	19	18	17	17	17	16	16	15	16	15	16	16
4	18	17	18	17	16	15	15	14	14	13	13	13
5	8	8	7	6	6	5	5	5	4	5	4	4
6 or more	6	5	4	2	2	2	2	2	2	2	2	2
*Weighted base (000's) = 100%**									24,450	24,845	24,592	24,529
*Unweighted sample**	*11988*	*12097*	*12006*	*9993*	*9955*	*9758*	*9158*	*8636*		*8221*	*8989*	*8620*
Average (mean) household size	2.91	2.78	2.70	2.56	2.48	2.40	2.43	2.36	2.32	2.30	2.33	2.31
					Percentage of persons in households of each size							
(b) Persons	%	%	%	%	%	%	%	%	%	%	%	%
1	6	7	8	10	11	12	11	12	13	14	13	13
2	22	23	23	26	27	29	28	30	30	30	29	30
3	20	19	19	20	20	20	20	19	20	20	20	20
4	25	25	27	27	25	24	25	24	23	22	23	22
5	15	14	14	12	11	10	10	10	9	10	9	9
6 or more	13	11	9	6	5	5	6	5	4	5	5	5
*Weighted base (000's) = 100%**									56,751	57,106	57,260	56,570
*Unweighted sample**	*34849*	*33579*	*32410*	*25555*	*24657*	*23385*	*22274*	*20396*		*19266*	*21180*	*20149*

* Trend tables show unweighted and weighted figures for 1998 to give an indication of the effect of the weighting. For weighted data (1998 to 2002) the weighted base (000's) is the base for percentages. Unweighted data (up to 1998) are based on the unweighted sample.

Table 3.2 Trends in household type: 1971 to 2002

(a) Households and (b) Persons   Great Britain

Household type	Unweighted								Weighted			
	1971	1975	1981	1985	1991	1995	1996	1998	1998	2000	2001	2002
	Percentage of households of each type											
(a) Households	%	%	%	%	%	%	%	%	%	%	%	%
1 adult aged 16-59	5	6	7	8	10	12	11	13	15	16	15	16
2 adults aged 16-59	14	14	13	15	16	17	15	16	16	17	17	16
Youngest person aged 0-4	18	15	13	13	14	13	13	13	12	11	11	11
Youngest person aged 5-15	21	22	22	18	16	16	17	16	16	15	16	16
3 or more adults	13	11	13	12	12	10	11	9	10	10	11	10
2 adults, 1 or both aged 60 or over	17	17	17	17	16	16	16	17	15	15	15	15
1 adult aged 60 or over	12	15	15	16	16	15	16	16	15	16	15	15
Weighted base (000's) = 100% *									24,450	24,845	24,592	24,529
Unweighted sample *	11934	12090	12006	9993	9955	9758	9158	8636		8221	8989	8620
	Percentage of persons in each type of household											
(b) Persons	%	%	%	%	%	%	%	%	%	%	%	%
1 adult aged 16-59	2	2	3	3	4	5	5	5	7	7	7	7
2 adults aged 16-59	10	10	10	12	13	14	13	14	14	14	14	14
Youngest person aged 0-4	27	23	21	21	22	20	21	21	19	19	18	18
Youngest person aged 5-15	31	34	33	28	25	26	27	25	26	25	25	26
3 or more adults	15	13	16	17	16	15	15	13	15	15	16	15
2 adults, 1 or both aged 60 or over	11	12	12	13	13	14	13	14	13	13	13	13
1 adult aged 60 or over	4	5	6	6	7	6	7	7	7	7	7	6
Weighted base (000's) = 100% *									56,751	57,106	57,260	56,570
Unweighted sample *	34720	33561	32410	25555	24657	23385	22274	20396		19266	21180	20149

* Trend tables show unweighted and weighted figures for 1998 to give an indication of the effect of the weighting. For weighted data (1998 to 2002) the weighted base (000's) is the base for percentages. Unweighted data (up to 1998) are based on the unweighted sample.

Table 3.3 Percentage living alone, by age: 1973 to 2002

All persons aged 16 and over   Great Britain

	Percentage who lived alone											
	Unweighted								Weighted†			
	1973	1983	1987	1991	1993	1995	1996	1998	1998	2000	2001	2002
16-24	2	2	3	3	4	5	4	4	4	5	5	5
25-44	2	4	6	7	8	9	8	10	12	12	12	12
45-64	8	9	10	11	11	12	11	14	15	16	15	15
65-74	26	28	28	29	28	27	31	27	28	29	27	27
75 and over	40	47	50	50	50	51	47	48	48	50	49	48
All aged 16 and over	9	11	12	14	14	15	14	16	17	17	16	17
Unweighted sample *												
16-24	3811	3498	3558	2819	2574	2318	2233	1885		1870	2064	2023
25-44	8169	7017	7418	7118	6875	6761	6489	5861		5393	6118	5579
45-64	7949	5947	5802	5493	5360	5615	5114	4892		4803	5147	5169
65-74	2847	2494	2389	2196	2303	2129	1943	1862		1672	1882	1766
75 and over	1432	1490	1596	1603	1581	1451	1485	1374		1344	1474	1435
All aged 16 and over	24208	20446	20763	19229	18693	18274	17264	15874		15082	16685	15972

* Trend tables show unweighted and weighted figures for 1998 to give an indication of the effect of the weighting. For weighted data (1998 to 2002) the weighted base (000's) is the base for percentages. Unweighted data (up to 1998) are based on the unweighted sample.

† Weighted bases are shown in Table 3.20.

Table 3.4   Percentage of men and women living alone, by age

All persons aged 16 and over   **Great Britain: 2002**

	Percentage who lived alone		
	Men	Women	Total
16-24	5	5	5
25-44	16	8	12
45-64	15	15	15
65-74	18	34	27
75 and over	29	60	48
All aged 16 and over	15	18	17
All persons*	12	15	13
Weighted base (000's) = 100%			
16-24	3,153	3,132	6,286
25-44	8,025	8,379	16,404
45-64	6,784	6,974	13,757
65-74	2,248	2,544	4,792
75 and over	1,536	2,456	3,993
All aged 16 and over	21,746	23,485	45,232
*All persons**	27,524	29,047	56,571
Unweighted sample			
16-24	982	1041	2023
25-44	2659	2920	5579
45-64	2494	2675	5169
65-74	849	917	1766
75 and over	595	840	1435
All aged 16 and over	7579	8393	15972
*All persons**	9706	10443	20149

* Including children.

Table 3.5   Type of household: 1979 to 2002

(a) Households and (b) Persons   **Great Britain**

Household type	Unweighted						Weighted			
	1979	1985	1991	1995	1996	1998	1998	2000	2001	2002
Percentage of households of each type										
(a) Households	%	%	%	%	%	%	%	%	%	%
1 person only	23	24	26	28	27	29	31	32	31	31
2 or more unrelated adults	3	4	3	2	3	2	3	3	3	2
Married/cohabiting couple										
with dependent children	31	28	25	24	25	23	22	21	22	21
with non-dependent children only	7	8	8	6	6	6	6	6	6	6
no children	27	27	28	29	28	30	28	28	29	28
Married couple										
with dependent children	..	..	..	..	23	20	19	18	18	18
with non-dependent children	..	..	..	..	6	5	6	6	6	6
no children	..	..	..	..	25	26	24	24	23	23
Cohabiting couple										
with dependent children	..	..	..	..	3	3	3	3	3	3
with non-dependent children	..	..	..	..	0	0	0	0	0	0
no children	..	..	..	..	4	4	5	5	5	5
Lone parent										
with dependent children	4	4	6	7	7	7	7	7	7	8
with non-dependent children only	4	4	4	3	3	3	3	2	2	2
Two or more families	1	1	1	1	1	1	1	1	1	1
Weighted base (000's) = 100%*†							24,389	24,787	24,493	24,449
Unweighted sample*	11454	9993	9955	9738	9138	8617		8204	8955	8594
Percentage of persons in each type of household										
(b) Persons	%	%	%	%	%	%	%	%	%	%
1 person only	9	10	11	12	11	12	13	14	13	13
2 or more unrelated adults	2	3	2	2	3	2	3	3	3	3
Married/cohabiting couple										
with dependent children	49	45	41	40	42	39	38	36	37	36
with non-dependent children only	9	11	11	9	9	8	9	9	9	9
no children	20	21	23	25	24	26	25	25	25	25
Married couple										
with dependent children	..	..	..	..	37	34	33	32	32	32
with non-dependent children	..	..	..	..	9	8	8	9	9	8
no children	..	..	..	..	21	22	21	21	20	21
Cohabiting couple										
with dependent children	..	..	..	..	4	5	5	5	5	5
with non-dependent children	..	..	..	..	0	0	0	0	0	0
no children	..	..	..	..	3	4	4	4	5	4
Lone parent										
with dependent children	5	5	7	8	8	9	8	9	8	9
with non-dependent children only	3	4	3	3	3	3	3	2	2	2
Two or more families	2	1	2	1	1	2	2	2	2	2
Weighted base (000's) = 100%*†							56,605	56,955	56,921	56,245
Unweighted sample*	30546	25454	24657	23325	22190	20350		19220	21065	20045

* See the footnote to Table 3.1.

See Appendix A for the definition of a household.

† Total includes a very small number of same sex cohabitees.

Table 3.6   Family type, and marital status of lone mothers: 1971 to 2002

Families with dependent children*   **Great Britain**

Family type	Unweighted								Weighted			
	1971	1975	1981	1985	1991	1995	1996	1998	1998	2000	2001	2002
	%	%	%	%	%	%	%	%	%	%	%	%
Married/cohabiting couple†	92	90	87	86	81	78	79	75	76	74	75	73
Lone mother	7	9	11	12	18	20	20	22	21	23	22	24
single	1	1	2	3	6	8	7	9	8	11	10	12
widowed	2	2	2	1	1	1	1	1	1	1	1	0
divorced	2	3	4	5	6	7	6	8	7	7	7	7
separated	2	2	2	3	4	5	5	5	5	5	4	5
Lone father	1	1	2	2	1	2	2	2	3	3	3	2
All lone parents	8	10	13	14	19	22	21	25	24	26	25	27
*Weighted base (000's) = 100%** *									7,182	7,105	7,146	7,206

* Dependent children are persons under 16, or aged 16-18 and in full-time education, in the family unit, and living in the household.

† Including married women whose husbands were not defined as resident in the household.

** Trend tables show unweighted and weighted figures for 1998 to give an indication of the effect of the weighting. For weighted data (1998 to 2002) the weighted base (000's) is the base for percentages. Unweighted data (up to 1998) are based on the unweighted sample.

Table 3.7   Family type and number of dependent children: 1972 to 2002

Dependent children*   **Great Britain**

	Percentage of all dependent children in each family type											
	Unweighted								Weighted			
	1972	1975	1981	1985	1991	1995	1996	1998	1998	2000	2001	2002
	%	%	%	%	%	%	%	%	%	%	%	%
Married/cohabiting couple with												
1 dependent child	16	17	18	19	17	16	17	15	17	17	17	18
2 or more dependent children	76	74	70	69	66	64	63	62	61	58	60	57
Lone mother with												
1 dependent child	2	3	3	4	5	5	5	6	6	7	6	8
2 or more dependent children	5	6	7	7	12	14	13	15	13	15	15	15
Lone father with												
1 dependent child	0	0	1	1	0	1	0	1	1	1	1	1
2 or more dependent children	1	1	1	1	1	1	1	1	1	2	1	1
Weighted base (000's) = 100%†									12,799	12,641	12,606	12,451
Unweighted sample†	9474	9293	8216	5966	5799	5559	5431	4897		4499	4846	4561

* Dependent children are persons under 16, or aged 16-18 and in full-time education, in the family unit, and living in the household.

† Trend tables show unweighted and weighted figures for 1998 to give an indication of the effect of the weighting. For weighted data (1998 to 2002) the weighted base (000's) is the base for percentages. Unweighted data (up to 1998) are based on the unweighted sample.

Table 3.8   **Average (mean) number of dependent children by family type: 1971 to 2002**

Families with dependent children*   **Great Britain**

Family type	Average (mean) number of children											
	Unweighted								Weighted††			
	1971	1975	1981	1985	1991	1995	1996	1998	1998	2000	2001	2002
Married/cohabiting couple†	2.0	2.0	1.9	1.8	1.9	1.9	1.9	1.9	1.8	1.8	1.8	1.8
Married couple	..	..	..	..	..	..	1.9	1.9	1.8	1.8	1.9	1.8
Cohabiting couple	..	..	..	..	..	..	1.7	1.7	1.7	1.7	1.6	1.6
Lone parent	1.8	1.7	1.6	1.6	1.7	1.7	1.7	1.7	1.6	1.7	1.7	1.6
Total: all families with dependent children	2.0	1.9	1.8	1.8	1.8	1.8	1.8	1.8	1.8	1.8	1.8	1.7
*Unweighted sample****												
Married/cohabiting couple	*4482*	*4299*	*3887*	*2890*	*2541*	*2358*	*2329*	*2004*		*1804*	*2004*	*1889*
Married couple	*..*	*..*	*..*	*..*	*..*	*..*	*2086*	*1753*		*1558*	*1720*	*1636*
Cohabiting couple	*..*	*..*	*..*	*..*	*..*	*..*	*243*	*251*		*246*	*284*	*253*
Lone parent	*382*	*477*	*558*	*458*	*595*	*658*	*635*	*652*		*660*	*682*	*679*
Total	*4864*	*4776*	*4445*	*3348*	*3136*	*3016*	*2964*	*2656*		*2464*	*2686*	*2568*

*   Dependent children are persons aged under 16, or aged 16-18 and in full-time education, in the family unit, and living in the household.

†   Including married women whose husbands were not defined as resident in the household.

**   Trend tables show unweighted and weighted figures for 1998 to give an indication of the effect of the weighting. For weighted data (1998 to 2002) the weighted base (000's) is the base for percentages. Unweighted data (up to 1998) are based on the unweighted sample.

††   Weighted bases are shown in Table 3.20.

Table 3.9   **Age of youngest dependent child by family type**

Families with dependent children*   **Great Britain: 2001 and 2002 combined**

Family type		Age of youngest dependent child					
		0-4	5-9	10-15	16 and over	*Unweighted sample***	Total
Married/cohabiting couple†	%	40	26	26	8	*3886*	% 74
Lone mother	%	35	30	29	5	*1239*	23
Lone father	%	13	27	45	14	*115*	3
All lone parents	%	33	30	31	6	*1354*	26
Total	%	38	27	27	7	*5240*	100

*   Dependent children are persons aged under 16, or aged 16-18 and in full-time education, in the family unit, and living in the household.

†   Including married women whose husbands were not defined as resident in the household.

**   Weighted base not shown for combined data sets.

Table 3.10   **Stepfamilies with dependent children by family type**

Stepfamilies with dependent children*
(Family head aged 16-59)   **Great Britain: 2002**

Type of stepfamily	
	%
Couple with child(ren) from the woman's previous marriage/ cohabitation	87
Couple with child(ren) from the man's previous marriage/ cohabitation	11
Couple with child(ren) from both partners' previous marriage/ cohabitation	3
Weighted base (000's) = 100%	*520*

*   Dependent children are persons under 16, or aged 16-18 and in full-time education, in the family unit, and living in the household.

Table 3.11   Usual gross weekly household income of families with dependent children by family type

Families with dependent children*

Great Britain: 2002

Family type		Usual gross weekly household income										Weighted base (000's) = 100%†	Unweighted sample
		£0.01- £100.00	£100.01 - £150.00	£150.01 - £200.00	£200.01 - £250.00	£250.01 - £300.00	£300.01 - £350.00	£350.01 - £400.00	£400.01 - £450.00	£450.01 - £500.00	£500.01 and over		
Married couple	%	3	2	2	2	2	3	4	5	5	71	4,097	1483
Cohabiting couple	%	2	5	4	2	5	4	6	12	10	50	682	236
Lone mother**	%	11	25	13	10	8	7	5	5	2	15	1,645	588
Single	%	15	34	11	10	7	6	4	1	1	11	832	292
Divorced	%	6	13	16	11	7	6	9	9	3	19	479	177
Separated	%	6	24	13	5	10	9	5	5	5	18	299	108
Lone father	%	10	17	7	3	3	6	11	6	8	28	174	56
All lone parents**	%	10	24	12	9	7	7	6	5	3	16	1,819	644

*   Dependent children are persons aged under 16, or aged 16-18 and in full-time education, in the family unit, and living in the household.
†   Bases exclude cases where income is not known.
** Includes eleven widowed lone mothers.

Table 3.12   The distribution of the population by sex and age: 1971 to 2002

All persons

Great Britain

Age	Unweighted								Weighted			
	1971	1975	1981	1985	1991	1995	1996	1998	1998	2000	2001	2002
	%	%	%	%	%	%	%	%	%	%	%	%
Males												
0- 4	9	8	7	7	8	7	7	8	6	6	6	6
5-15*	19	18	18	16	15	16	16	16	15	15	15	15
16-44*	39	40	41	42	41	39	39	38	42	42	42	41
45-64	24	23	22	22	22	24	23	24	23	24	24	25
65-74	7	8	8	9	8	9	8	9	8	8	8	8
75 and over	3	3	4	4	5	5	6	5	5	5	5	6
Weighted base (000's) = 100%†									27,921	28,134	28,212	27,524
Unweighted sample†	16908	16242	15735	12551	11913	11376	10781	9831		9322	10166	9706
	%	%	%	%	%	%	%	%	%	%	%	%
Females												
0- 4	8	6	6	6	7	6	7	7	6	6	6	5
5-15*	16	17	16	15	14	14	15	14	14	14	14	14
16-44*	37	38	39	41	39	39	39	38	40	40	40	40
45-64	24	24	22	21	22	24	23	24	23	24	24	24
65-74	9	10	10	10	10	9	9	9	9	9	9	9
75 and over	5	6	7	8	8	8	7	8	8	8	8	8
Weighted base (000's) = 100%†									28,828	28,973	29,048	29,047
Unweighted sample†	17871	17328	16675	13522	12744	12009	11493	10564		9944	11014	10443
	%	%	%	%	%	%	%	%	%	%	%	%
Total												
0- 4	8	7	6	6	7	7	7	7	6	6	6	6
5-15*	17	17	17	15	15	15	15	15	14	14	14	14
16-44*	38	39	40	42	40	39	39	38	41	41	41	40
45-64	24	23	22	21	22	24	23	24	23	24	24	24
65-74	8	9	9	9	9	9	9	9	8	8	8	8
75 and over	4	4	5	6	7	6	7	7	7	7	7	7
Weighted base (000's) = 100%†									56,749	57,106	57,260	56,571
Unweighted sample†	34779	33570	32410	26073	24657	23385	22274	20395		19266	21180	20149

*   5-14 and 15-44 in 1971 and 1975.
†   Trend tables show unweighted and weighted figures for 1998 to give an indication of the effect of the weighting. For weighted data (1998 to 2002) the weighted base (000's) is the base for percentages. Unweighted data (up to 1998) are based on the unweighted sample.

Table 3.13   **Percentage of males and females by age**

All persons   Great Britain: 2002

Age		Males	Females	Weighted base (000's) = 100%	Unweighted sample
0- 4	%	52	48	3,286	1168
5-15	%	50	50	8,052	3009
16-19	%	51	49	2,776	922
20-24	%	49	51	3,510	1101
25-29	%	49	51	3,366	1106
30-34	%	49	51	4,262	1451
35-39	%	49	51	4,482	1560
40-44	%	49	51	4,293	1462
45-49	%	48	52	3,740	1372
50-54	%	50	50	3,739	1390
55-59	%	50	50	3,599	1349
60-64	%	50	50	2,680	1058
65-69	%	46	54	2,599	964
70-74	%	48	52	2,193	802
75 and over	%	38	62	3,994	1435
Total	%	49	51	56,571	20149

Table 3.14   **Socio-economic classification based on own current or last job by sex and age**

All persons aged 16 and over   Great Britain: 2002

Socio-economic classification*	Age group						
	16-24	25-34	35-44	45-54	55-64	65 and over	All
	%	%	%	%	%	%	%
Males							
Higher managerial and professional	6	18	17	19	14	13	15
Lower managerial and professional	11	23	25	25	18	19	21
Intermediate	10	8	5	5	4	5	6
Small employers and own account	3	9	13	15	15	10	11
Lower supervisory and technical	16	13	14	15	15	19	15
Semi-routine	22	10	10	9	13	13	12
Routine	22	14	13	11	18	20	16
Never worked and long-term unemployed	11	4	2	2	2	1	3
Weighted base (000's) = 100%	1,973	3,469	4,210	3,551	3,118	3,769	20,090
Unweighted sample	616	1127	1425	1289	1167	1439	7063
	%	%	%	%	%	%	%
Females							
Higher managerial and professional	3	9	8	6	3	1	5
Lower managerial and professional	13	29	25	27	22	16	22
Intermediate	20	19	19	18	18	20	19
Small employers and own account	1	3	5	7	6	4	5
Lower supervisory and technical	5	5	6	6	6	7	6
Semi-routine	28	18	22	21	27	25	23
Routine	16	8	11	11	15	20	14
Never worked and long-term unemployed	14	8	5	3	4	7	6
Weighted base (000's) = 100%	1,937	3,746	4,341	3,772	3,103	4,980	21,879
Unweighted sample	637	1297	1524	1421	1218	1750	7847
	%	%	%	%	%	%	%
Total							
Higher managerial and professional	4	13	12	12	9	6	10
Lower managerial and professional	12	26	25	26	20	17	22
Intermediate	15	14	12	11	11	13	13
Small employers and own account	2	6	9	11	10	7	8
Lower supervisory and technical	11	9	10	10	10	12	10
Semi-routine	25	14	16	15	20	20	18
Routine	19	11	12	11	17	20	15
Never worked and long-term unemployed	12	6	4	3	3	5	5
Weighted base (000's) = 100%	3,907	7,214	8,551	7,322	6,222	8,750	41,966
Unweighted sample	1253	2424	2949	2710	2385	3189	14910

* From April 2001 the National Statistics Social-economic Classification (NS-SEC) was introduced for all official statistics and surveys. It replaced Social Class based on Occupation and Socio-economic Groups (SEG). Full-time students and persons in inadequately described occupations are excluded (see Appendix A).

Table 3.15   **Ethnic group: 2001 to 2002**

All persons   Great Britain

Ethnic group	2001		2002	
	%		%	
White British	89	92	88	91
Other White	3		3	
Mixed background	1		1	
Indian	2		2	
Pakistani and Bangladeshi	2	4	2	4
Other Asian background	0		1	
Black Caribbean	1	2*	1	2*
Black African	1		1	
Other ethnic group	1		1	
Weighted base (000's) = 100%	*57,034*		*56,302*	
Unweighted sample	*21102*		*20053*	

* Including other Black groups not shown separately.

Table 3.16   **Age by ethnic group**

All persons   Great Britain: 2001 and 2002 combined

Age	Ethnic group										
	White British	Other White	Mixed background	Indian	Pakistani and Bangladeshi	Other Asian background	Black Caribbean	Black African	Other Black background	Other ethnic group	All
	%	%	%	%	%	%	%	%	%	%	%
0-15	20	11	50	20	37	20	21	32	33	20	20
16-24	10	13	14	16	19	18	12	12	15	18	11
25-44	29	45	28	38	29	42	34	41	45	38	30
45-64	25	21	7	21	12	16	19	13	7	18	24
65 and over	16	11	2	5	4	3	15	2	0	6	15
*Unweighted sample**	*36694*	*1106*	*447*	*625*	*771*	*200*	*406*	*372*	*67*	*467*	*41155*

* Weighted bases not shown for combined data sets.

Table 3.17   **Sex by ethnic group**

All persons   Great Britain: 2001 and 2002 combined

Ethnic group		Male	Female	*Unweighted sample**
White British	%	49	51	*36694*
Other White	%	48	52	*1106*
Mixed background	%	53	47	*447*
Indian	%	50	50	*625*
Pakistani and Bangladeshi	%	48	52	*771*
Other Asian background	%	46	54	*200*
Black Caribbean	%	48	52	*406*
Black African	%	44	56	*372*
Other Black background	%	45	55	*67*
Other ethnic group	%	52	48	*467*

* Weighted bases not shown for combined data sets.

Table 3.18   Government Office Region by ethnic group

All persons   Great Britain: 2001 and 2002 combined

Government Office Region	Ethnic group										
	White British	Other White	Mixed background	Indian	Pakistani and Bangladeshi	Other Asian background	Black Caribbean	Black African	Other Black background	Other ethnic group	Total
	%	%	%	%	%	%	%	%	%	%	%
England											
North East	5	2	1	0	3	3	0	1	0	3	5
North West	12	5	11	14	12	7	4	3	4	5	12
Yorkshire and the											
Humber	9	4	6	5	15	6	3	1	2	4	8
East Midlands	8	4	7	12	4	3	11	3	7	7	7
West Midlands	8	5	12	19	21	10	14	3	3	10	9
East of England	10	8	8	3	8	5	5	3	0	5	9
London	9	39	32	36	25	49	56	80	80	46	13
South East	15	17	11	8	5	11	4	3	1	9	14
South West	10	4	7	1	1	1	0	2	1	6	9
Wales	5	3	3	1	4	2	1	0	0	2	5
Scotland	10	8	3	0	2	3	2	1	1	4	9
*Unweighted sample**	*36694*	*1106*	*447*	*625*	*771*	*200*	*406*	*372*	*67*	*467*	*41155*

* Weighted bases not shown for combined data sets.

Table 3.19   Average household size by ethnic group of household reference person

Households   Great Britain: 2001 and 2002 combined

Ethnic group	Average (mean) household size	Unweighted sample*
White British	2.27	*15969*
Other White	2.32	*506*
Mixed background	2.38	*104*
Indian	3.40	*191*
Pakistani and Bangladeshi	4.28	*181*
Other Asian background	3.07	*68*
Black Caribbean	2.24	*198*
Black African	2.56	*147*
Other Black background	2.77	*24*
Other ethnic group	2.47	*178*
Total	2.32	*17566*

* Weighted bases not shown for combined data sets.

Table 3.20   **Weighted bases for Tables 3.3 and 3.8**

(a) Persons aged 16 and over   **Great Britain**

	1998	**2000**	**2001**	**2002**	**Table Reference**
16-24	6,139	6,191	6,192	6,286	3.3
25-44	17,117	17,130	17,275	16,404	3.3
45-64	13,226	13,519	13,540	13,757	3.3
65-74	4,767	4,719	4,727	4,792	3.3
75 and over	3,836	3,888	3,898	3,993	3.3
All aged 16 and over	45,085	45,447	45,632	45,232	3.3

(b) Families with dependent children*   **Great Britain**

	1998	**2000**	**2001**	**2002**	**Table Reference**
Married/cohabiting couple†	5,465	5,232	5,366	5,244	3.8
Married couple	4,765	4,496	4,580	4,515	3.8
Cohabiting couple	700	736	785	729	3.8
Lone parent	1,717	1,861	1,739	1,922	3.8
Total	7,182	7,093	7,105	7,166	3.8

*   Dependent children are persons aged under 16, or aged 16-18 and in full-time education, in the family unit, and living in the household.
†   Including married women whose husbands were not defined as resident in the household.

Chapter 4 Housing and consumer durables

The General Household Survey (GHS) has included questions on housing and the availability of consumer durables since 1971. Periodically, new consumer items have been added and some older items have been dropped depending on their availability. This year the GHS asked questions for the first time about access to DVD players and separately identified satellite, cable and digital television access. This chapter looks at trends as well as data from the 2002 GHS.

- Trends in housing tenure

- Trends in accommodation

- Characteristics of tenure groups

  Owner occupiers

  Social sector tenants

  Private renters

- National Statistics Socio-economic Classification (NS-SEC) by tenure

- Persons per room

- Cars and vans

- Consumer durables – trends over time

- Consumer durables and socio-economic classification (NS-SEC)

- Consumer durables and gross weekly income

- Consumer durables and household type

- Consumer durables and lone-parent families

Trends in housing tenure

Over the past 32 years there has been an increase in home ownership. In 2002, 69% of all households owned their own home (either outright or being purchased with a mortgage) compared with 49% in 1971. Most of the increase in home ownership occurred during the 1980s (from 54% in 1981 to 67% in 1991). This was when the 'right to buy' scheme was introduced, which has allowed local authority tenants to buy their own homes. The proportion of households owning their own home has remained relatively constant since the 1990s (between 67% and 69%).

The proportion of households renting from a housing association[1] has increased in every decade (from 1% in 1971, 2% in 1981 and 3% in 1991 to 7% in 2002). This has resulted from the transfer of housing stock from local authorities to Registered Social Landlords (RSLs) since the mid-1990s and the concentration of public funding for new housing in the RSL sector.

- Just under a third (31%) of households were renting council homes in 1971. The proportion increased to 34% in 1981 and then gradually declined to 14% in 2002.
- Just over one in ten (11%) of all households were private renters in 2002 compared with 20% in 1971.
- The proportion of private renter households has remained relatively constant from 1998 to 2002 (between 10% and 11%). **Table 4.1, Figure 4A**

Trends in accommodation

There has been little change in the proportion of households living in each type of accommodation over the period 1971 to 2002. The greatest change has been in the proportion of households living in detached homes (from 16% in 1971 to 22% in 2002) although it has remained relatively constant over the last five years.

In every decade semi-detached and terraced houses were the most common types of accommodation lived in. The proportion of households living in a house, whether detached, semi-detached or terraced, has remained relatively constant over the last five years (between 79% and 80%). **Table 4.2**

Characteristics of tenure groups

In 2002, as in previous years, there was variation in the characteristics of the households in each tenure group and the type of accommodation in which they lived.

Owner occupiers

Owner occupiers include households who own their home outright and those who are purchasing their home with a mortgage.

- In 2002 the majority (91%) of owner occupiers lived in a house. Just under two thirds (65%) of owner occupiers lived in a detached or semi-detached house, just over a quarter (26%) lived in a terraced house and the remainder (9%) lived in a flat.

Figure 4A   **Tenure, 1971 to 2002**

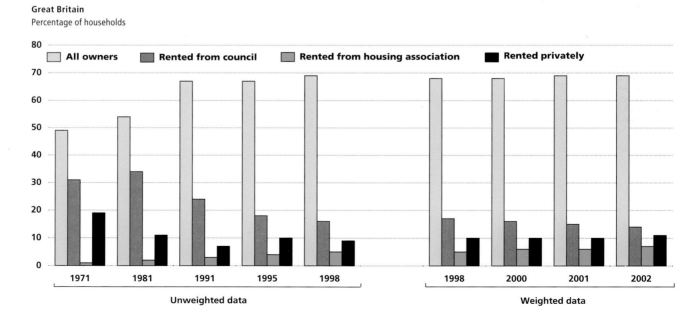

Great Britain
Percentage of households

- Over a quarter (28%) of homes were owned outright and 40% were being purchased with a mortgage. The latter group was more likely than those who owned their home outright to be living in a terraced house (29% compared with 20%).
- Almost two thirds (65%) of homes owned outright contained at least one adult aged 60 or over. Of those being bought with a mortgage, over half (52%) were occupied by either a small family or by two adults aged 16 to 59.
- Unlike other tenure groups, owner-occupied accommodation was fairly evenly spread between properties regardless of when they were built.
- Lone-parent families with dependent children were less likely to be owner occupier households with a mortgage (28%) than other families with dependent children (71%).
- Of all households, those buying with a mortgage had on average the highest gross weekly household income at £763.
- Outright home-ownership was more likely in the older age groups. For homes owned outright, the median age group of the household reference person was 65 to 69 compared with 30 to 44 for homes purchased with a mortgage.

Tables 4.4, 4.6, 4.7, 4.9, 4.10

Social sector tenants

Social sector tenants include households renting from either the council or a housing association.

- Over half (56%) of the households who were social sector tenants lived in a house. Just over a quarter (27%) lived in a detached or semi-detached house and 29% lived in a terraced house. The remainder (44%) lived in a flat - 42% in a purpose-built flat or maisonette and 2% in a converted flat/maisonette or room.
- A higher proportion of social sector tenants lived in purpose-built flats or maisonettes than any other tenure group (42% compared with 24% of private renter households and 7% of owner occupiers).
- Households renting from housing associations were more likely to live in newer accommodation than other tenure groups. Almost a third (31%) of households renting from a housing association were living in accommodation built in 1985 or later, compared with 15% of owner-occupier households

and 14% of private renters. This results from the introduction of the 'New Start' scheme in 1988 when housing associations became responsible for building the majority of social rented accommodation.
- A quarter (25%) of households that were renting from the social sector contained only one adult who was aged 60 or over.
- Lone-parent families were more likely to be social sector households than other families with dependent children (51% compared with 14% of other families with dependent children).
- Households renting from the social sector had on average the lowest gross weekly income of all households (£231 gross household income per week for council tenants and £223 gross household income per week for housing association tenants).

Tables 4.4–4.7,4.9

Private renters

Private renter households include those who rent either furnished or unfurnished properties from a private landlord.

- Over half (52%) of converted flats or maisonettes were lived in by private renter households.
- Private renter households were more likely to live in accommodation built before 1919 than any other tenure group (39% compared with 21% of owner occupiers and 6% of households renting from the social sector).
- Over a quarter (28%) of private renter households contained only one adult aged under 60. This was higher than among owner-occupier (13%) and social sector households (18%).
- In 2002 the average gross weekly household income of private renters was £461.
- Private renter households were more likely to be in the younger age groups. For private renter households, the median age group of the household reference person was 30 to 44 years compared with 45 to 59 for both owner-occupier and social sector households.
- Private renter households had the lowest median length of tenure of the household reference person (12 months but less than 2 years compared with 5 years or more for owner-occupier and social sector households). **Tables 4.4–4.6, 4.9–4.10, 4.13**

National Statistics Socio-economic Classification (NS-SEC) by tenure

Based on the National Statistics socio-economic classification (NS-SEC), tenure varied by socio-economic status.

- Households living in social sector accommodation were more likely to have an economically inactive household reference person than any other tenure group (63% compared with just under a third (31%) of households in both the owner occupier and private renter groups).
- Home ownership was less likely in households headed by people in the semi-routine and routine occupational groups (60% to 62%) compared with, households headed by people within a large employer and higher managerial group (90%).
- Among households headed by an economically active person, those in the semi-routine or routine occupational groups were more likely than other households to be social sector tenants (27% and 29% respectively) compared with between 1% to 13% of other households. **Table 4.12**

Persons per room

In 2002 the mean number of persons per room was 0.45. The mean number of persons per room decreased from 0.57 in 1975 to 0.48 in 1995. It has remained relatively stable in the last five years (0.46 to 0.45).

Among the tenure groups, occupancy rates of one or more persons per room were more likely in private renter households (6%) or those renting from the social sector (8%) than those that were owner occupied (3%). The majority (96%) of all households had an occupancy rate of less than one person per room.

Overall, only 2% of all households were below the 'bedroom standard' (see Appendix A). This proportion was lowest among owner occupiers (1%), with 5% of households renting from the social sector and 3% of private renter households living in accommodation below the bedroom standard. Over half (55%) of households that owned their home outright had two or more bedrooms above the bedroom standard, compared with 9% to 35% of households in the other tenure groups. As outright ownership was more prevalent among the older age groups, older households were less likely than others to be overcrowded. **Tables 4.14–4.16**

Cars and vans

Over the last 30 years, the proportion of households with access to a car or van has increased but this increase has slowed since the early 1990s. In 2002 nearly three quarters (73%) of households had access to a car or van. The proportion of households with access to a car or van increased from 52% in 1972 to 71% in 1995 and has remained relatively stable in the last five years. Just under a quarter (22%) of households had access to two cars or vans in 2002 compared with 8% of households in 1972. Over the same period, the proportion of households with access to three or more cars or vans rose from 1% in 1972 to 5% in 2002.

Table 4.17, Figure 4B

The availability of a car or van varied by the socio-economic classification of the household reference person.

- Households headed by an economically active person in the large employer and higher managerial group were more likely to have access to two or more cars or vans (61% compared with 21% to 24% of the semi-routine and routine households and 43% to 50% of other managerial and professional households).
- Around a quarter (24% to 28%) of households headed by an economically active person in the routine and semi-routine groups had no access to a car or van

Figure 4B **Households with access to a car or van, 1972 to 2002**

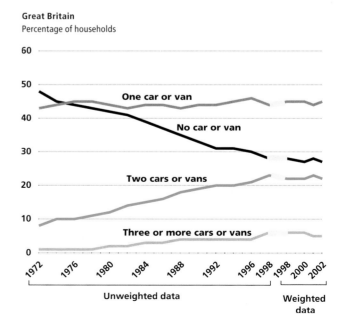

Great Britain
Percentage of households

compared with 4% to 15% of other economically active households.

- Of the households headed by an economically inactive household reference person, 50% had no access to a car or van.   **Table 4.18**

Consumer durables – trends over time

The GHS has shown the steady increase in the ownership of consumer durables since the early 1970s. This trend has continued in 2002, with a few exceptions. By the mid 1990s most homes had central heating and had access to a television, telephone, washing machine and freezer.

Between 1998 and 2002 there has been an increase in the proportion of households with access to more recent forms of consumer durables:

- video recorders – 85% in 1998 to 89% in 2002;
- CD players – 69% in 1998 to 83% in 2002;
- home computers – 34% in 1998 to 54% in 2002;
- microwave ovens – 78% in 1998 to 87% in 2002.

In addition to the increase in home computers, access to the Internet at home has also increased. In 2002, 44% of households had access to the Internet at home compared with 40% in 2001 and 33% in 2000. Nearly all of those households (42% of all households) accessed the Internet from their home computer rather than through another medium such as a digital television, mobile phone or games console.

In 2002, almost all (99%) households had a telephone (either fixed or mobile phone). Over the period 2000 to 2002, fixed telephone access has been relatively constant (between 92% and 93%). Over the same period, the proportion of households with access to a mobile phone has increased each year from 58% in 2000, to 70% in 2001 and to 75% in 2002. Just under a quarter (24%) of households had a fixed telephone only and 6% of households had a mobile phone only.

Table 4.19, Figures 4C, 4D

Consumer durables and socio-economic classification (NS-SEC)

Based on the National Statistics socio-economic classification (NS-SEC), access to consumer durables varied by the socio-economic status of the household reference person.

- Just under a fifth (19%) of households headed by an economically inactive person had access to the Internet at home compared with 70% to 83% of those in the professional and managerial groups, and 36% and 33% of semi-routine and routine groups.
- Households where the household reference person was economically active were more likely to have access to a mobile phone (between 81% and 93% compared with 52% of households headed by someone who was economically inactive).
- Households headed by a person in the professional and managerial groups were more likely to have

Figure 4C   **Percentage of households with consumer durables, 1972 to 2002**

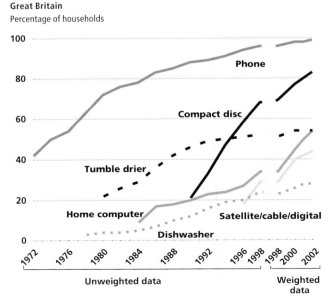

Figure 4D   **New technology, 2000 to 2002**

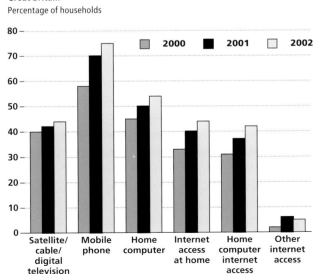

access to more recent items of consumer durables (43% to 60% had access to a DVD player compared with 36% to 37% of semi-routine and routine groups). Just over one in ten (12%) of households headed by an economically inactive person had access to a DVD player. **Table 4.20**

Consumer durables and gross weekly income

As one might expect, the higher the household income the more likely it was for a household to have consumer durables.

In 2002, households with a gross weekly income of over £500 were more likely than households with an income of £100 or less to have a:
- home computer (81% compared with 25%);
- mobile phone (92% compared with 47%);
- dishwasher (48% compared with 9%);
- tumble drier (67% compared with 36%);
- CD player (96% compared with 61%);
- DVD player (48% compared with 16%);
- digital television (37% compared with 14%); or,
- access to the Internet at home (71% compared with 17%).

Colour television was the main exception with access ranging from 96% to 100% across all the income groups. **Table 4.21**

Consumer durables and household type

Larger households were more likely to have items of consumer durables, in particular home entertainment items and home computers. In general, single-person households were the least likely to have any of the consumer durables listed, with older (aged 60 or over) single-person households being the least likely to have the more recent items such as DVD players and home computers. The differences in these household types may be explained by the pooled resources and the increased chance in larger households of any one member of the household having a listed item.

In 2002, the differences between single-person and other household types were:
- washing machine (80% of older and 87% of younger single-person households compared with 96% to 99% of other households);

- freezer (90% of older and 89% of younger single-person households compared with almost all, 97% to 99%, of other households);
- tumble drier (34% of older and 37% of younger single-person households compared with 54% to 74% of other households).

There were differences between older and younger person households.
- Households with adults aged 60 or over were the least likely to have a home computer, with a third (33%) of older two-person households and 9% of older single-person households having a home computer. This compares with 69% of younger two-person households and 79% of large adult households.
- Access to the Internet at home was least common among older households (6% of older single-person households and 27% of older two-person households compared with 57% to 59% of family households, 60% of younger two-person households and 66% of large adult households).
- Single-person and older households were least likely to have satellite, cable or digital television. For example, 9% of older and 21% of younger single-person households had digital television compared with 40% of both large family and large adult households.
- Older households were the least likely to have access to a mobile phone (28% of older single-person households and 65% of older two-person households compared with 76% of younger single-person households and 88% to 91% of other households). **Table 4.22**

Consumer durables and lone-parent families

Lone-parent families were less likely than other families with dependent children to have access to more recent consumer durables such as home computers and mobile phones. Lone-parent families were also less likely to have access to the Internet.

Differences between lone-parent families and other families with dependent children were:
- satellite/cable/digital television (48% compared with 62%);
- home computer (56% compared with 80%);
- access to the Internet (36% compared with 68%);
- tumble drier (58% compared with 72%);

- dishwasher (17% compared with 46%);
- mobile phone (83% compared with 91%);
- car or van (52% compared with 93%).

Table 4.23, Figure 4E

Figure 4E   **Consumer durables: Lone parent families compared with couples with dependent children, 2002**

Great Britain
Percentage of families

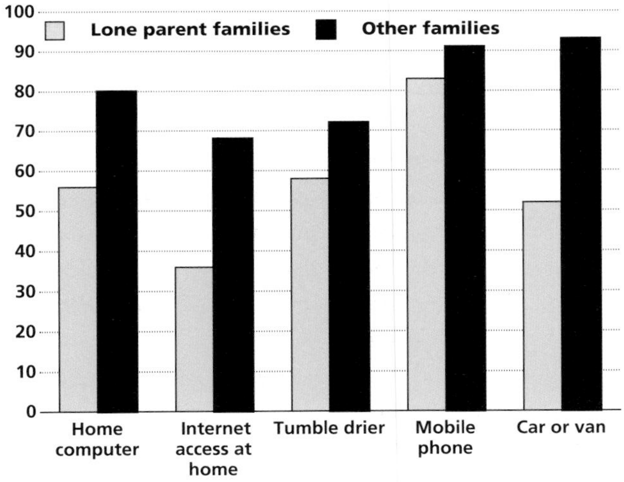

Notes

1   Since 1996, housing associations are described as Registered Social Landlords (RSLs). RSLs are not-for-profit organisations which include: charitable housing associations; industrial and provident societies, and companies registered under the Companies Act 1985.

Table 4.1　　**Tenure: 1971 to 2002**

Households　　　　　　　　　　　　　　　　　　　　　　　　　　　　　Great Britain

Tenure	Unweighted								Weighted			
	1971	1975	1981	1985	1991	1995	1996	1998	1998	2000	2001	2002
	%	%	%	%	%	%	%	%	%	%	%	%
Owner occupied, owned outright	22	22	23	24	25	25	26	28	26	27	27	29
Owner occupied, with mortgage	27	28	31	37	42	42	41	41	42	41	41	40
Rented from council*	31	33	34	28	24	18	19	16	17	16	15	14
Rented from housing association	1	1	2	2	3	4	5	5	5	6	6	7
Rented with job or business	5	3	2	2	1	2	**	**	**	**	**	**
Rented privately, unfurnished†	12	10	6	5	4	5	7	7	7	7	7	8
Rented privately, furnished	3	3	2	2	2	3	3	2	3	3	3	3
Weighted base (000's) = 100%††									24,436	24,838	24,592	24,508
Unweighted sample††	11936	11970	11939	9933	9922	9723	9155	8631		8219	8989	8613

* Council includes local authorities, New Towns and Scottish Homes from 1996.
† Unfurnished includes the answer 'partly furnished'.
** From 1996 all tenants whose accommodation goes with the job of someone in the household have been allocated to 'rented privately'. Squatters are also included in the privately rented category.
†† Trend tables show unweighted and weighted figures for 1998 to give an indication of the effect of the weighting. For the weighted data (1998, and 2000 to 2002) the weighted base (000's) is the base for percentages. Unweighted data (up to 1998) are based on the unweighted sample.

Table 4.2　　**Type of accommodation: 1971 to 2002**

Households　　　　　　　　　　　　　　　　　　　　　　　　　　　　　Great Britain

Type of accommodation*	Unweighted								Weighted			
	1971	1975	1981	1985	1991	1995	1996	1998	1998	2000	2001	2002
	%	%	%	%	%	%	%	%	%	%	%	%
Detached house	16	15	16	19	19	22	21	23	21	21	21	22
Semi-detached house	33	34	32	31	32	31	32	33	32	31	31	32
Terraced house	30	28	31	29	29	28	27	26	26	28	28	27
Purpose-built flat or maisonette	13	14	15	15	14	15	15	15	17	16	16	16
Converted flat or maisonette/rooms	6	8	5	5	4	4	5	4	4	4	4	4
With business premises/other	2	1	1	1	1	1	0	0	0	0	0	0
Weighted base (000's) = 100%†									24,398	24,806	24,520	24,421
Unweighted sample†	11846	12041	11978	9890	9917	9730	9128	8615		8207	8963	8581

* Tables for type of accommodation exclude households living in caravans.
† Trend tables show unweighted and weighted figures for 1998 to give an indication of the effect of the weighting. For the weighted data (1998, and 2000 to 2002) the weighted base (000's) is the base for percentages. Unweighted data (up to 1998) are based on the unweighted sample.

Table 4.3   **Type of accommodation occupied by households renting from a council compared with other households: 1981 to 2002**

Households   **Great Britain**

Type of accommodation	Unweighted						Weighted data			
	1981	**1987**	**1991**	**1995**	**1996**	**1998**	**1998**	**2000**	**2001**	**2002**
	%	%	%	%	%	%	%	%	%	%
Renting from council										
Detached house	1	1	1	0	1	1	1	1	1	0
Semi-detached house	30	28	28	28	28	29	27	27	26	27
Terraced house	34	35	34	33	31	28	27	31	29	30
Purpose-built flat or maisonette	33	34	35	38	38	40	43	40	42	40
Converted flat or maisonette	2	2	3	1	2	2	2	2	2	2
*Weighted base (000's) = 100%**							4,021	3,870	3,695	3,404
*Unweighted sample**	*4007*	*2600*	*2339*	*1770*	*1748*	*1410*		*1240*	*1325*	*1138*
	%	%	%	%	%	%	%	%	%	%
Other households										
Detached house	24	25	25	26	25	27	26	24	25	25
Semi-detached house	33	33	33	32	33	33	32	32	32	32
Terraced house	29	28	28	27	27	26	26	27	27	26
Purpose-built flat or maisonette	6	7	8	10	9	10	11	11	11	12
Converted flat or maisonette	7	5	5	4	5	4	5	5	4	4
*Weighted base (000's) = 100%**							20,328	20,896	20,808	20,980
*Unweighted sample**	*7904*	*7511*	*7578*	*7953*	*7379*	*7189*		*6954*	*7632*	*7431*
	%	%	%	%	%	%	%	%	%	%
All households										
Detached house	16	18	19	22	21	23	22	21	21	22
Semi-detached house	32	32	32	31	32	33	32	31	31	32
Terraced house	31	30	29	28	27	26	26	28	28	27
Purpose-built flat or maisonette	15	14	14	15	15	15	17	16	16	16
Converted flat or maisonette	5	5	4	4	5	4	4	4	4	4
*Weighted base (000's) = 100%**							24,349	24,766	24,503	24,384
*Unweighted sample**	*11911*	*10111*	*9917*	*9723*	*9127*	*8599*		*8194*	*8957*	*8569*

*   Trend tables show unweighted and weighted figures for 1998 to give an indication of the effect of the weighting. For the weighted data (1998, and 2000 to 2002) the weighted base (000's) is the base for percentages. Unweighted data (up to 1998) are based on the unweighted sample.

Table 4.4   **(a) Type of accommodation by tenure**
             (b) Tenure by type of accommodation

Households   Great Britain: 2002

Tenure		Type of accommodation*							Weighted base (000's) = 100%	Unweighted sample
		Detached house	Semi-detached house	Terraced house	All houses	Purpose-built flat or maisonette	Converted flat or maisonette/ rooms	All flats/ rooms		
(a)										
Owner occupied, owned outright	%	35	37	20	92	6	1	8	6,928	2576
Owner occupied, with mortgage	%	26	34	29	90	8	3	10	9,864	3472
All owners	%	30	35	26	91	7	2	9	16,793	6048
Rented from council†	%	0	27	30	58	40	2	42	3,404	1138
Rented from housing association	%	1	23	28	52	45	3	48	1,591	539
Social sector tenants	%	1	26	29	56	42	2	44	4,995	1677
Rented privately, unfurnished**	%	15	20	30	65	19	16	35	1,868	630
Rented privately, furnished	%	4	11	23	39	37	24	61	728	214
Private renters††	%	12	18	28	58	24	19	42	2,596	844
Total	%	22	32	27	80	16	4	20	24,384	8569
(b)									Total	
		%	%	%	%	%	%	%	%	
Owner occupied, owned outright		45	33	22	33	11	8	11	28	
Owner occupied, with mortgage		48	44	44	45	20	27	21	40	
All owners		93	77	66	78	31	36	32	69	
Rented from council†		0	12	16	10	35	7	29	14	
Rented from housing association		0	5	7	4	18	5	16	7	
Social sector tenants		1	17	23	14	53	12	45	20	
Rented privately, unfurnished**		5	5	9	6	9	33	13	8	
Rented privately, furnished		1	1	3	1	7	19	9	3	
Private renters††		6	6	11	8	16	52	23	11	
Weighted base (000's) =100%		*5,332*	*7,708*	*6,503*	*19,542*	*3,918*	*924*	*4,842*	*24,384*	
Unweighted sample		*2024*	*2764*	*2279*	*7067*	*1199*	*303*	*1502*	*8569*	

*   Tables for type of accommodation exclude households living in caravans.
†   Council includes local authorities, New Towns and Scottish Homes from 1996.
**   Unfurnished includes the answer 'partly furnished'.
††   All tenants whose accommodation goes with the job of someone in the household have been allocated to 'rented privately'. Squatters are also included in the privately rented category.

Table 4.5   **Age of building by tenure**

Households   Great Britain: 2002

Age of building* containing household's accommodation	Tenure									
	Owners			Social sector tenants			Private renters			Total
	Owned outright	With mortgage	All owners	Council†	Housing association	Social sector tenants	Unfurnished private**	Furnished private	Private Renters††	
(a)	%	%	%	%	%	%	%	%	%	%
Before 1919	21	22	21	4	10	6	39	36	39	20
1919-1944	22	18	20	20	14	18	19	22	20	19
1945-1964	24	17	20	39	16	31	13	9	12	21
1965-1984	24	25	24	33	29	32	15	15	15	25
1985 or later	10	18	15	5	31	13	13	17	14	15
Weighted base (000's) = 100%	*6,871*	*9,721*	*16,592*	*3,170*	*1,512*	*4,682*	*1,784*	*663*	*2,447*	*23,722*
Unweighted sample	*2555*	*3424*	*5979*	*1062*	*514*	*1576*	*603*	*195*	*798*	*8353*

*   For an assessment of the reliability of age of building estimates, see Birch F, Age of buildings (OPCS Social Survey Division, GHS Series No.7, 1974).

†   Council includes local authorities, New Towns and Scottish Homes from 1996.

**   Unfurnished includes the answer 'partly furnished'.

††   All tenants whose accommodation goes with the job of someone in the household have been allocated to 'rented privately'. Squatters are also included in the privately rented category.

Table 4.6 **(a) Household type by tenure**
         (b) Tenure by household type

Households   Great Britain: 2002

Tenure		Household type							Weighted base (000's) = 100%	Unweighted sample
		1 adult aged 16-59	2 adults aged 16-59	Small family	Large family	Large adult household	2 adults, 1 or both aged 60 or over	1 adult aged 60 or over		
(a)										
Owner occupied, owned outright	%	7	10	4	1	12	37	28	7,027	2612
Owner occupied, with mortgage	%	17	24	28	6	18	5	2	9,871	3474
All owners	%	13	18	18	4	15	18	13	16,898	6086
Rented from council*	%	16	7	23	8	9	11	25	3,404	1138
Rented from housing association	%	22	7	22	8	5	10	25	1,600	542
Social sector tenants	%	18	7	22	8	8	11	25	5,004	1680
Rented privately, unfurnished†	%	26	22	23	4	7	8	11	1,878	633
Rented privately, furnished	%	34	26	9	2	23	2	3	728	214
Private renters**	%	28	23	19	3	11	6	9	2,606	847
Total	%	16	16	19	5	13	15	15	24,507	8613

b)								**Total**	
	%	%	%	%	%	%	%	%	
Owner occupied, owned outright	14	18	6	8	25	69	54	29	
Owner occupied, with mortgage	44	59	59	50	53	12	5	40	
All owners	58	77	66	58	79	81	59	69	
Rented from council*	14	6	16	23	10	10	24	14	
Rented from housing association	9	3	8	11	3	4	11	7	
Social sector tenants	24	9	24	34	12	15	35	20	
Rented privately, unfurnished†	13	10	9	6	4	4	6	8	
Rented privately, furnished	6	5	1	1	5	0	1	3	
Private renters**	19	15	10	7	9	4	6	11	
Weighted base (000's) =100%	*3,884*	*4,019*	*4,715*	*1,204*	*3,275*	*3,768*	*3,642*	*24,507*	
Unweighted sample	*1177*	*1445*	*1684*	*457*	*1106*	*1450*	*1294*	*8,613*	

*   Council includes local authorities, New Towns and Scottish Homes from 1996.
†   Unfurnished includes the answer 'partly furnished'.
** From 1996 all tenants whose accommodation goes with the job of someone in the household have been allocated to 'rented privately'. Squatters are also included in the privately rented category.

Table 4.7   Housing profile by family type: lone-parent families compared with other families

Families with dependent children*

	Lone-parent families	Other families
	%	%
Tenure		
Owner occupied, owned outright	5	8
Owner occupied, with mortgage	28	71
Rented from council or from housing association	51	14
Rented privately unfurnished	13	6
Rented privately furnished	2	1
Central heating	%	%
Yes	92	96
No	8	4
Type of accommodation	%	%
Detached house	8	28
Semi-detached house	30	37
Terraced house	40	28
Purpose-built flat or maisonette	18	6
Converted flat or maisonette/rooms	4	1
Bedroom standard	%	%
2 or more below standard	0	0
1 below standard	9	4
Equals standard	54	34
1 above standard	33	44
2 or more above standard	4	18
Persons per room	%	%
Under 0.5	24	8
0.5-0.99	70	77
1.0-1.49	6	15
1.5 or above	0	0
Unweighted sample†	*1283*	*3873*

* Dependent children are persons aged under 16, or aged 16-18 and in full-time education, in the family unit, and living in the household.
† Weighted base not shown for combined data sets.

Table 4.8   Type of accommodation by household type

Households

Household type		Type of accommodation*							*Weighted base (000's) = 100%*	*Unweighted sample*
		Detached house	Semi-detached house	Terraced house	All houses	Purpose-built flat or maisonette	Converted flat or maisonette/ rooms	All flats/ rooms		
One adult aged 16-59	%	9	23	27	58	32	10	42	3,861	1170
Two adults aged 16-59	%	24	30	28	82	13	5	18	4,010	1442
Small family	%	21	36	30	86	11	2	14	4,713	1683
Large family	%	21	36	34	92	8	1	8	1,204	457
Large adult household	%	30	33	29	92	6	1	8	3,273	1105
Two adults, one or both aged 60 or over	%	32	37	20	89	9	2	11	3,740	1439
One adult aged 60 or over	%	17	29	23	69	28	3	31	3,604	1280
Total	%	22	32	27	80	16	4	20	24,405	8576

* Tables for type of accommodation exclude households living in caravans.

Table 4.9   **Usual gross weekly income by tenure**

Households   Great Britain: 2002

Usual gross weekly income (£)	Tenure									Total
	Owners			Social sector tenants			Private renters			
	Owned outright	With mortgage	All owners	Council*	Housing association	Social sector	Unfurnished private†	Furnished private	Private renters**	
Income of household reference person										
Mean	328	535	452	165	172	168	328	344	333	380
Lower quartile	128	300	194	92	89	91	115	95	111	134
Median	222	433	350	129	132	130	247	223	233	282
Upper quartile	377	623	542	209	220	212	400	415	407	479
Income of household reference person and partner										
Mean	409	711	590	205	205	205	410	392	405	490
Lower quartile	160	381	246	98	96	98	128	109	123	164
Median	271	584	458	150	150	150	283	241	273	355
Upper quartile	480	843	733	265	262	263	522	531	523	630
Total household income										
Mean	444	763	635	231	223	229	449	494	461	531
Lower quartile	171	412	267	101	97	98	136	132	136	178
Median	291	624	492	167	153	161	307	353	313	392
Upper quartile	524	898	780	306	278	295	573	692	597	685
Weighted base (000's) =100%	*7,027*	*9,871*	*16,898*	*3,404*	*1,600*	*5,004*	*1,874*	*728*	*2,606*	*24,508*
Unweighted sample	*2612*	*3474*	*6086*	*1138*	*542*	*1680*	*632*	*214*	*847*	*8613*

*   Council includes local authorities, New Towns and Scottish Homes from 1996.
†   Unfurnished includes the answer 'partly furnished'.
**   From 1996 all tenants whose accommodation goes with the job of someone in the household have been allocated to 'rented privately'. Squatters are
     also included in the privately rented category.

Table 4.10 (a) **Age of household reference person by tenure**
       (b) **Tenure by age of household reference person**

Household reference persons   *Great Britain: 2002*

Tenure		Age of household reference person*								*Weighted base (000's) = 100%*	*Unweighted sample*
		Under 25	25-29	30-44	45-59	60-64	65-69	70-79	80 and over		
(a)											
Owner occupied, owned outright	%	0	1	5	25	13	15	26	14	*7,027*	*2612*
Owner occupied, with mortgage	%	3	8	49	34	4	1	1	0	*9,871*	*3474*
All owners	%	2	5	31	30	8	7	12	6	*16,898*	*6086*
Rented from council†	%	8	6	27	21	5	8	15	10	*3,404*	*1138*
Rented from housing association	%	7	6	31	19	5	6	17	9	*1,600*	*542*
Social sector tenants	%	8	6	29	20	5	7	16	10	*5,004*	*1680*
Rented privately, unfurnished**	%	11	15	37	17	3	3	7	6	*1,878*	*633*
Rented privately, furnished	%	32	24	31	7	2	1	2	1	*728*	*214*
Private renters††	%	17	18	35	15	2	3	6	5	*2,606*	*847*
Total	%	4	6	31	27	6	7	12	7	*24,507*	*8613*
(b)										Total	
		%	%	%	%	%	%	%	%	%	
Owner occupied, owned outright		3	2	5	27	57	65	63	61	29	
Owner occupied, with mortgage		22	50	64	52	23	9	4	2	40	
All owners		25	52	69	79	81	73	68	63	69	
Rented from council†		24	12	12	11	11	17	18	21	14	
Rented from housing association		10	6	7	5	5	6	10	8	7	
Social sector tenants		34	19	19	15	16	22	27	30	20	
Rented privately, unfurnished**		19	18	9	5	3	4	5	7	8	
Rented privately, furnished		21	11	3	1	1	0	0	0	3	
Private renters††		41	30	12	6	4	4	5	7	11	
Weighted base (000's) =100%		*1,103*	*1,581*	*7,524*	*6,545*	*1,588*	*1,607*	*2,919*	*1,640*	*24,507*	
Unweighted sample		*331*	*510*	*2537*	*2384*	*614*	*599*	*1054*	*584*	*8613*	

* Boxed figures indicate median age-groups.
† Council includes local authorities, New Towns and Scottish Homes from 1996.
** Unfurnished includes the answer 'partly furnished'.
†† From 1996 all tenants whose accommodation goes with the job of someone in the household have been allocated to 'rented privately'. Squatters are
   also included in the privately rented category.

Table 4.11   Tenure by sex and marital status of household reference person

Household reference persons   **Great Britain: 2002**

Tenure	Males						Females						Total
	Married	Cohabiting	Single	Widowed	Divorced/ separated	All males	Married	Cohabiting	Single	Widowed	Divorced/ separated	All females	
	%	%	%	%	%	%	%	%	%	%	%	%	%
Owner occupied, owned outright	34	7	16	56	18	29	29	10	10	56	19	28	29
Owner occupied, with mortgage	50	64	37	6	37	46	51	48	31	6	35	30	40
All owners	84	71	53	63	54	75	79	57	41	62	54	58	69
Rented from council*	7	7	13	22	20	10	10	20	25	23	23	21	14
Rented from housing association	3	3	7	10	11	5	5	8	13	10	12	10	7
Social sector tenants	10	11	20	32	31	14	15	28	38	32	35	30	20
Rented privately, unfurnished†	5	13	12	4	12	7	4	12	15	5	10	9	8
Rented privately, furnished	1	5	14	1	3	3	1	3	7	0	1	2	3
Private renters**	5	19	26	5	15	10	6	15	22	5	11	11	11
Weighted base (000's) =100%	*9,938*	*1,298*	*2,077*	*731*	*1,217*	*15,260*	*1,809*	*744*	*2,173*	*2,461*	*2,061*	*9,247*	*24,508*
Unweighted sample	*3679*	*439*	*605*	*277*	*387*	*5387*	*667*	*263*	*723*	*845*	*728*	*3226*	*8613*

*   Council includes local authorities, New Towns and Scottish Homes from 1996.

†   Unfurnished includes the answer 'partly furnished'.

**   From 1996 all tenants whose accommodation goes with the job of someone in the household have been allocated to 'rented privately'. Squatters are also included in the privately rented category.

Table 4.12 (a) Socio-economic classification and economic activity status of household reference person by tenure
(b) Tenure by socio-economic classification and economic activity status of household reference person

Household reference persons

Great Britain: 2002

Socio-economic classification and economic activity status of household reference person*		Tenure										
		Owners			Social sector tenants			Private renters			Total	
		Owned outright	With mortgage	All owners	Council†	Housing associa-tion	Social sector tenants	Unfurn-ished private**	Furnished private	Private Renters††		
(a)		%	%	%	%	%	%	%	%	%	%	
Economically active HRP:												
Large employers and higher managerial		2	8	5	0	0	0	4	4	4	4	
Higher professional		3	12	8	0	0	0	7	12	8	6	
Lower managerial and professional		9	29	21	4	6	4	17	21	18	17	
Intermediate		3	8	6	4	4	4	7	9	7	6	
Small employers and own account		6	9	8	2	3	3	6	4	5	7	
Lower supervisory and technical		4	12	9	4	4	4	9	6	9	8	
Semi-routine		4	8	6	9	11	9	11	4	9	7	
Routine		4	7	6	10	9	9	6	6	6	7	
Never worked and long-term unemployed		0	0	0	3	3	3	2	2	2	1	
Economically inactive HRP		64	7	31	64	61	63	31	32	31	37	
Weighted base (000's) =100%		*6,986*	*9,698*	*16,684*	*3,380*	*1,584*	*4,964*	*1,824*	*655*	*2,479*	*24,126*	
Unweighted sample		*2598*	*3418*	*6016*	*1130*	*537*	*1667*	*616*	*194*	*810*	*8493*	

		Owned outright	With mortgage	All owners	Council†	Housing associa-tion	Social sector tenants	Unfurn-ished private**	Furnished private	Private Renters††	*Weighted base (000's) = 100%*	*Un-weighted sample*
(b)												
Economically active HRP:												
Large employers and higher managerial	%	16	74	90	0	1	1	7	3	9	*1,014*	*370*
Higher professional	%	14	72	86	1	0	1	8	5	13	*1,564*	*559*
Lower managerial and professional	%	15	68	84	3	2	5	8	3	11	*4,110*	*1449*
Intermediate	%	17	57	74	9	4	13	9	4	13	*1,375*	*470*
Small employers and own account	%	27	57	84	5	3	8	6	2	8	*1,600*	*566*
Lower supervisory and technical	%	13	64	78	8	3	11	9	2	11	*1,863*	*628*
Semi-routine	%	17	43	60	17	10	27	11	2	13	*1,754*	*606*
Routine	%	18	44	62	20	9	29	7	3	9	*1,607*	*546*
Never worked and long-term unemployed	%	12	5	16	43	22	65	13	5	18	*250*	*81*
Economically inactive HRP	%	50	7	57	24	11	35	6	2	9	*8,988*	*3218*
Total	%	29	40	69	14	7	21	8	3	10	*24,126*	*8493*

* From April 2001 the National Statistics Socio-economic classification (NS-SEC) was introduced for all official statistics and surveys. It replaced Social Class based on occupation and Socio-economic Groups (SEG). Excludes full-time students and persons in inadequately described occupations.

† Council includes local authorities, New Towns and Scottish Homes from 1996.

** Unfurnished includes the answer 'partly furnished'.

†† From 1996 all tenants whose accommodation goes with the job of someone in the household have been allocated to 'rented privately'. Squatters are also included in the privately rented category.

Table 4.13   (a)   Length of residence of household reference person by tenure
(b)   Tenure by length of residence of household reference person

Household reference persons   **Great Britain: 2002**

Length of residence* (years)	Tenure									Total	
	Owners			Social sector tenants			Private renters				
	Owned outright	With mortgage	All owners	Council†	Housing associa-tion	Social sector tenants	Unfurn-ished private**	Furnished private	Private Renters††		
(a)	%	%	%	%	%	%	%	%	%	%	
Less than 12 months	3	9	7	9	12	10	31	[51]	36	10	
12 months but less than 2 years	2	7	5	6	7	6	14	18	[15]	6	
2 years but less than 3 years	3	8	6	8	12	9	[12]	11	12	7	
3 years but less than 5 years	5	14	10	11	17	13	11	7	10	11	
5 years but less than 10 years	10	[22]	17	[21]	[24]	[22]	11	7	10	[17]	
10 years or more	[77]	39	[55]	45	29	40	21	6	16	48	
Weighted base (000's) =100%	7,024	9,871	16,895	3,400	1,600	5,001	1,878	724	2,601	24,497	
Unweighted sample	2611	3474	6085	1137	542	1679	633	213	846	8610	

										Weighted base (000's) = 100%	Un-weighted sample
b)											
Less than 12 months	%	8	36	44	12	7	19	23	14	37	2,548	835
12 months but less than 2 years	%	11	45	55	12	7	20	17	8	25	1,583	528
2 years but less than 3 years	%	11	47	58	14	11	25	13	5	17	1,777	603
3 years but less than 5 years	%	13	52	65	14	10	24	8	2	10	2,627	908
5 years but less than 10 years	%	16	51	68	17	9	26	5	1	6	4,277	1511
10 years or more	%	46	33	79	13	4	17	3	0	4	11,684	4225
Total	%	29	40	69	14	7	20	8	3	11	24,497	8610

*   Boxed figures indicate median length of residence.
†   Council includes local authorities, New Towns and Scottish Homes from 1996.
** Unfurnished includes the answer 'partly furnished'.
†† From 1996 all tenants whose accommodation goes with the job of someone in the household have been allocated to 'rented privately'. Squatters are also included in the privately rented category.

Table 4.14   **Persons per room: 1971 to 2002**

Households   **Great Britain**

Persons per room	Unweighted								Weighted			
	1971	1975	1981	1985	1991	1995	1996	1998	1998	2000	2001	2002
	%	%	%	%	%	%	%	%	%	%	%	%
Under 0.5	37	39	42	45	50	52	51	55	55	57	57	57
0.5 to 0.65	25	25	25	26	24	25	24	23	23	22	22	23
0.66 to 0.99	24	23	23	21	19	18	19	18	18	16	16	16
1	9	8	7	6	5	5	5	4	4	4	4	3
Over 1 to 1.5	4	3	2	1	1	1	1	1	1	1	1	1
Over 1.5	1	0	0	0	0	0	0	0	0	0	0	0
*Weighted base (000's) = 100%**									24,450	24,845	24,592	24,529
*Unweighted sample**	11990	12096	12002	9982	9646	9754	9154	8636		8221	8989	8620
Mean persons per room	..	0.57	0.56	0.52	0.50	0.48	0.49	0.47	0.46	0.45	0.46	0.45

*   Trend tables show unweighted and weighted figures for 1998 to give an indication of the effect of the weighting. For the weighted data (1998 and 2000 to 2002) the weighted base (000's) is the base for percentages. Unweighted data (up to 1998) are based on the unweighted sample.

Table 4.15   **Persons per room and mean household size by tenure**

Households   Great Britain: 2002

Persons per room*	Tenure									Total
	Owners			Social sector tenants			Private renters			
	Owned outright	With mortgage	All owners	Council†	Housing association	Social sector tenants	Unfur- nished private**	Furnished private	Private renters††	
	%	%	%	%	%	%	%	%	%	%
Under 0.5	78	48	61	51	51	51	52	35	48	57
0.5 to 0.65	16	27	23	23	24	23	25	30	26	23
0.66 to 0.99	5	21	14	17	18	17	18	25	20	16
1	0	3	2	6	5	6	4	8	5	3
Over 1	0	1	1	3	2	2	1	2	1	1
Weighted base (000's) =100%	7,027	9,871	16,898	3,404	1,600	5,004	1,878	728	2,606	24,508
Unweighted sample	2612	3474	6086	1138	542	1680	633	214	847	8613
Mean persons per room	0.35	0.49	0.43	0.50	0.49	0.50	0.46	0.54	0.49	0.45
Mean household size	1.91	2.71	2.38	2.19	2.07	2.15	2.09	2.25	2.13	2.31

*   Boxed figures indicate median density of occupation.
†   Council includes local authorities, New Towns and Scottish Homes from 1996.
**   Unfurnished includes the answer 'partly furnished'.
††   From 1996 all tenants whose accommodation goes with the job of someone in the household have been allocated to 'rented privately'.
    Squatters are also included in the privately rented category.

Table 4.16   **Closeness of fit relative to the bedroom standard by tenure**

Households   Great Britain: 2002

Difference from bedroom standard (bedrooms)	Tenure									Total
	Owners			Social sector tenants			Private renters			
	Owned outright	With mortgage	All owners	Council*	Housing association	Social sector tenants	Unfur- nished private†	Furnished private	Private renters**	
	%	%	%	%	%	%	%	%	%	%
1 or more below standard	1	2	1	6	4	5	3	4	3	2
Equals standard	9	21	16	48	58	51	37	53	41	26
1 above standard	35	43	40	34	29	32	41	29	38	38
2 or more above standard	55	35	43	13	9	11	19	13	18	34
Weighted base (000's) = 100%	7,027	9,871	16,898	3,404	1,600	5,004	1,878	728	2,606	24,508
Unweighted sample	2612	3474	6086	1138	542	1680	633	214	847	8613

*   Council includes local authorities, New Towns and Scottish Homes.
†   Unfurnished includes the answer 'partly furnished'.
**   From 1996 all tenants whose accommodation goes with the job of someone in the household have been allocated to 'rented privately'. Squatters are also included in the
    privately rented category.

Table 4.17   Cars or vans: 1972 to 2002

Households   Great Britain

Cars or vans	Unweighted								Weighted			
	1972	1975	1981	1985	1991	1995	1996	1998	1998	2000	2001	2002
	%	%	%	%	%	%	%	%	%	%	%	%
Households with:												
no car or van	48	44	41	38	32	29	30	28	28	27	28	27
one car or van	43	45	44	45	44	45	46	44	45	45	44	45
two cars or vans	8	10	12	14	19	22	21	23	22	22	23	22
three or more cars or vans	1	1	2	3	4	4	4	6	6	6	5	5
*Weighted base (000's) = 100%**									24,450	24,845	24,592	24,529
*Unweighted sample**	11624	11929	11989	9963	9910	9758	9158	8636		8221	8989	8620

*   Trend tables show unweighted and weighted figures for 1998 to give an indication of the effect of the weighting. For the weighted data (1998 and 2000 to 2002) the weighted
    base (000's) is the base for percentages. Unweighted data (up to 1998) are based on the unweighted sample.

Table 4.18   Availability of a car or van by socio-economic classification of household reference person

Households   Great Britain: 2002

Socio-economic classification of household reference person*		Number of cars or vans available to household			Weighted base (000's) = 100%	Unweighted sample
		None	1	2 or more		
Economically active HRP						
Large employers and higher managerial	%	4	35	61	1,014	370
Higher professional	%	9	41	50	1,566	560
Lower managerial and professional	%	9	47	43	4,110	1449
Intermediate	%	15	54	31	1,375	470
Small employers and own account	%	5	41	54	1,603	567
Lower supervisory and technical	%	11	57	32	1,863	628
Semi-routine	%	28	52	21	1,754	606
Routine	%	24	52	24	1,607	546
Never worked and long-term unemployed	%	62	36	1	250	81
Economically inactive HRP	%	50	42	9	9,001	3222
Total	%	27	45	27	24,144	8499

*   From April 2001 the National Statistics Socio-economic Classification (NS-SEC) was introduced for all official statistics and surveys. It replaced Social Class based on Occupation and
    Socio-economic Groups (SEG). Excludes full-time students and persons in inadequately described occupations.

Table 4.19   Consumer durables, central heating and cars: 1972 to 2002

Households   **Great Britain**

	Unweighted								Weighted			
	1972	1975	1981	1985	1991	1995	1996	1998	1998	2000	2001	2002
Percentage of households with:												
Television (total)	93	96	97	98	98	98	99	98	98	99	99	99
colour	93	96	74	86	95	97	97	98	97	98	98	99
black and white only			23	11	4	2	2	1	1	1	0	0
satellite/cable/digital	..	..	..	..	..	..	18	29	29	40	42	44
satellite*	..	..	..	..	..	..	..	..	..	..	..	26
cable*	..	..	..	..	..	..	..	..	..	..	..	14
digital*	..	..	..	..	..	..	..	..	..	..	..	28
Video recorder	..	..	..	31	68	79	82	85	85	88	88	89
DVD player	..	..	..	..	..	..	..	..	..	..	..	32
CD player	..	..	..	..	27	52	58	68	69	77	80	83
Home computer	..	..	..	13	21	25	27	34	34	45	50	54
Access to internet at home	..	..	..	..	..	..	..	..	..	33	40	44
Access from home computer	..	..	..	..	..	..	..	..	..	31	37	42
Other access	..	..	..	..	..	..	..	..	..	2	6	5
Microwave oven	..	..	..	..	55	70	74	79	78	83	85	87
Refrigerator†	73	88	93	95	..	..	..	..	..	..	..	..
Deep freezer†	..	..	49	66	83	89	91	93	92	93	94	95
Washing machine	66	71	78	81	87	90	90	92	91	93	92	93
Tumble drier	..	..	23	33	48	51	51	52	51	54	54	54
Dishwasher	..	..	4	6	14	20	20	24	23	26	28	28
Telephone (fixed or mobile)	42	54	75	81	88	93	94	96	96	98	98	99
fixed telephone**	..	..	..	..	..	..	..	..	..	93	93	92
mobile telephone**	..	..	..	..	..	..	..	..	..	58	70	75
Central heating	37	43	59	69	82	86	88	90	90	92	92	93
A car or van (total)	52	56	59	62	67	71	70	72	72	73	72	73
A car or van	43	45	44	45	44	45	46	44	45	45	44	45
- more than 1	9	11	14	17	23	26	24	28	27	28	28	28
Weighted base (000's) = 100%††									24,450	24,575	24,592	24,529
Unweighted sample††	11663	11929	11718	9993	9955	9757	9156	8636		8213	8984	8618

*   Data only available from 2002.
†   Fridge freezers are attributed to both 'refrigerator' and 'deep freezer' from 1979 on.
**   Data only available from 2000. Percentages for fixed and mobile phones sum to greater than 100% because some households owned both.
††   Trend tables show unweighted and weighted figures for 1998 to give an indication of the effect of the weighting. For the weignted data (1998 and 2000 to 2002) the weighted base (000's) is the base for percentages. Unweighted data (up to 1998) are based on the unweighted sample.

Table 4.20   **Consumer durables, central heating and cars by socio-economic classification of household reference person**

Household reference persons   Great Britain: 2002

Consumer durables	Socio-economic classification of household reference person*								Economically inactive	Total
	Economically active									
	Large employers and higher managerial	Higher professional	Lower managerial and professional	Intermediate	Small employers and own account	Lower supervisory and technical	Semi-routine	Routine		
Percentage of households with:										
Television										
colour	100	98	99	99	99	100	98	99	98	99
black and white only	0	0	0	0	0	0	1	1	1	0
satellite	30	28	32	28	39	36	28	30	16	26
cable	17	17	16	16	15	17	17	14	11	14
digital	37	30	35	30	38	41	32	30	17	28
Video recorder	97	94	95	95	93	96	93	93	81	89
DVD player	60	49	43	38	43	48	36	37	12	32
CD player	98	97	96	94	90	93	87	88	65	82
Home computer	89	88	80	65	67	64	53	47	26	53
Access to internet at home	82	83	70	52	58	51	36	33	19	44
Microwave oven	93	89	89	89	91	94	93	92	82	87
Deep freezer/fridge freezer	97	96	97	96	96	97	96	96	94	95
Washing machine	99	98	98	95	97	96	95	96	88	93
Tumble drier	71	63	61	60	63	59	56	56	45	54
Dishwasher	59	56	43	28	44	27	19	18	15	28
Telephone (fixed or mobile)	100	100	100	99	100	100	99	99	98	99
fixed telephone†	98	98	96	92	95	93	89	85	92	92
mobile telephone†	93	92	91	90	88	88	83	81	52	75
Central heating	100	98	95	92	93	92	92	90	92	93
Car or van - more than one	61	50	43	31	54	32	21	24	9	27
Weighted base (000's)										
=100%	1,014	1,566	4,110	1,375	1,603	1,863	1,754	1,607	8,996	24,139
Unweighted sample	370	560	1449	470	567	628	606	546	3220	8497

* From April 2001 the National Statistics Socio-economic classification (NS-SEC) was introduced for all official statistics and surveys. It replaced Social Class based on Occupation and Socio-economic Groups (SEG). Excludes full-time students and persons in inadequately described occupations.

† Percentages for fixed and mobile telephones sum to greater than 100 because some households owned both.

Table 4.21   **Consumer durables, central heating and cars by usual gross weekly household income**

Households   Great Britain: 2002

Consumer durables	Usual gross weekly household income (£)										
	0.01-100	100.01-150	150.01-200	200.01-250	250.01-300	300.01-350	350.01-400	400.01-450	450.01-500	500.01 or more	Total*
Percentage of households with:											
Television											
colour	96	98	98	98	99	99	99	98	99	100	99
black and white only	1	1	1	1	0	0	0	0	0	0	0
satellite	13	16	15	16	20	23	27	31	30	35	26
cable	10	10	13	13	10	13	19	15	18	16	14
digital	14	15	16	19	24	25	34	33	34	37	28
Video recorder	73	78	83	85	91	93	92	95	94	97	89
DVD player	16	14	15	18	20	25	34	34	37	48	32
CD player	61	62	66	70	79	86	88	90	94	96	83
Home computer	25	22	26	35	35	47	52	55	58	81	54
Access to internet at home	17	15	16	24	25	33	40	42	45	71	44
Microwave oven	78	82	83	87	87	88	90	89	90	92	87
Deep freezer/fridge freezer	88	93	94	94	96	97	96	96	98	98	95
Washing machine	80	88	89	91	92	93	96	97	98	99	93
Tumble drier	36	39	44	44	50	49	55	57	60	67	54
Dishwasher	9	8	9	15	15	19	18	26	26	48	28
Telephone	95	97	99	99	99	99	99	99	99	100	99
fixed telephone†	81	85	91	90	91	93	91	94	95	97	92
mobile telephone†	47	49	53	65	70	71	80	83	90	92	75
Central heating	89	90	91	91	90	91	93	92	93	96	93
Car or van - more than one	6	3	4	7	10	17	15	22	26	51	28
Weighted base (000's)											
=100%	2,059	2,018	1,708	1,430	1,278	1,149	1,077	1152	1,041	8,666	24,529
Unweighted sample	692	705	607	505	447	400	373	409	364	3087	8618

*   Total includes no answers to income.

†   Percentages for fixed and mobile telephones sum to greater than 100 because some households owned both.

Table 4.22   **Consumer durables, central heating and cars by household type**

Households   **Great Britain: 2002**

Consumer durables	Household type							
	1 adult aged 16-59	2 adults aged 16-59	Small family	Large family	Large adult household	2 adults, 1 or both aged 60 or over	1 adult aged 60 or over	Total
Percentage of households with:								
Television								
colour	96	99	99	99	100	99	98	99
black and white only	1	0	0	0	0	0	1	0
satellite	17	31	35	42	37	20	7	26
cable	15	15	17	16	20	11	8	14
digital	21	35	35	40	40	21	9	28
Video recorder	84	94	97	96	97	92	68	89
DVD player	28	43	46	49	54	11	2	32
CD player	84	95	96	92	95	76	42	83
Home computer	49	69	73	76	79	33	9	54
Access to internet at home	39	60	59	57	66	27	6	44
Microwave oven	84	89	93	94	94	87	75	87
Deep freezer/fridge freezer	89	97	98	99	99	98	90	95
Washing machine	87	98	98	99	99	96	80	93
Tumble drier	37	58	65	74	72	54	34	54
Dishwasher	15	35	35	37	45	27	9	28
Telephone (fixed or mobile)	97	100	99	99	100	100	97	99
fixed telephone*	82	92	91	88	96	99	96	92
mobile telephone*	76	90	90	88	91	65	28	75
Central heating	90	94	95	92	95	93	91	93
Car or van - more than one	4	44	36	39	58	19	1	28
Weighted base (000's)								
=100%	*3,891*	*4,019*	*4,715*	*1,204*	*3,281*	*3,774*	*3,645*	*24,529*
Unweighted sample	*1179*	*1445*	*1684*	*457*	*1108*	*1451*	*1294*	*8618*

*   Percentages for fixed and mobile telephones sum to greater than 100 because some households owned both.

Table 4.23   **Consumer durables, central heating and cars: lone-parent families
compared with other families**

Families with dependent children*   **Great Britain: 2001 and 2002 combined**

Consumer durables	Lone-parent families	Other families
Percentage of households with:		
Television		
colour	99	100
black and white	0	0
satellite/cable/digital	48	62
Video recorder	94	98
CD player	92	96
Home computer	56	80
Access to internet at home	36	68
Microwave oven	90	94
Deep freezer/fridge freezer	96	99
Washing machine	97	99
Tumble drier	58	72
Dishwasher	17	46
Telephone (fixed or mobile)	98	100
fixed telephone†	81	95
mobile telephone†	83	91
Central heating	92	96
Car or van - one or more	52	93
*Unweighted sample***	*1283*	*3873*

*   Dependent children are persons aged under 16, or aged 16-18 and in full time education, and living in the household.

†   Percentages for fixed and mobile telephones sum to greater than 100 because some households owned both.

**   Weighted base not shown for combined data sets.

Chapter 5 Marriage and cohabitation

The General Household Survey (GHS) is a key source of information on marriage and cohabitation. It has collected details on current marital status since 1971, since when the questions have been extended to include cohabitation. This chapter looks at the characteristics of the marital status groups (i.e. single, married, cohabiting, divorced, separated or widowed) in 2002 and the trends in the data.

- Development of the marriage and cohabitation questions

- De facto marital status

- Current cohabitation

  Current cohabitation by legal marital status and age

  Current cohabitation and trends over time among women

- Women aged 16 to 59 and dependent children in the household

  Dependent children in the household and current cohabitation among women

- Past cohabitations not ending in marriage

  Past cohabitations by current marital status

  First cohabitations not ending in marriage

  Duration of past cohabitations not ending in marriage

Development of the marriage and cohabitation questions

The marriage and cohabitation questions on the GHS have been developed over time to reflect changes in society.

- In 1979 questions on marital history were introduced for both men and women. Questions were also introduced for women aged 18 to 49 relating to pre-marital cohabitation before the current or most recent marriage.
- In 1986 these questions[1] on pre-marital cohabitation were extended to both men and women aged 16 to 59 and every marriage past and present.
- In 1998 a further question[2] was added to find out the number of past cohabitations not ending in marriage.
- In 2000 new questions[3] were included on the length of past cohabitations not ending in marriage. These included the start and end dates of cohabitations and what people perceived to be the end of the cohabitation (the end of the relationship, the end of sharing the accommodation or both).

We collect marital status information in two stages. First, the marital status of all adults aged 16 and over is collected from the person answering the household questionnaire (usually the household reference person or their partner). At the second stage, each household member aged 16 to 59 is asked detailed questions about their marriage and cohabitation history. If interviewers judge that lack of privacy may affect reporting, they can offer respondents a self-completion questionnaire. In 2002, 4% of respondents chose this option (unweighted data).

De facto marital status

De facto marital status (that is, including cohabitation) has been taken as the legal marital status of the respondent as recorded at the beginning of the interview unless the respondent was currently cohabiting in which case cohabiting is the de facto status. Cohabiting couples are people who live together as a couple in a household without being married to each other. We have here

defined single, widowed, divorced or separated respondents who were cohabiting as cohabiting rather than by their legal marital status. Those who were not cohabiting have been classified by their marital status.

In 2002, the de facto marital status of adults aged 16 and over was as follows:
- 54% of men and 50% of women were married;
- 10% of men and 9% of women were cohabiting;
- 26% of men and 20% of women were single;
- 4% of men and 11% of women were widowed;
- 6% of men and 9% of women were either divorced or separated.   **Table 5.1**

There were differences in the de facto marital status groups by sex and age.
- In 2002, women aged 16 to 24 were more likely to be living together as a couple (married or cohabiting) than men in the same age group. Among those aged 16 to 24, 7% of women were married compared with 2% of men, and 16% of women were cohabiting compared with 10% of men.
- Among people aged 35 to 44, there was no statistically significant difference in the percentage of men and women who were married (64%) or who were cohabiting (12% of men and 10% of women).
- Women aged 75 and over were more likely to be widowed than men (62% compared with 28% of men aged 75 and over).   **Table 5.2**

Current cohabitation

Individuals who described themselves as 'separated' were, strictly speaking, legally married. However, because men and women who are separated can cohabit, they have been included in the 'non-married' category in the following analyses.

In 2002:
- overall 12% of both men and women aged 16 to 59 were cohabiting[4];
- among women aged 16 to 59, those in their twenties were more likely to be cohabiting than any other age group (25% of women aged 20 to 24 and 26% of women aged 25 to 29 compared with 4% to 16% in the other age groups);
- a quarter (25%) of men aged 25 to 29 were cohabiting compared with 3% to 20% in the other age groups;
- overall among non-married people aged 16 to 59, a quarter (25%) of both men and women were cohabiting.   **Table 5.3**

Current cohabitation by legal marital status and age

> Data from 2001 and 2002 were combined to provide a large enough sample to analyse cohabitation by legal marital status[5] and age.

- Among men aged 16 to 59, those who were divorced were more likely than those who were single or separated to be cohabiting (33% compared with 23% of single men and 21% of separated men).
- Of women aged 16 to 59 there was little difference between the proportions of divorced and single women who were cohabiting (29% and 28% respectively).   **Table 5.4**

Analysis of people aged 16 to 59 who were cohabiting showed:

- about three-quarters of all cohabiters were single (76% of men and 72% of women) and about a fifth were divorced (19% of men and 22% of women);
- older cohabiters were more likely to be divorced (among those aged 50 to 59, 64% of men and 76% of women);
- the majority of young people – those aged 16 to 24 - who were cohabiting were single (99% of both men and women).   **Tables 5.5–5.6**

Current cohabitation and trends over time among women

The development of the marriage and cohabitation questions on the GHS means that the longest time series is for women aged 18 to 49. Tables 5.7 and 5.8 look at the trends for this group.

Among women aged 18 to 49:
- since 1979 the proportion of women who were married has continuously declined (from 74% in 1979, to 61% in 1991 and to 49% in 2002);
- the proportion of women who were single has more than doubled from 18% in 1979 to 38% in 2002.
                                                      Table 5.7

In terms of current cohabitation, among women aged 18 to 49:
- the percentage of non-married women who were cohabiting increased from 11% in 1979 to 29% in 2002;
- the percentage of single women cohabiting increased from 8% in 1979 to 23% in 1991 and 31% in 2002.
                                           Table 5.8, Figure 5.A

Figure 5A   **Percentage of single, divorced and separated women* aged 18–49 cohabiting by legal marital status, 1979 to 2002**

Great Britain
Percentage of woman aged 18-49

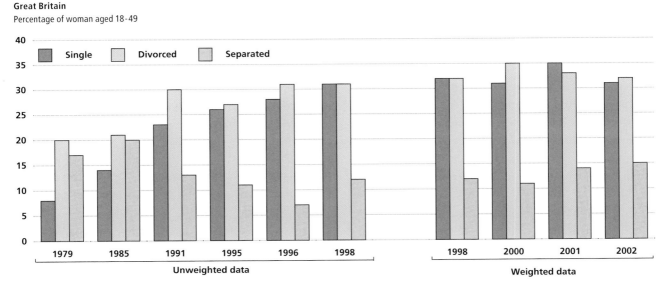

* Widows have not been included because their numbers are so small

Women aged 16 to 59 and dependent children in the household

In 2002 49% of women aged 16 to 59 had no children living with them, 42% had at least one dependent child and 9% had only non-dependent children living with them. The proportion of women who had dependent children living with them varied by marital status.

Among women aged 16 to 59:
- over half of both married women (52%) and separated women (60%) had at least one dependent child living with them compared with 11% of widows, 23% of single women, 36% of cohabiting women and 44% of divorced women;
- 76% of single women and 71% of widows had no children living with them compared with 61% of cohabiting women, 45% of divorced women, 35% of married women and 29% of separated women;
- 18% of widows and 11% of both divorced and separated women had at least one non-dependent child living with them;
- nearly two thirds (64%) of women with at least one dependent child were married, 13% were single and 11% were cohabiting.   **Table 5.9**

Dependent children in the household and current cohabitation among women

In 2002, as in previous years, non-married women aged 16 to 59 who had dependent children living with them were more likely than those without dependent children to be cohabiting.

Among women aged 16 to 59:
- 30% of non-married women who had at least one dependent child living with them were cohabiting, compared with 23% of non-married women without dependent children;
- over a third (36%) of single women who had dependent children living with them were cohabiting, compared with 24% of single women without dependent children.   **Table 5.10**

Among women aged 16 to 59 who were currently cohabiting:
- of those who had at least one dependent child living with them, over two-thirds (69%) were single, 25% were divorced and the remainder (6%) were either separated or widowed;
- of those with no dependent children, 75% were single and 20% were divorced.   **Table 5.11**

Past cohabitations not ending in marriage

In 1998 the GHS for the first time asked men and women aged 16 to 59 about the number of past cohabitations they had had which did not end in marriage. Since 2000, the questions were extended to include the length of these past cohabitations[6] not ending in marriage. These periods of completed cohabitation did not include the current relationship of a respondent living as a couple at the time of interview.

With the exception of those who chose the self-completion option, the majority of married and cohabiting respondents were interviewed in the presence of their partner. Therefore, it is possible that previous cohabitations may be under-reported for these groups.

In 2002 among adults aged 16 to 59:
- 15% had at least one completed cohabitation not ending in marriage;
- 12% had one of these relationships, 3% had two and 1% had three or more;
- there was no statistically significant difference between the proportions of men and women who had at least one completed cohabitation not ending in marriage (15% of men and 16% of women);
- the highest proportions reporting at least one such relationship were among adults aged 25 to 39 (between 22% and 27% compared with between 5% and 13% of those aged 45 to 59).   **Table 5.12**

Past cohabitations by current marital status

As in previous years, the proportions reporting past cohabitations not ending in marriage varied by current marital status for both men and women. Married people were less likely than other respondents to report such relationships.

Among adults aged 16 to 59:
- 85% of men and 84% of women had never had a past cohabitation that did not end in marriage;
- 9% of married men said they had had at least one past cohabitation not ending in marriage, compared with 23% of currently cohabiting men (for women, these proportions were 7% and 18% respectively).**Table 5.13**

First cohabitations not ending in marriage

Overall, 70% of first cohabitations not ending in marriage began by the age of 25.

Among adults aged 16 to 59 who have cohabited:
- first cohabitations not ending in marriage were most likely to begin when people were aged 20 to 24 (43%);
- women were twice as likely as men to have started a first cohabitation which did not end in marriage at a younger age (35% of women began such a relationship by the age of 19 compared with 17% of men).

The age at which people started their first completed cohabitation also varied with the year in which the cohabitation began. Adults starting their first cohabitation in the 1960-70s tended to be younger than those who started their first completed cohabitation in later decades. For cohabitations started in the 1960s to 1970s, 38% of adults started a first cohabiting union by the age of 19 compared with 25% between 1990 and 2002. **Table 5.14**

Duration of past cohabitations not ending in marriage

Respondents were asked the duration of each completed cohabitation and gave either the start date or the end date. The date that had not been given was then calculated and checked with the respondent, who could disagree and correct it. Length of cohabitation was based on the corrected dates where required.

Among adults aged 16 to 59 who have cohabited:
- the mean length of time for the first cohabitation not ending in marriage was 39 months compared with 33 months for the second cohabitation;
- first cohabitations that were the only past cohabitation tended to be longer than those that were the first of two or more (the mean length of the first and only completed cohabitation was 41 months compared with 31 months for a cohabitation that was the first of two or more);
- women were more likely to have a longer mean length of first cohabitation than men (41 months compared with 36 months for a cohabitation that was the first of at least one such relationship). **Table 5.15**

Notes and references

1 These were developed in conjunction with John Haskey, Office for National Statistics. The results of this work were reported in *Population Trends, 68*. The Stationery Office (London, Summer 1992).

2 This question was added following John Haskey's work on a module of questions on cohabitional and marriage histories for the Omnibus Survey. The results of this work were reported in *Population Trends, 96*. The Stationery Office (London, Summer 1999).

3 These were developed in conjunction with John Haskey, Office for National Statistics. The results of this work were reported in *Survey Methodology Bulletin*. No 46. January 2000 (Lilly R. Developing questions on cohabitation histories) and *Population Trends, 103*. The Stationery Office (London, Spring 2001).

4 'Cohabiting' includes same sex cohabitees.

5 The section on marital history identified cases where the current or most recent marriage was in fact a cohabitation and also ascertained for cases where the spouse was not listed as a household member whether the marriage had broken down. This additional information was used to derive a modified version of marital status and it is this variable which is used in the rest of the chapter. Less than 1% of respondent's aged 16 to 59 were classified differently as a result of this exercise.

6 In 1998 the question about previous cohabitations not ending in marriage specified cohabitations with 'someone of the opposite sex'. This part of the wording was dropped in 2000 and same sex couples were not specifically identified in the analysis.

Table 5.1   **Sex by marital status**

All persons aged 16 and over   Great Britain: 2002

Marital status*	Men	Women
	%	%
Married	54	50
Cohabiting	10	9
Single	26	20
Widowed	4	11
Divorced	5 ⎤ 6	7 ⎤ 9
Separated	2 ⎦	3 ⎦
Weighted base (000's) = 100%	*21,746*	*23,488*
Unweighted sample†	*7579*	*8393*

* Marital status as recorded at the beginning of the interview.
† Total includes a very small number of same sex cohabitees.

Table 5.2 (a) Age by sex and marital status
(b) Marital status by sex and age

Persons aged 16 and over
Great Britain: 2002

(a)

Age	Married	Cohabiting	Single	Widowed	Divorced	Separated	Total
	%	%	%	%	%	%	%
Men							
16-24	1	15	49	0	0	3	15
25-34	12	39	25	1	2	18	17
35-44	23	25	13	1	21	26	20
45-54	21	12	7	2	36	23	17
55-64	20	6	3	14	23	16	14
65-74	14	2	2	27	12	9	10
75 and over	8	1	1	55	5	4	7
Weighted base (000's) =100%	*11,844*	*2,080*	*5,623*	*783*	*1,009*	*330*	*21,746*
Unweighted sample	*4379*	*711*	*1739*	*297*	*327*	*102*	*7579*
	%	%	%	%	%	%	%
Women							
16-24	2	24	51	0	0	4	13
25-34	14	37	25	0	6	24	17
35-44	24	22	11	1	26	31	19
45-54	23	11	5	4	26	23	16
55-64	19	4	3	11	22	11	13
65-74	12	1	2	27	14	4	11
75 and over	6	1	3	58	6	2	10
Weighted base (000's) =100%	*11,854*	*2,088*	*4,687*	*2,638*	*1,575*	*601*	*23,488*
Unweighted sample	*4383*	*714*	*1599*	*908*	*558*	*213*	*8393*
	%	%	%	%	%	%	%
Total							
16-24	1	20	50	0	0	4	14
25-34	13	38	25	0	5	22	17
35-44	24	23	12	1	24	29	19
45-54	22	11	6	3	30	23	17
55-64	20	5	3	11	23	13	14
65-74	13	2	2	27	13	6	11
75 and over	7	1	2	57	6	3	9
Weighted base (000's) =100%	*23,697*	*4,167*	*10,309*	*3,423*	*2,582*	*929*	*45,230*
Unweighted sample†	*8762*	*1425*	*3338*	*1205*	*885*	*315*	*15972*

(b)

Age		Married	Cohabiting	Single	Widowed	Divorced	Separated	Weighted base (000's) = 100%	Unweighted sample†
Men									
16-24	%	2	10	87	0	0	0	*3,154*	*982*
25-34	%	37	22	38	0	1	2	*3,721*	*1204*
35-44	%	64	12	16	0	5	2	*4,304*	*1455*
45-54	%	70	7	10	0	10	2	*3,640*	*1318*
55-64	%	77	4	6	3	8	2	*3,143*	*1176*
65-74	%	76	2	5	9	6	1	*2,248*	*849*
75 and over	%	63	1	4	28	3	1	*1,536*	*595*
Total	%	54	10	26	4	5	2	*21,746*	*7579*
Women									
16-24	%	7	16	76	0	0	1	*3,133*	*1041*
25-34	%	44	20	30	0	2	4	*3,909*	*1353*
35-44	%	64	10	12	0	9	4	*4,471*	*1567*
45-54	%	71	6	6	3	10	4	*3,837*	*1444*
55-64	%	71	3	4	9	11	2	*3,137*	*1231*
65-74	%	58	1	4	28	8	1	*2,544*	*917*
75 and over	%	28	1	5	62	4	1	*2,457*	*840*
Total	%	50	9	20	11	7	3	*23,488*	*8393*
Total									
16-24	%	4	13	82	0	0	1	*6,285*	*2023*
25-34	%	40	21	34	0	2	3	*7,628*	*2557*
35-44	%	64	11	14	0	7	3	*8,775*	*3022*
45-54	%	71	6	8	1	10	3	*7,477*	*2762*
55-64	%	74	3	5	6	9	2	*6,279*	*2407*
65-74	%	66	1	5	19	7	1	*4,792*	*1766*
75 and over	%	41	1	4	49	4	1	*3,994*	*1435*
Total	%	52	9	23	8	6	2	*45,230*	*15972*

* Marital status as recorded at the beginning of the interview.
† Total includes a very small number of same sex cohabitees.

Table 5.3   Percentage currently cohabiting by sex and age

Men and women aged 16-59   Great Britain: 2002

Age	All	Non-married*		Weighted base (000's) = 100%		Unweighted sample	
				All	Non-married*	All	Non-married*
	Percentage cohabiting						
Men							
16-19	3	3		1,088	1,088	348	348
20-24	20	21		1,362	1,306	413	395
25-29	25	35		1,423	1,031	449	318
30-34	20	38		1,837	987	603	306
35-39	13	33		1,939	768	656	238
40-44	10	32		1,884	615	630	188
45-49	8	23		1,578	550	567	180
50-54	6	21		1,650	450	604	148
55-59	4	17		1,694	431	626	140
Total	12	25		14,455	7,226	4896	2261
Women							
16-19	7	7		1,116	1,109	386	384
20-24	25	28		1,550	1,366	498	440
25-29	26	39		1,601	1,049	545	355
30-34	16	33		2,044	1,010	717	351
35-39	10	27		2,161	793	768	275
40-44	11	31		2,056	735	709	247
45-49	8	25		1,859	582	695	210
50-54	4	16		1,768	458	670	168
55-59	4	14		1,693	514	642	183
Total	12	25		15,848	7,616	5630	2613

* Men and women describing themselves as 'separated' were, strictly speaking, legally married. However, because the separated can cohabit, they have been included in the 'non-married' category.

Table 5.4   Percentage currently cohabiting by legal marital status and age

Men and women aged 16-59   Great Britain: 2001 and 2002 combined

Legal marital status*	16-24	25-34	35-49	50-59	Total	Unweighted sample†				
						16-24	25-34	35-49	50-59	Total
	Percentage cohabiting									
Men										
Married	-	-	-	-	-	41	903	2581	1852	5377
Non-married										
Single	11	36	31	10	23	1488	1250	698	206	3642
Widowed	0 [11]	** [36]	** [30]	[10] [18]	16 [24]	0	3	14	44	61
Divorced	**	51	34	27	33	2	57	374	262	695
Separated	**	[27]	20	19	21	4	42	108	66	220
Total	11	23	11	5	12	1535	2255	3775	2430	9995
Women										
Married	-	-	-	-	-	123	1208	2896	1973	6200
Non-married										
Single	19	41	32	10	28	1647	1208	596	123	3574
Widowed	** [19]	** [39]	13 [28]	7 [15]	8 [26]	1	3	51	148	203
Divorced	**	41	30	22	29	3	158	591	366	1118
Separated	**	17	13	5	12	13	117	241	88	459
Total	18	22	9	4	13	1787	2694	4375	2698	11554

* Men and women describing themselves as 'separated' were, strictly speaking, legally married. However, because the separated can cohabit they have been included in the 'non-married' category.

† Weighted bases not shown for combined data sets.

** Base too small to enable reliable analysis to be made.

Table 5.5   Cohabiters: age by legal marital status

Cohabiting persons aged 16-59

Great Britain: 2001 and 2002 combined

Legal marital status*	16-24	25-34	35-49	50-59	Total
Men	%	%	%	%	%
Married	-	-	-	-	-
Non-married					
Single	99	92	59	20	76
Widowed	0	0	1	4	1
Divorced	1	6	34	64	19
Separated	1	2	6	12	4
Unweighted sample†	*163*	*536*	*411*	*126*	*1236*
Women					
Married	-	-	-	-	-
Non-married					
Single	99	86	47	11	72
Widowed	0	0	2	9	1
Divorced	1	11	43	76	22
Separated	0	3	8	4	4
Unweighted sample†	*292*	*554*	*405*	*118*	*1369*

* Men and women describing themselves as 'separated' were, strictly speaking, legally married. However, because the separated can cohabit they have been included in the 'non-married' category.

† Weighted bases not shown for combined data sets.

Table 5.6   Cohabiters: age by sex

Cohabiting persons aged 16-59

Great Britain: 2002

Age	Men	Women
	%	%
16-19	2	4
20-24	15	20
25-29	20	21
30-34	21	17
35-39	14	11
40-44	11	12
45-49	7	7
50-54	5	4
55-59	4	4
Weighted base (000's) = 100%	*1,782*	*1,937*
Unweighted sample	*601*	*658*

Table 5.7   Legal marital status of women aged 18-49: 1979 to 2002

Women aged 18-49   **Great Britain**

Legal marital status*	Unweighted									Weighted			
	1979	1983	1985	1989	1991	1993	1995	1996	1998	1998	2000	2001	2002
	%	%	%	%	%	%	%	%	%	%	%	%	%
Married	74	70	68	63	61	59	58	57	53	53	51	50	49
Non-married													
Single	18	21	22	26	26	28	28	29	30	32	35	36	38
Widowed	1	1	1	1	1	1	1	1	1	1	0	1	0
Divorced	4	6	6	7	8	9	9	9	11	10	9	9	9
Separated	3	2	3	3	3	4	4	4	5	4	5	4	4
Weighted base (000's) = 100%†										11,827	11,946	11,689	11,752
Unweighted sample†	6006	5285	5364	5483	5359	5171	4953	4695	4181		3979	4325	4092

* Men and women describing themselves as 'separated' were, strictly speaking, legally married. However, because the separated can cohabit they have been included in the 'non-married' category.

† Trend tables show unweighted and weighted figures for 1998 to give an indication of the effect of the weighting. For the weighted data (1998 and 2000 to 2002) the weighted base (000's) is the base for percentages. Unweighted data (up to 1998) are based on the unweighted sample.

Table 5.8   Percentage of women aged 18-49 cohabiting by legal marital status: 1979 to 2002

Women aged 18-49   **Great Britain**

Legal marital status*	Unweighted							Weighted			
	1979	1985	1991	1993	1995	1996	1998	1998	2000	2001	2002
					Percentage cohabiting						
Married	-	-	-	-	-	-	-	-	-	-	-
Non-married											
Single	8	14	23	23	26	28	31	32	31	35	31
Widowed	0	5	2	[8]	[8]	[5]	[8]	[11]	[15]	[11]	**
Divorced	20	21	30	25	27	31	31	32	35	33	32
Separated	17	20	13	11	11	11	7	12	11	14	15
	11	16	23	22	25	26	29	30	30	32	29
Total	3	5	9	9	10	11	13	14	15	16	15
Weighted bases (000's) = 100%†											
Married								6,212	6,051	5,899	5,727
Non-married											
Single								3,760	4,176	4,155	4,458
Widowed								99	55	99	52
Divorced								1,229	1,120	1,023	1,029
Separated								528	544	513	486
Total								11,828	11,946	11,689	11,752
Unweighted sample†											
Married	4461	3653	3265	3053	2864	2683	2234	2032	2176	2050
Non-married											
Single	1061	1175	1416	1431	1405	1361	1268	1342	1523	1490
Widowed	61	55	55	49	40	44	36	20	37	18
Divorced	256	338	448	453	437	421	443	393	389	363
Separated	167	143	175	185	206	186	200	192	200	171
Total	6006	5364	5359	5171	4952	4695	4181	3979	4325	4092

* Men and women describing themselves as 'separated' were, strictly speaking, legally married. However, because the separated can cohabit they have been included in the 'non-married' category.

† Trend tables show unweighted and weighted figures for 1998 to give an indication of the effect of the weighting. For the weighted data (1998 and 2000 to 2002) the weighted base (000's) is the base for percentages. Unweighted data (up to 1998) are based on the unweighted sample.

** Base too small for reliable analysis.

Table 5.9   **Women aged 16–59:**
             (a) Whether has dependent children in the household by marital status
             (b) Marital status by whether has dependent children in the household

Women aged 16-59   Great Britain: 2002

Marital status		Children			Weighted base (000's) =100%	Unweighted sample
		Dependent children	Non dependent children only	No children		
(a)						
Married	%	52	13	35	8,167	2994
Non-married						
Cohabiting	%	36	3	61	2,018	686
Single	%	23	1	76	3,824	1299
Widowed	%	11	18	71	219	77
Divorced	%	44	11	45	1,077	381
Separated	%	60	11	29	509	180
Total	%	42	9	49	15,814	5617
					Total	
(b)		%	%	%	%	
Married		64	79	36	52	
Non-married						
Cohabiting		11	4	16	13	
Single		13	1	38	24	
Widowed		0	3	2	1	
Divorced		7	8	6	7	
Separated		5	4	2	3	
Weighted base (000's) =100%		6,685	1,390	7,739	15,814	
Unweighted sample		2402	480	2735	5617	

Table 5.10   **Women aged 16-59: percentage cohabiting by legal marital status and whether**
              has dependent children in the household

Women aged 16-59   Great Britain: 2002

Legal marital status	Has dependent children	No dependent children	Total	Weighted bases (000's) = 100%			Unweighted sample		
				Has dependent children	No dependent children	Total*	Has dependent children	No dependent children	Total*
		Percentage cohabiting							
Married	-	-	-	4,250	3,917	8,214	1542	1452	3010
Non-married									
Single	36	24	27	1,353	3,861	5,259	468	1292	1776
Widowed	†	8	9	26	212	237	9	75	84
Divorced	28	29	28	658	845	1,532	240	294	545
Separated	11	16	13	343	242	590	123	83	208
Total	11	13	12	6,630	9,077	15,832	2382	3196	5623

(Widowed/Divorced/Separated grouped: 30   23   25)

* Totals with dependent children and without dependent children do not sum to the total because the dependency of some children could not be established.
† Base too small for reliable analysis.

Table 5.11   **Cohabiting women aged 16-59: whether has dependent children in the household by legal marital status**

Cohabiting women aged 16-59   Great Britain: 2002

Legal marital status	Has dependent children	No dependent children	Total
	%	%	%
Non-married			
Single	69	75	73
Widowed	1	1	1
Divorced	25	20	22
Separated	5	3	4
Weighted base (000's) = 100%	*713*	*1,212*	*1,925*
Unweighted sample	*246*	*407*	*653*

Table 5.12   **Number of past cohabitations not ending in marriage by sex and age**

Men and women aged 16-59   Great Britain: 2002

Age		Number of completed cohabitations*					Weighted base (000's) = 100%	Unweighted sample
		None	One	Two	Three or more	Total at least one		
Men								
16-19	%	99	1	0	0	1	*1,088*	*348*
20-24	%	91	7	1	0	9	*1,361*	*413*
25-29	%	79	16	4	0	21	*1,421*	*448*
30-34	%	73	20	6	2	27	*1,828*	*600*
35-39	%	77	15	6	2	23	*1,926*	*652*
40-44	%	83	14	2	2	17	*1,878*	*628*
45-49	%	84	10	4	2	16	*1,569*	*564*
50-54	%	92	6	1	1	8	*1,651*	*604*
55-59	%	96	3	1	0	4	*1,687*	*624*
Total	%	85	11	3	1	15	*14,409*	*4881*
Women								
16-19	%	97	3	0	0	3	*1,116*	*386*
20-24	%	81	16	2	1	19	*1,550*	*498*
25-29	%	72	23	5	1	28	*1,598*	*544*
30-34	%	74	20	5	1	26	*2,041*	*716*
35-39	%	79	14	6	1	21	*2,160*	*768*
40-44	%	84	12	4	0	16	*2,057*	*709*
45-49	%	90	8	2	0	10	*1,853*	*693*
50-54	%	94	5	1	0	6	*1,763*	*668*
55-59	%	95	4	1	0	5	*1,685*	*639*
Total	%	84	12	3	0	16	*15,823*	*5621*
All								
16-19	%	98	2	0	0	2	*2,204*	*734*
20-24	%	86	12	2	0	14	*2,911*	*911*
25-29	%	75	20	4	0	25	*3,019*	*992*
30-34	%	73	20	5	1	27	*3,869*	*1316*
35-39	%	78	15	6	1	22	*4,086*	*1420*
40-44	%	83	13	3	1	17	*3,934*	*1337*
45-49	%	87	9	3	1	13	*3,423*	*1257*
50-54	%	93	5	1	1	7	*3,414*	*1272*
55-59	%	95	4	1	0	5	*3,372*	*1263*
Total	%	85	12	3	1	15	*30,232*	*10502*

* Excludes current cohabitations.

Table 5.13   **Number of past cohabitations not ending in marriage by current marital status and sex**

Men and women aged 16-59   **Great Britain: 2002**

Number of cohabitations*	Marital status						
	Married	Non-married					
		Cohabiting	Single	Widowed	Divorced	Separated	Total†
	%	%	%	%	%	%	%
Men							
None	91	77	80	[100]	72	86	85
One	7	17	13	[0]	18	8	11
Two	1	4	5	[0]	5	4	3
Three or more	0	2	2	[0]	4	2	1
Total at least one	9	23	20	[0]	28	14	15
Weighted base (000's) = 100%	*7,213*	*1,765*	*4,334*	*90*	*693*	*248*	*14,408*
Unweighted sample	*2629*	*596*	*1322*	*28*	*213*	*73*	*4881*
Women							
None	93	82	69	86	81	84	84
One	6	15	24	8	16	12	12
Two	1	3	6	6	2	3	3
Three or more	0	0	1	0	0	1	0
Total at least one	7	18	31	14	19	16	16
Weighted base (000's) = 100%	*8,223*	*1,920*	*3,821*	*218*	*1,104*	*499*	*15,822*
Unweighted sample	*3013*	*652*	*1297*	*77*	*391*	*176*	*5621*

* Excludes current cohabitations.
† Total includes a small number of same sex cohabitees.

Table 5.14   **Age at first cohabitation which did not end in marriage by year cohabitation began and sex**

Persons aged 16-59 who have cohabited   **Great Britain: 2002**

Age at first cohabitation	Year first cohabitation began			
	1960-79	1980-89	1990-2002	All
	%	%	%	%
Men				
16-19	39	16	14	17
20-24	52	50	45	47
25-29	6	19	23	20
30-34	3	9	10	9
35-59	0	5	8	6
Weighted base (000's) =100%	*160*	*653*	*1,004*	*1,817*
Unweighted sample	*55*	*215*	*311*	*581*
Women				
16-19	38	38	33	35
20-24	40	36	41	39
25-29	16	18	14	15
30-34	6	6	6	6
35-59	0	3	7	5
Weighted base (000's) =100%	*204*	*707*	*1,297*	*2,208*
Unweighted sample	*76*	*252*	*434*	*762*
All				
16-19	38	27	25	27
20-24	45	43	43	43
25-29	12	19	18	18
30-34	5	8	8	7
35-59	0	4	7	6
Weighted base (000's) =100%	*364*	*1,361*	*2,300*	*4,025*
Unweighted sample	*131*	*467*	*745*	*1343*

Table 5.15   **Duration of past cohabitations which did not end in marriage by number of past cohabitations and sex**

Persons aged 16-59 who have cohabited   Great Britain: 2002

Duration of cohabitation	First cohabitation			Second cohabitation
	One only	One of two or more	All	All
	%	%	%	%
Men				
Less than 1 year	23	29	25	32
1 year, less than 2	21	22	21	22
2 years, less than 3	14	19	16	14
3 years, less than 5	19	17	18	17
5 years or more	23	13	20	14
Mean length in months	38	29	36	29
Weighted base (000's) =100%	*1,347*	*476*	*1,824*	*483*
Unweighted sample	*434*	*149*	*583*	*151*
Women				
Less than 1 year	20	21	20	22
1 year, less than 2	18	24	19	19
2 years, less than 3	14	16	15	22
3 years, less than 5	21	21	21	17
5 years or more	27	18	25	20
Mean length in months	43	33	41	36
Weighted base (000's) =100%	*1,776*	*474*	*2,250*	*497*
Unweighted sample	*612*	*165*	*777*	*173*
All persons				
Less than 1 year	21	25	22	27
1 year, less than 2	19	23	20	20
2 years, less than 3	14	18	15	18
3 years, less than 5	20	19	20	17
5 years or more	25	15	23	17
Mean length in months	41	31	39	33
Weighted base (000's) =100%	*3,123*	*950*	*4,074*	*980*
Unweighted sample	*1046*	*314*	*1360*	*324*

Chapter 6 Occupational and personal pension schemes

The General Household Survey (GHS) has included questions on occupational pensions on a regular basis since 1981 and on personal pensions since 1987. This chapter provides information on occupational and personal pensions for employees and then considers the pension arrangements of the self-employed. The analyses show trends on key variables and detailed data for 2002.

- Pension provision

- Pension arrangements for employees

- Membership of current employer's pension scheme

- Trends in membership of an occupational pension scheme

- Variation in pension scheme membership by employee characteristics

    Socio-economic classification (NS-SEC)

    Earnings

    Length of time with current employer

    Number of employees in the establishment

    Industry

- Personal pension arrangements among the self-employed

    Trends in personal pension scheme membership: self-employed men working full time

    Membership of a personal pension scheme and length of time in self-employment

Pension provision

At present all working people (both employees and the self-employed) are required to pay National Insurance contributions towards the basic state pension. Employers and their employees are also required either to contribute through National Insurance contributions to the second-tier state pension, S2P (State Second Pension), or to make alternative provision through an occupational scheme or a personal pension arrangement. Self-employed people cannot contribute to the S2P, so the only second pension choice for them is a personal pension.

The following changes in pension provision have recently been introduced.

- April 2001 - the stakeholder pension was introduced. It is a type of personal pension suitable for those who do not belong to an occupational pension scheme. Charges on stakeholder pensions are low and they offer greater flexibility than traditional personal pensions. For example, individuals may make intermittent contributions.
- October 2001 - certain types of employer were now required to provide access to a stakeholder pension scheme.
- April 2002 - the S2P was introduced. The new pension reformed the State Earnings Related Pension Scheme (SERPS) to provide a more generous additional

pension for low and moderate earners, certain groups of carers and people with a long-standing illness or disability.

The majority of tables in this chapter show data for men working full time. This is because the sample sizes for self-employed men working part time and for male employees working part time are too small to give reliable estimates.

Pension arrangements for employees[1]

In 2002:

- Two thirds of full-time employees (66% of both men and women) were currently members of a pension scheme, as were 39% of women who were part-time employees;
- over half of full-time employees were members of an occupational pension scheme (55% of men and 60% of women) but only a third (33%) of women working part time were;
- membership of personal pension schemes was less common than membership of occupational pension schemes[2]. (For example, 12% of female employees who were working full time had personal pensions, while 60% belonged to occupational pension schemes.)

Younger employees (those under the age of 25) were the least likely to be a member of a current pension scheme.

- Among male employees aged 18 to 24 working full time, 27% were in a current pension scheme compared with 62% or more in the older age groups.
- Among employees working full time, women in all age groups were more likely than men to belong to an occupational pension scheme. This difference was greatest in the younger age groups.
- Conversely, in all but the youngest age group, men were more likely than women to have a personal pension.   **Table 6.1**

Membership of current employer's pension scheme

In 2002, about seven in ten employees (71% of men and 69% of women) said their present employer had a pension scheme. Membership of such schemes varied by sex and work status (full or part-time).

Figure 6A   **Membership of current employer's pension scheme by sex and whether working full time or part time, 2002**

Great Britain
Percentage

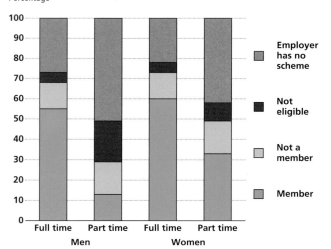

Employees who worked full time were more likely than those working part time to say that their present employer had a pension scheme (78% compared with 58% of women working part time, and 73% compared with 49% among men). **Table 6.2, Figure 6A**

Employees who belong to an occupational pension scheme may make further contributions known as additional voluntary contributions (AVCs). If these payments are made to a fund with their employer's scheme then these are AVCs. If the contributions are to a fund separate to their employer's scheme, usually with an insurance company, then these are free-standing AVCs (FSAVCs). In 2002, 10% of employees who belonged to an occupational pension scheme were also paying either AVCs or FSAVCs (table not shown).

Trends in membership of an occupational pension scheme

The analysis of trends in occupational pension scheme membership concentrates on the period since 1988 when important changes were introduced in pension provision for employees.

Since 1978, employers have been allowed to contract out of the S2P (formerly SERPS) if their pension scheme satisfied certain specific requirements. Individual employees who either belong to an occupational scheme which is not contracted out of the S2P or who do not belong to an occupational scheme have had the option since 1987 of contracting out of the S2P by starting their own personal pension plan. In such cases, the employee receives a rebate on his or her National Insurance contribution from the Inland Revenue (formerly from the DSS). In September 1994 a European Court of Justice ruling required parity of access to occupational pension schemes for full-time and part-time employees.

Trends in membership of the current employer's pension scheme differed for men and women and for employees working full and part time.
- Among men working full time, current membership of an occupational pension scheme has been stable at 55% in the four years to 2002 following a decline from 64% in 1989..
- The percentage of women working full time who were members of an occupational scheme showed a steady growth, rising from 55% in 1989 to 60% in 2002.

- The percentage of women working part time who were members of an occupational scheme rose from 15% in 1989 to 33% in 2002.

Since 2001, there has been no change in the percentage of both men and women full-time workers who reported their current employer did not have a pension scheme. **Table 6.3**

Variation in pension scheme membership by employee characteristics

Pension scheme membership is associated with a number of characteristics of employees and their type of employment, as shown in Tables 6.4 to 6.11. The following analysis mainly focuses on membership of an occupational pension scheme, as this is the most common type of pension arrangement for employees.

Socio-economic classification (NS-SEC)

National Statistics Socio-economic Classification (NS-SEC) was introduced for all official statistics and surveys in April 2001. When the results are shown separately for men and women working full and part time, they are only shown by the three broad categories of NS-SEC. This is because the sample bases are too small for more detailed analysis.

As in previous years, there were in 2002 variations in the current pension arrangements of respondents in different occupational groups.
- The percentages of female employees working full time who said that their current employer ran a pension scheme were: managerial and professional, 85%; intermediate occupations, 80%; routine and manual occupations, 64%.
- Among women working full time, 71% of those in the managerial or professional group belonged to an occupational pension scheme compared with 60% of those in the intermediate group and 41% of those in the routine and manual groups.
- The percentages for male full-time employees showed a similar variation. **Tables 6.4 and 6.5**

Earnings

Employees in the higher usual gross weekly earnings groups were more likely both to belong to their employer's pension scheme and to have a personal pension scheme than employees with lower earnings.

- Among employees working full time, 75% of men and 81% of women with usual gross weekly earnings of more than £600 belonged to their occupational pension scheme. Among their counterparts earning between £100 and £200 per week, the percentages were 20% and 26%.

- Almost a quarter (24%) of men working full time with usual gross weekly earnings over £600 belonged to a personal pension scheme compared with 12% earning between £100 and £200 per week.

- Earnings variations in membership of occupational and personal pension schemes for female full and part-time employees showed the same pattern.

Table 6.6

Length of time with current employer

The likelihood of belonging to the current employer's pension scheme increased with the length of time respondents had worked for that employer. This variation reflected differences in two factors – whether the employer ran a pension scheme and whether the respondent was eligible.

- Among men working full time, around a quarter (28%) who had worked for their current employer for less than two years belonged to an occupational pension scheme. This compared with 71% of those who had been with their employer for five years or more.

- For these two groups of men working full time, the percentages who said that their employer ran a pension scheme were 59% and 81% respectively.

- Among men who had worked full time for their current employer for less than two years, 12% said that they were not eligible to belong to their employer's pension scheme. This compared with 2% of those who had been with their employer for five years or more.

Female full and part-time employees showed a similar pattern to men in terms of both the availability of an occupational scheme and whether they belonged to a scheme.   **Tables 6.7 and 6.8**

Number of employees in the establishment

As in previous years, membership of an occupational pension scheme was most common in large establishments, which were more likely than small establishments to be owned or managed by an employer offering an occupational scheme.

- Among male employees working full time, the percentages belonging to their employer's pension scheme ranged from 30% of those in establishments with between three and 24 employees to 81% of those in establishments with 1,000 or more employees.

- The corresponding percentages for those who said that their employer ran an occupational scheme were 48% and 91% respectively.

There was a similar pattern for women working full and part time.

- Among female employees working full time, the percentages belonging to their employer's pension scheme ranged from 25% of those in establishments with one or two employees to 84% of those in establishments with 1,000 or more employees.

- The corresponding percentages who said that their employer ran an occupational scheme were 40% and 95%.   **Tables 6.9 and 6.10**

Industry

To provide a sufficiently large sample to analyse membership of occupational pension schemes by industry groups, data from the most recent three years of the GHS have been combined. The sample sizes of some groups are still small, however, and the results for these industries should be treated with caution.

There was wide variation in membership of occupational pension schemes between industry groups.

- Occupational pension scheme membership for male employees working full time ranged from 22% in the agriculture, forestry and fishing industry to 77% in the public and personal services.

- The relatively high level of occupational pension scheme membership among employees in public and personal services was also recorded for both full-time and part-time female employees (72% and 45%).

- The percentage of full-time employees belonging to their employer's pension scheme was below average for workers in construction (36% for men) and distribution, hotels, catering and repairs (34% for both men and women).   **Table 6.11**

Personal pension arrangements among the self-employed

Self-employed people, like employees, currently have to pay National Insurance contributions towards a basic state pension but they cannot contribute to the S2P. The second pension choice for them is either a personal pension or, since April 2001, a stakeholder pension. However, many self-employed people make provision for their retirement through other savings and investments.

In 2002, as among employees, self-employed men were more likely than women to have personal pension arrangements. Likewise, personal pension-scheme membership was more common among full-time than part-time workers.

Among self-employed people aged 16 and over:
- 48% of self-employed men and 32% of self-employed women were currently in a personal pension scheme;
- over a third (36%) of self-employed men had never had a personal pension scheme compared with over a half (55%) of women;
- around a quarter of self-employed men and women who worked part time were currently members of a personal pension scheme (28% of men and 24% of women);
- among self-employed people who worked part time, the percentage who had never had a personal pension scheme was higher among women than men (63% compared with 44%);
- more than one in ten (men, 16%; women, 13%) had been in a personal pension scheme but no longer belonged to one in 2002. (At 28%, this percentage was particularly high among self-employed men working part time.)   **Table 6.12**

Trends in personal pension scheme membership: self-employed men working full time

The GHS provides trend data since 1991 on the personal pension arrangements among self-employed men working full time. The possession of a current personal pension among self-employed men working full time remained fairly stable between 1991 and 1998 at around two thirds. Between 1998 and 2002 the percentage with a personal pension decreased from 64% to 52%.

Table 6.13

Membership of a personal pension scheme and length of time in self-employment

Data from the most recent three years of GHS have been combined to provide a sufficiently large sample to analyse membership of personal pension schemes by the length of time spent in self-employment.

The likelihood of having a personal pension increased with the length of time spent in self-employment.
- Men working full time who had been self-employed for five years or more were over twice as likely to have a current personal pension as those who had been self-employed for less than two years (62% compared with 26%).
- There was a similar pattern for women. Of women working full time who had been self-employed for five years or more, 47% had a current personal pension compared with 25% who had been self-employed for less than two years. For women working part time, the percentages were 37% and 11%.

Table 6.14

Notes

1   The GHS questions about pension arrangements for employees are asked only about the respondent's current employer. Therefore, some people may have held entitlements in the occupational pension scheme of a previous employer.

2   Individuals may have both a personal pension and an occupational pension. These options are not mutually exclusive.

Table 6.1   **Current pension scheme membership by age and sex**

Employees aged 16 and over excluding YT and ET   Great Britain: 2002

Pension scheme members	Age						
	16-17	18-24	25-34	35-44	45-54	55 and over	Total
				Percentages			
Men full time							
Occupational pension*	[8]	23	50	65	68	54	55
Personal pension	[0]	6	19	23	24	16	19
Any pension	[8]	27	62	78	80	64	66
Women full time							
Occupational pension*	[4]	34	62	67	71	56	60
Personal pension	[0]	6	13	17	11	9	12
Any pension	[4]	37	68	76	77	61	66
Women part time							
Occupational pension*	1	11	39	41	43	26	33
Personal pension	0	1	11	12	9	9	9
Any pension	1	12	46	50	50	32	39
Weighted base (000's) =100%							
Men full time	129	1,272	2,654	3,015	2,287	1,441	10,798
Women full time	91	900	1,667	1,597	1,494	612	6,361
Women part time	262	547	881	1,426	1,017	828	4,961
Unweighted sample							
Men full time	41	396	866	1023	838	537	3701
Women full time	30	298	565	546	564	232	2235
Women part time	94	179	311	508	383	319	1794

* Including a few people who were not sure if they were in a scheme but thought it possible.

Table 6.2   **Membership of current employer's pension scheme by sex and whether working full time or part time**

Employees aged 16 and over excluding YT and ET   Great Britain: 2002

Pension scheme coverage	Men			Women		
	Working full time	Working part time	Total*	Working full time	Working part time	Total*
	%	%	%	%	%	%
Current employer has a pension scheme						
Member†	55	13	51	60	33	48
Eligible but not a member	13 / 73	16 / 49	13 / 71	13 / 78	16 / 58	14 / 69
Not eligible to belong	5	20	7	5	9	7
Does not know if a member	0	0	0	0	0	0
Current employer does not have a pension scheme	26	45	28	21	40	29
Not known if present employer has a pension scheme	1	7	1	1	2	1
Weighted base (000's) = 100%	*10,820*	*1,141*	*11,989*	*6,362*	*4,963*	*11,351*
Unweighted sample	*3709*	*390*	*4109*	*2236*	*1795*	*4040*

* Including a few people whose hours of work were not known.
† Including a few people who were not sure if they were in a scheme but thought it possible.

Table 6.3   Membership of current employer's pension scheme by sex: 1983 to 2002

Employees aged 16 and over excluding YT and ET*   Great Britain

Pension scheme coverage	Unweighted								Weighted			
	1983	1987	1989	1991	1993	1995	1996	1998	1998	2000	2001	2002
	%	%	%	%	%	%	%	%	%	%	%	%
Men full time												
Current employer has a pension scheme												
Member†	66	63	64	61	60	58	58	57	55	54	54	55
Not a member	10	12	14	16	16	16	16	15	15	16	19	18
Does not know if a member	1	0	0	1	0	0	0	0	0	0	0	0
(total)	77	74	79	77	76	74	74	72	71	70	73	73
Current employer does not have a pension scheme	22	22	19	21	22	25	25	28	29	29	26	26
Not known if present employer has a pension scheme	2	3	2	2	2	1	1	1	1	1	1	1
Weighted base (000's) = 100%**									11,009	11,323	11,220	10,820
Unweighted sample**	5087	5129	4906	4563	3976	4062	3937	3697		3558	3881	3709
	%	%	%	%	%	%	%	%	%	%	%	%
Women full time												
Current employer has a pension scheme												
Member†	55	52	55	55	54	55	53	56	55	58	58	60
Not a member	17	16	21	21	22	20	20	17	18	17	20	18
Does not know if a member	0	1	0	0	0	0	0	0	0	0	0	0
(total)	72	68	76	77	77	76	73	73	73	75	78	78
Current employer does not have a pension scheme	24	28	21	20	22	24	26	26	27	25	21	21
Not known if present employer has a pension scheme	4	4	3	3	2	1	1	0	0	1	1	1
Weighted base (000's) = 100%**									6,429	6,353	6,465	6,362
Unweighted sample**	2256	2562	2602	2484	2239	2331	2143	2244		2089	2384	2236
	%	%	%	%	%	%	%	%	%	%	%	%
Women part time												
Current employer has a pension scheme												
Member†	13	11	15	17	19	24	26	27	26	31	33	33
Not a member	39	34	37	34	35	32	28	26	26	25	29	25
Does not know if a member	0	0	0	1	0	0	0	0	0	0	0	0
(total)	53	46	52	52	55	55	53	53	52	56	63	58
Current employer does not have a pension scheme	40	44	40	39	38	42	44	45	46	42	36	40
Not known if present employer has a pension scheme	7	10	7	8	7	3	2	2	3	2	2	2
Weighted base (000's) = 100%**									4,628	5,059	4,990	4,963
Unweighted sample**	1638	2126	2102	1977	1938	2038	1908	1674		1732	1878	1795

*   Prior to 1985 full-time students are excluded. Figures since 1987 include full-time students who were working but exclude those on Government schemes.

†   Including a few people who were not sure if they were in a scheme but thought it possible.

**   Trend tables show unweighted and weighted figures for 1998 to give an indication of the effect of the weighting. For the weighted data (1998 and 2000 to 2002) the weighted base (000's) is the base for percentages. Unweighted data (up to 1998) are based on the unweighted sample.

Table 6.4   **Current pension scheme membership by sex and socio-economic classification**

Employees aged 16 and over excluding YT and ET   Great Britain: 2002

Pension scheme members	Socio-economic classification			
	Managerial and professional	Intermediate	Routine and manual	Total[†]
	Percentages			
Men full time				
Occupational pension*	67	64	42	55
Personal pension	23	11	17	19
Any pension	78	70	54	66
Women full time				
Occupational pension*	71	60	41	60
Personal pension	12	13	11	12
Any pension	77	68	47	66
Women part time				
Occupational pension*	57	44	26	33
Personal pension	11	16	7	9
Any pension	63	56	31	39
Weighted base (000's) =100%				
Men full time	4,862	764	5,069	10,798
Women full time	3,119	1,438	1,707	6,361
Women part time	857	1,037	2,560	4,961
Unweighted sample				
Men full time	1713	253	1703	3701
Women full time	1104	504	594	2235
Women part time	317	383	919	1794

* Including a few people who were not sure if they were in a scheme but thought it possible.

† From April 2001 the National Statistics Socio-economic Classification (NS-SEC) was introduced for all official statistics and surveys. It replaced Social Class based on Occupation and Socio-economic Groups (SEG). Full-time students, persons in inadequately described occupations, persons who have never worked and the long term unemployed are not shown as separate categories, but are included in the figure for all persons (see Appendix A).

Table 6.5   **Membership of current employer's pension scheme by sex and socio-economic classification**

Employees aged 16 and over excluding YT and ET   **Great Britain: 2002**

Pension scheme coverage	Socio-economic classification			
	Managerial and professional	Intermediate	Routine and manual	Total*
	%	%	%	%
Men full time				
Current employer has a pension scheme				
Member†	67 ⎤ 81	64 ⎤ 84	42 ⎤ 64	55 ⎤ 73
Not a member **	14 ⎦	21 ⎦	23 ⎦	18 ⎦
Current employer does not have a pension scheme	19	15	34	26
Not known if employer has a pension scheme	0	0	1	1
Weighted base (000's) =100%	*4,895*	*758*	*5,065*	*10,820*
Unweighted sample	*1725*	*251*	*1702*	*3709*
Women full time				
Current employer has a pension scheme				
Member†	71 ⎤ 85	60 ⎤ 80	41 ⎤ 64	60 ⎤ 78
Not a member **	14 ⎦	20 ⎦	23 ⎦	18 ⎦
Current employer does not have a pension scheme	15	19	34	21
Not known if employer has a pension scheme	0	2	2	1
Weighted base (000's) =100%	*3,119*	*1,441*	*1,707*	*6,362*
Unweighted sample	*1104*	*505*	*594*	*2236*
Women part time				
Current employer has a pension scheme				
Member†	57 ⎤ 78	44 ⎤ 68	26 ⎤ 52	33 ⎤ 58
Not a member **	21 ⎦	24 ⎦	26 ⎦	25 ⎦
Current employer does not have a pension scheme	21	31	46	40
Not known if employer has a pension scheme	1	1	2	2
Weighted base (000's) =100%	*858*	*1,038*	*2,561*	*4,963*
Unweighted sample	*318*	*383*	*919*	*1795*

* From April 2001 the National Statistics Socio-economic Classification (NS-SEC) was introduced for all official statistics and surveys. It replaced Social Class based on Occupation and Socio-economic Groups (SEG). Full-time students, persons in inadequately described occupations, persons who have never worked and the long term unemployed are not shown as separate categories, but are included in the figure for all persons (see Appendix A).

† Including a few people who were not sure if they were in a scheme but thought it possible.

** Including people who were not eligible and a few people who did not know if they were a member.

Table 6.6   Current pension scheme membership by sex and usual gross weekly earnings

Employees aged 16 and over excluding YT and ET   Great Britain: 2002

Pension scheme members	Usual gross weekly earnings (£)							
	0.01-100.00	100.01-200.00	200.01-300.00	300.01-400.00	400.01-500.00	500.01-600.00	600.01 or more	Total†
					Percentages			
Men full time								
Occupational pension*	37	20	38	50	65	71	75	55
Personal pension	13	12	16	20	21	22	24	19
Any pension	41	27	50	64	77	83	86	66
Women full time								
Occupational pension*	[32]	26	52	66	78	81	81	60
Personal pension	[10]	9	11	14	10	15	15	12
Any pension	[42]	31	59	74	82	86	88	66
Women part time								
Occupational pension*	15	41	60	66	[68]	**	**	33
Personal pension	3	12	13	19	[15]	**	**	9
Any pension	18	49	68	76	[71]	**	**	39
Weighted base (000's) =100%								
Men full time	163	496	1,830	2,080	1,505	1,052	2,045	10,798
Women full time	105	796	1,762	1,164	764	529	601	6,361
Women part time	1,984	1,681	614	155	55	20	52	4,961
Unweighted sample								
Men full time	54	162	608	697	519	368	725	3701
Women full time	37	275	617	408	268	190	213	2235
Women part time	713	605	224	57	21	8	19	1794

*   Including a few people who were not sure if they were in a scheme but thought it possible.

†   Totals include no answers to income.

** Base too small for reliable analysis to be made.

Table 6.7   **Current pension scheme membership by sex and length of time with current employer**

Employees aged 16 and over excluding YT and ET   **Great Britain: 2002**

Pension scheme members	Length of time with current employer			
	Less than 2 years	2 years, but less than 5 years	5 years or more	Total[†]
	Percentages			
Men full time				
Occupational pension*	28	49	71	55
Personal pension	18	20	19	19
Any pension	42	61	81	66
Women full time				
Occupational pension*	37	59	77	60
Personal pension	11	13	12	12
Any pension	45	65	82	66
Women part time				
Occupational pension*	17	31	49	33
Personal pension	7	10	10	9
Any pension	23	38	55	39
Weighted base (000's) = 100%				
Men full time	*2,980*	*2,343*	*5,470*	*10,798*
Women full time	*2,056*	*1,396*	*2,895*	*6,361*
Women part time	*1,888*	*1,039*	*2,030*	*4,961*
Unweighted sample				
Men full time	*988*	*783*	*1928*	*3701*
Women full time	*700*	*486*	*1044*	*2235*
Women part time	*666*	*377*	*750*	*1794*

* Including a few people who were not sure if they were in a scheme but thought it possible.
† Including a few where length of time in job was not known.

Table 6.8   **Whether or not current employer has a pension scheme by sex and length of time with current employer**

Employees aged 16 and over excluding YT and ET

Great Britain: 2002

Pension scheme coverage	Length of time with current employer			
	Less than 2 years	2 years, but less than 5 years	5 years or more	Total*
	%	%	%	%
Men full time				
Current employer has a pension scheme				
Member†	28	49	71	55
Not a member **	19 _59_	19 _73_	8 _81_	13 _73_
Not eligible to belong	12	5	2	5
Current employer does not have a pension scheme	39	26	19	26
Not known if employer has a pension scheme	2	1	0	1
Weighted base (000's) =100%	*2,969*	*2,355*	*5,492*	*10,820*
Unweighted sample	*984*	*787*	*1936*	*3709*
Women full time				
Current employer has a pension scheme				
Member†	37	59	77	60
Not a member **	19 _67_	16 _78_	7 _86_	13 _78_
Not eligible to belong	12	4	1	5
Current employer does not have a pension scheme	30	22	14	21
Not known if employer has a pension scheme	3	0	0	1
Weighted base (000's) =100%	*2,056*	*1,396*	*2,896*	*6,362*
Unweighted sample	*700*	*486*	*1045*	*2236*
Women part time				
Current employer has a pension scheme				
Member†	17	31	49	33
Not a member **	17 _47_	19 _60_	13 _68_	16 _58_
Not eligible to belong	12	10	6	9
Current employer does not have a pension scheme	50	38	32	40
Not known if employer has a pension scheme	4	2	0	2
Weighted base (000's) =100%	*1,890*	*1,041*	*2,031*	*4,963*
Unweighted sample	*667*	*377*	*750*	*1795*

*   Including a few whose length of time in job was not known.
†   Including a few people who were not sure if they were in a scheme but thought it possible.
**   Including a few people who did not know if they were a member.

Table 6.9   **Current pension scheme membership by sex and number of employees in the establishment**

Employees aged 16 and over excluding YT and ET

Great Britain: 2002

Pension scheme members	Number of employees at establishment					
	1-2	3-24	25-99	100-999	1000 or more	Total†
				Percentages		
Men full time						
Occupational pension*	40	30	52	68	81	55
Personal pension	26	21	21	18	12	19
Any pension	56	47	65	76	84	66
Women full time						
Occupational pension*	25	41	60	71	84	60
Personal pension	14	15	10	12	9	12
Any pension	39	51	64	76	87	66
Women part time						
Occupational pension*	8	20	39	51	71	33
Personal pension	7	9	9	10	5	9
Any pension	15	28	45	56	72	39
Weighted base (000's) = 100%						
Men full time	381	2,628	2,819	3,602	1,200	10,798
Women full time	174	1,705	1,712	1,976	711	6,361
Women part time	292	2,131	1,242	929	298	4,961
Unweighted sample						
Men full time	135	895	964	1234	415	3701
Women full time	62	604	605	689	247	2235
Women part time	105	768	452	336	108	1794

* Including a few people who were not sure if they were in a scheme but thought it possible.
† Includes a few people for whom the number of employees at establishment was not known.

81

Table 6.10   **Membership of current employer's pension scheme by sex and number of employees at the establishment**

Employees aged 16 and over excluding YT and ET   Great Britain: 2002

Pension scheme coverage	Number of employees at establishment					
	1-2	3-24	25-99	100-999	1000 or more	Total*
	%	%	%	%	%	%
Men full time						
Current employer has a pension scheme						
Member†	40 ⎤ 50	30 ⎤ 48	52 ⎤ 74	68 ⎤ 88	81 ⎤ 91	55 ⎤ 73
Not a member **	10 ⎦	18 ⎦	22 ⎦	20 ⎦	10 ⎦	18 ⎦
Current employer does not have a pension scheme	50	51	25	11	9	26
Not known if employer has a pension scheme	1	1	1	1	1	1
Weighted base (000's) =100%	*380*	*2,634*	*2,823*	*3,600*	*1,211*	*10,820*
Unweighted sample	*135*	*898*	*966*	*1233*	*419*	*3709*
Women full time						
Current employer has a pension scheme						
Member†	25 ⎤ 40	41 ⎤ 58	60 ⎤ 80	71 ⎤ 91	84 ⎤ 95	60 ⎤ 78
Not a member **	15 ⎦	17 ⎦	21 ⎦	20 ⎦	11 ⎦	18 ⎦
Current employer does not have a pension scheme	59	41	19	9	5	21
Not known if employer has a pension scheme	1	2	1	1	0	1
Weighted base (000's) =100%	*174*	*1,706*	*1,711*	*1,985*	*711*	*6,362*
Unweighted sample	*62*	*604*	*604*	*692*	*247*	*2236*
Women part time						
Current employer has a pension scheme						
Member†	8 ⎤ 22	20 ⎤ 40	39 ⎤ 70	51 ⎤ 86	71 ⎤ 92	33 ⎤ 58
Not a member **	14 ⎦	19 ⎦	31 ⎦	35 ⎦	21 ⎦	25 ⎦
Current employer does not have a pension scheme	76	58	28	13	8	40
Not known if employer has a pension scheme	1	2	2	1	0	2
Weighted base (000's) =100%	*292*	*2,132*	*1,243*	*928*	*299*	*4,963*
Unweighted sample	*105*	*768*	*453*	*336*	*108*	*1795*

*   Includes a few people for whom the number of employees at establishment was not known.
†   Including a few people who were not sure if they were in a scheme but thought it possible.
** Including people who were not eligible and a few people who did not know if they were a member.

Table 6.11　　**Current pension scheme membership by sex and industry group**

Employees aged 16 and over excluding YT and ET　　　　　　　　　　　　　　　　Great Britain: 2000, 2001 and 2002 combined

Pension scheme members	Industry group*										
	Agriculture, forestry, fishing	Coal mining, energy and water supply	Mining (excl coal), manufact-ure of metals, minerals and chemicals	Metal goods, engineer-ing and vehicle	Other manufact-uring	Construc-tion	Distribu-tion, hotels, catering repairs	Transport and commun-ications	Banking, finance, insurance business services	Public and other personal services	Total
						Percentages					
Men full time											
Occupational pension†	22	73	[63]	59	52	36	34	60	54	77	54
Personal pension	39	17	[15]	23	25	24	25	17	26	12	21
Any pension	53	82	[75]	71	67	52	51	70	67	81	67
Women full time											
Occupational pension†	[20]	[80]	**	62	47	54	34	52	55	72	59
Personal pension	[22]	[18]	**	14	19	10	13	13	19	9	13
Any pension	[42]	[81]	**	68	58	61	42	59	65	76	66
Women part time											
Occupational pension†	[7]	**	**	27	29	18	16	37	34	45	33
Personal pension	[11]	**	**	8	14	20	7	10	14	8	9
Any pension	[19]	**	**	32	38	35	21	44	42	50	38
Unweighted sample††											
Men full time	*130*	*190*	*34*	*1661*	*1435*	*1073*	*1610*	*1090*	*1725*	*2125*	*11073*
Women full time	*23*	*44*	*8*	*333*	*561*	*109*	*993*	*347*	*1310*	*2952*	*6680*
Women part time	*26*	*8*	*1*	*113*	*222*	*82*	*1687*	*152*	*616*	*2472*	*5379*

* Standard Industrial Classification, 1992.
† Including a few people who were not sure if they were in a scheme but thought it possible.
** Base too small for reliable analysis to be made.
†† Weighted bases not shown for combined data sets.

Table 6.12　　**Membership of personal pension scheme by sex and whether working full time or part time: self-employed persons**

Self-employed persons aged 16 and over　　　　　　　　　　　　　　　　　　Great Britain: 2002

Pension scheme coverage	Men			Women		
	Working full time	Working part time	Total*	Working full time	Working part time	Total*
	%	%	%	%	%	%
Informant belongs to a personal pension scheme	52	28	48	42	24	32
Informant no longer has a personal pension scheme	14	28	16	12	14	13
Informant has never had a personal pension scheme	34	44	36	46	63	55
Weighted base (000's) = 100%	*1,926*	*369*	*2,337*	*425*	*498*	*927*
Unweighted sample	*695*	*136*	*844*	*156*	*186*	*343*

* Including a few people whose hours of work were not known.

Table 6.13   **Membership of a personal pension scheme for self-employed men working full time: 1991 to 2002**

Self-employed persons aged 16 and over   **Great Britain**

Pension scheme coverage	Unweighted							Weighted			
	1991	1992	1993	1994	1995	1996	1998	1998	2000	2001	2002
						Percentages					
Men working full time											
Informant belongs to a personal pension scheme	66	65	67	60	61	64	65	64	54	54	52
Informant no longer has a personal pension scheme	7	9	9	10	11	11	10	10	12	15	14
Informant has never had a personal pension scheme	27	26	24	30	28	25	26	27	34	32	34
*Weighted base (000's) =100%**								1,958	1,904	1,942	1,926
*Unweighted sample**	929	869	852	842	879	696	683		625	700	695

* Trend tables show unweighted and weighted figures for 1998 to give an indication of the effect of the weighting. For the weighted data (1998 and 2000 to 2002) the weighted base (000's) is the base for percentages. Unweighted data (up to 1998) are based on the unweighted sample.

Table 6.14   **Membership of personal pension scheme by sex and length of time in self-employment**

Self-employed persons aged 16 and over   **Great Britain: 2000, 2001 to 2002 combined**

	Length of time in self-employment			
	Less than 2 years	2 years, but less than 5 years	5 years or more	Total
	Percentage of self-employed who belong to a personal pension scheme			
Men full time	26	39	62	53
Women full time	25	25	47	38
Women part time	11	18	37	26
*Unweighted sample**				
Men full time	294	256	1461	2011
Women full time	76	103	269	448
Women part time	134	97	273	504

* Weighted bases not shown for combined data sets.

Chapter 7 General health and use of health services

The GHS has asked a series of questions about health and the use of health services since the beginning of the survey in 1971. Although periodic changes have been made to the content of the health section, it is possible to monitor changes in self-reported health and the use of health services over 30 years. This chapter looks at these trends and provides a summary of the main findings.

- Self-reported health

- Chronic sickness

- Acute sickness

- Self-reported sickness and socio-economic classification

- Self-reported sickness and economic activity status

- Self-reported sickness and regional variations

- Details of longstanding conditions

- Use of health services

- NHS General Practitioner (GP) consultations

- NHS GP consultations and economic activity status

- NHS GP consultations and prescriptions

- NHS and private GP consultations

- Practice nurse consultations

- Children's use of other health services

- Hospital visits

Reports such as the *Independent Inquiry into Inequalities in Health*[1] and *Our Healthier Nation*[2] indicate that health inequalities exist. *The NHS Plan*[3] gives priority to dealing with this problem. Following its publication, the Secretary of State set two national targets in February 2001 to reduce health inequalities in infant mortality and life expectancy. The series of *Tackling Health Inequalities*[4] reports build on *The NHS Plan*, to develop further ways of reducing health inequalities. The World Health Organisation recently published *Social determinants of health. The solid facts*[5], which provides a summary of what it considers to be the main social factors associated with health inequalities. Analysis of GHS data makes it possible to monitor changes over time and in health measures alongside other socio-economic factors.

Many of the tables here refer to data on adults and children. Questions on health and use of health services are asked of all adults in the household aged 16 and over. For the majority of questions, information is also collected from a responsible adult about all children in the household. (See Appendix E for the full questionnaire.)

Self-reported health

The GHS has asked a question on self-perception of health since 1977. Past research[6] has indicated that self-assessed health is a good predictor of premature mortality, even after controlling for other socio-economic variables. In 2002 the GHS found that 56% of adults said they had good health, 30% reported they had fairly good health and 14% said their health was not good. Although the proportion reporting good health has decreased from 59% in 2001 to 56% in 2002 there has been no change in the proportion saying their health was not good.

In 2000 this question was introduced for children aged less than 16. There is no evidence of change between 2000 and 2002.
- 81% of children were reported by their parent or guardian as having good health and 15% were reported as having fairly good health. For 4% it was

reported that their health was not good (table not shown).   **Table 7.1**

Chronic sickness

Since 1972, respondents have been asked whether they have a longstanding illness, disability or infirmity[7]. Those who report a longstanding illness are then asked if this limits their activities in any way. This question has been asked since 1973. Limiting longstanding illness is widely used as a measure of health status and has been shown to be an accurate predictor of early mortality, psychological health and hospital utilisation[8]. Data on longstanding illness and limiting longstanding illness include adults and children.

The prevalence of reported longstanding illness has increased from 21% in 1972 to 35% in 2002. The proportion of people reporting a longstanding illness increased steadily through the 1970s and early 1980s, and then ranged from 30% in 1985 to 35% in 1996, with no clear pattern over time. It is too early to say whether this increase is part of an upward trend. However, between 2001 and 2002, the proportion reporting longstanding illness has increased from 32% to 35%. These trends are partly explained by the ageing population over this period.

The prevalence of limiting longstanding illness has shown a similar trend, although the overall increase has been smaller (15% in 1975 reported a limiting condition compared with 21% in 2002). In 1975, the proportion of children who were reported to have a limiting longstanding illness was 2% for under fives and 5% for 5 to 15 year-olds. By 1995 these proportions had increased to 4% and 8% respectively and have remained at this level.

In 2002, as in all previous years, the likelihood of reporting a chronic condition, whether limiting or otherwise, increased with age.
- The prevalence of longstanding illness, disabilities or infirmities increased from 15% of those aged under five to 72% of those aged 75 and over.
- The increase in prevalence was particularly marked among those aged 45 or over. Just over one in five

(22%) of respondents aged under 45 reported a longstanding illness compared with just over half (53%) of older respondents[9].

- Between 2001 and 2002 the proportion of those aged 75 and over with a longstanding illness increased from 63% to 72%.
- The proportion reporting a limiting condition increased from 4% of children aged under five to 53% of adults in the 75 and over age group.
- Approximately one in ten (11%) respondents under the age of 45 reported a limiting longstanding illness, compared with just over a third (35%) of older respondents[10].

It should be noted that these reports of chronic sickness are based on the respondent's own assessment. A higher prevalence may therefore reflect people's increased expectations about their health as well as changes in the actual prevalence of sickness. Another possible contributory factor is an increase in the absolute numbers of people with severe chronic conditions who are surviving now compared with the past (perhaps due to new treatments or the wider application of successful treatments). It should also be remembered both that people vary in how much they are troubled by the same symptoms, and that their need to limit activities will depend on what they usually do.

A report from the 2001 GHS 'People aged 65 and over' discussed how GHS data were likely to under-estimate the incidence of longstanding conditions among older people in three ways[10]. First, Census data has shown that the prevalence of chronic conditions is highest among those living in communal establishments, whereas the GHS only samples private households. Second, evidence from other sources indicates that among those aged 65 and over some who are assessed as having a disability nevertheless say that they have no chronic illness or disability. Third, it pointed to research showing that older people are likely to under-report chronic conditions because they regard limitations in their daily activities as a normal part of growing old and not as evidence of illness or disability.

Overall, there were no statistically significant differences between men and women in the reported prevalence of longstanding or limiting longstanding illness. There was a higher reported prevalence of longstanding illness among boys compared with girls, however. Among boys, 17% of those aged under five and 21% aged 5 to 15 were reported to have a longstanding illness compared with 12% and 19% of girls respectively.   **Table 7.2, Figure 7A**

Figure 7A   **Percentage of males and females reporting (a) longstanding illness (b) limiting longstanding illness (c) restricted activity in the 14 days before interview: 1972 to 2002**
(data not available for 1977, 1978, 1997 and 1999)

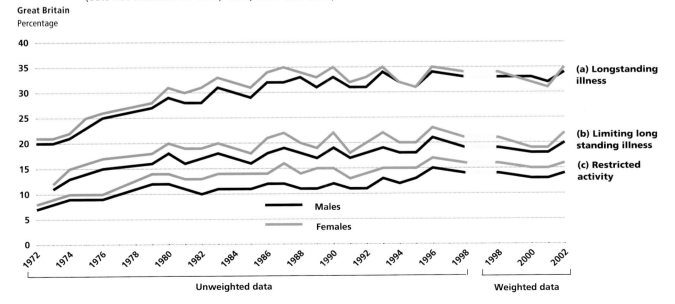

Acute sickness

Respondents were asked whether they had to cut down on their normal activities in the two weeks prior to interview as a result of illness or injury.

From 1972 to 1996, the proportion of those reporting restricted activity increased from 8% to 16%. It has remained relatively stable since then, and was 15% in 2002. Reported acute sickness increased with age but not as sharply as for chronic conditions. About one in ten (11%) of those aged under 45 had an acute sickness which restricted their activities during the reference period, increasing to just over a quarter (26%) of those aged 75 and over.

In 2002, 14% of males and 16% of females reported restricted activity due to illness or injury during the two weeks prior to interview. The largest difference in reported acute sickness between men and women was in the 75 and over age group where 28% of women reported restricted activity compared with 24% of men. Women aged 75 and over had on average 82 days' restricted activity per year whereas men in the same age group had 72.   **Tables 7.2–7.3**

Self-reported sickness and socio-economic classification

Tackling Health Inequalities – consultation on a plan for delivery[4] states: 'at the turn of the 21st century, opportunity for a healthy life is still linked to social circumstances'. The GHS provides several social and economic measures including socio-economic classification (NS-SEC) data, which confirms that social circumstances are still related to health.

Tables 7.4 to 7.6 present data using NS-SEC and are also grouped into the following three main classes:
* managerial and professional occupations;
* intermediate occupations;
* routine and manual occupations.

There were differences between each of the three main NS-SEC groups in the proportion of respondents reporting longstanding illness. Respondents living in households whose reference person was in the routine and manual group had the highest prevalence of longstanding illness (40% of males and 41% of females). They were followed by the intermediate group (32% of

males and 34% of females) and the managerial and professional group (30% of both males and females). The difference between respondents whose household reference person was in the intermediate group and managerial and professional group was only statistically significant for females.

A similar trend was evident among respondents who reported limiting longstanding illness. Females whose household reference person was in the routine and manual group were the most likely to suffer from a limiting longstanding illness (26%), followed by those in the intermediate group (21%) and those in the managerial and professional group (17%). Among males, those living in households in the routine and manual group were also the most likely to have reported a limiting longstanding illness. Unlike females, however, there was no statistically significant difference between the managerial and professional and intermediate groups. A quarter (25%) of males in routine and manual group households reported limiting longstanding illness compared with 18% in the intermediate group and 16% in managerial and professional households.

Respondents living in households headed by someone in the routine and manual group had the highest incidence of restricted activity in the two weeks prior to interview (16% of males and 17% of females). There were no significant differences between respondents whose household reference person was in the managerial and professional or intermediate group.

Of the three main NS-SEC groups, therefore, male and female respondents living in households headed by someone in the routine and manual group were the most likely to report a longstanding illness, a limiting longstanding illness or restricted activity in the last two weeks. Females whose household reference person was in the managerial and professional group were the least likely to report a longstanding illness or limiting longstanding illness.

There were no statistically significant differences in reported chronic sickness between respondents whose household reference person was in the routine and manual group and those respondents whose reference person had never worked or was long-term unemployed. In terms of those reporting acute sickness, there was a significant difference between females living in households in the routine and manual occupations group and those in the never worked or long-term unemployed group (17% and 24% respectively). This difference was not significant among males, however.

Tables 7.4–7.6

Self-reported sickness and economic activity status

Social determinants of health. The solid facts[5] states that unemployment puts people's health at risk and increases their likelihood of premature death. Furthermore, it argues that joblessness has detrimental effects on people's mental health, self-reported ill-health and physical health because of its psychological and financial impact. However, it also points out that job quality is important; merely having a job does not always protect physical and mental health, particularly if the nature of the employment is insecure or unsatisfactory. Poor health may also lead to job loss and long-term unemployment.

Among the GHS respondents, unemployed people were more likely than those in work to report longstanding illness, but the difference was only significant for men[11]. However, both unemployed men and unemployed women reported significantly higher levels of limiting longstanding illness than those who were working.

- 38% of men who were unemployed reported a longstanding illness compared with 27% of those who were working.
- 21% of unemployed women and 20% of unemployed men reported a limiting longstanding illness, compared with 13% and 12% of their working counterparts.

The prevalence of longstanding illness, limiting illness and restricted activity was highest among economically inactive respondents. The difference between economically inactive and working respondents who reported chronic and acute sickness was significant across all age groups but was more pronounced among men than women.

- 27% of working men and women reported a longstanding illness compared with 64% of men and 54% of women who were economically inactive.

Tables 7.7–7.9

Self-reported sickness and regional variations

People living in Wales had the highest prevalence of limiting longstanding illness (24%), followed by people living in England (21%) and people living in Scotland (18%). A similar trend was found in the prevalence of longstanding illness and restricted activity but not all differences were significant. When analysed by sex, these regional differences in the prevalence of limiting longstanding illness were only significant for males living in Wales (25% compared with 20% in England and 17% in Scotland).

Males living in Scotland had the lowest incidence of longstanding illness (29%) and restricted activity (11%). There were no statistically significant differences between regions in reported chronic or acute sickness among females.

Within England, by Government Office Region, people in London had the lowest incidence of longstanding illness (27% compared with 34% to 42% for other regions). When analysed by sex, however, this was only significant among males.

Table 7.10

With a few exceptions, similar results were found for NHS Regional Office areas (shown in Table 7.11). People in the Northern and Yorkshire region reported significantly higher amounts of longstanding illness (41% compared with 37% to 27% in other regions), limiting longstanding illness (26% compared with 23% to 17% in other regions) and restricted activity (20% compared with 16% to 13% in other regions). When analysed by sex, however, this trend was only significant among females reporting restricted activity (22% compared with 17% to 13% in other regions). **Table 7.11**

Details of longstanding conditions

Respondents aged 16 and over who reported a longstanding illness or condition were asked 'What is the matter with you?' Details of the illness were recorded by the interviewer and coded during the interview using a computer-assisted coding frame[12]. The categories into which respondents' replies were coded were later collapsed into broad groups approximate to the chapter headings of the International Classification of Diseases (ICD10).

Studies of the validity of self-reported data have shown that there is a high level of agreement between incidence based on self-reporting and on medical examinations[13], and between self-reporting and doctor diagnosis of specific conditions[14]. The level of agreement is highest for those conditions that require ongoing treatment, have commonly recognised names and are salient to respondents because they cause discomfort or worry[15].

Similar to previous years of the GHS, the most common conditions reported by respondents were musculoskeletal problems and conditions of the heart and circulatory system. There has been no change in the ranking order of frequency of conditions since 1998.

Table 7.12

For the majority of conditions a higher prevalence was found among older people than among young people. The difference was more marked for some complaints than others. A condition of the musculoskeletal system was reported at a rate of 76 per 1000 for those aged 16 to 44, compared with a rate of 363 per 1000 among people aged 75 and over. While 18 per 1000 people in the 16 to 44 age group reported a heart and circulatory system condition, the corresponding rate among those aged 75 and over was 386 per 1000.

Skin complaints were one of a few conditions that did not increase with age. For example, people aged 16 to 44 were just as likely as those aged 45 to 64 to have a skin condition (10 per 1000). Similarly, there was a higher prevalence of mental disorders among 45 to 64 year olds (38 per 1000) than among the two older age groups (16 and 24 per 1000). (The GHS does not collect data on people living in institutions because the sample is based on private households. Therefore it is possible that rates of some disorders are an underestimation of the true population rates, particularly among those aged 75 and over.)

Table 7.13

In line with the GHS results from previous years, overall, women were more likely than men to report musculoskeletal problems (179 compared to 156 per 1000). However, among those aged 16 to 44 this trend was reversed and the rate of reported musculoskeletal problems was higher among men than among women (82 per 1000 for men compared with 70 per 1000 for women).

Overall there was no difference between men and women in the reporting of heart and circulatory problems. For all age groups above the age of 44, however, rates of reported heart and circulatory problems were higher for men than for women. Among people aged 65 to 74, 330 men per 1000 reported heart and circulatory problems compared with 291 women per 1000.

Table 7.14

Table 7.15 shows the major disease groups separated into their component parts. This shows that the higher levels of older women compared with older men who reported musculoskeletal problems was mainly explained by the higher rate of arthritis and rheumatism among women (219 compared with 153 per 1000 among those aged 65 to 74 and 268 compared with 194 per 1000 among those aged 75 and over). Conversely, the higher levels of men compared with women who reported musculoskeletal problems in the youngest age group (16 to 44) can be explained by a higher rate of bone and joint problems among men.

Perhaps unsurprisingly, heart attacks, strokes and other heart complaints all increased with age among both men and women. Hay fever was noticeably higher among young men. The rates for men aged 16 to 44 were 10 per 1000 compared with 4 per 1000 for all the older age groups.

Table 7.15

Tables 7.16 and 7.17 look at the rate of reporting selected longstanding conditions by socio-economic classification of the household reference person.

- Respondents whose household reference person was in the managerial and professional group had the lowest incidence of musculoskeletal (130 per 1000) and heart and circulatory (97 per 1000) problems.

- Respondents whose household reference person was in the semi-routine and routine group were the most likely to report a musculoskeletal condition (223 per 1000).

- Out of all those with a longstanding illness, respondents whose household reference person was in the managerial and professional group reported the lowest average number of conditions (1.42). Respondents whose household reference person was in the semi-routine and routine group, meanwhile, reported the highest average (1.63).   **Table 7.16–7.17**

Use of health services

The GHS provides data about the use of health services among children and adults in the general population. It complements other sources of data which refer to those who have made use of health services, as it also includes those who make little or no use of these services.

The topics covered include:
- whether they have seen a General Practitioner (GP) in the two weeks before interview;
- whether they have seen a practice nurse in the two weeks before interview;
- whether they have attended an outpatient or casualty department in the three months before interview;
- whether they have been a day patient in the last 12 months;
- whether they have been an inpatient in the last 12 months.

Overall, females were more likely than males to have made use of any of these services. Use was highest among children aged less than five and adults in the older age groups.

NHS General Practitioner (GP) consultations

There has been little change in the proportion of respondents reporting an NHS GP consultation for 2002 compared with those reported since 1998. In 2002, 13% of males reported consulting a GP during the 14 days prior to interview – this is an increase on 2001 when it was 11%. The proportion of females consulting a GP has remained at around the same level since 1998 (between 16% and 17%).

As one would expect, the likelihood of having consulted a GP generally increases with age. For example, 14% of people aged 16 to 44 consulted a GP compared with a quarter of people aged 75 and over. Similar to previous years, the difference between men and women was particularly marked in the 16 to 44 age group (9% compared with 18%). Many consultations by women of this age are likely to be associated with birth control or pregnancy, which could account for some of the difference. Differences between the sexes were also most apparent among the youngest and oldest age groups.
- Boys aged under five were more likely than girls of the same age to have been reported to have had a

consultation with a NHS GP during the 14 days prior to interview (19% compared with 14%). Over the years since 1972, the proportion of girls aged 5 to 15 reported to have consulted a NHS GP show greater fluctuation than that of boys in the same age band.
- Women in the oldest age group (75 and over) were more likely than men in the same group to have consulted a GP (27% compared with 21%[16]).
- Females had an average of six consultations per year whereas males had four.
- The average number of NHS GP consultations per year has remained relatively stable, between four and five, since 1972.　　　　　**Tables 7.18–7.19, Figure 7B**

In 2002, 86% of consultations took place in a surgery. Trend data from 1971 shows that there has been an overall reduction in the number of NHS GP consultations taking place at home and an increase in surgery and telephone consultations. The proportion of consultations taking place in respondents' homes has fallen from 22% in 1971 to only 5% in 2002. In 1971 73% of consultations took place in the surgery, but by 2002 this figure had increased to 86%. In 1971 4% of NHS GP consultations took place over the telephone (at a time when less than half of households owned a telephone). During the late 1990s and early 2000s, the proportion of NHS GP consultations on the telephone peaked at 10% and has since fallen to 9% in 2002.

Table 7.20

Figure 7B　**Percentage of males and females consulting an NHS GP in the 14 days before interview: 1971 to 2002**

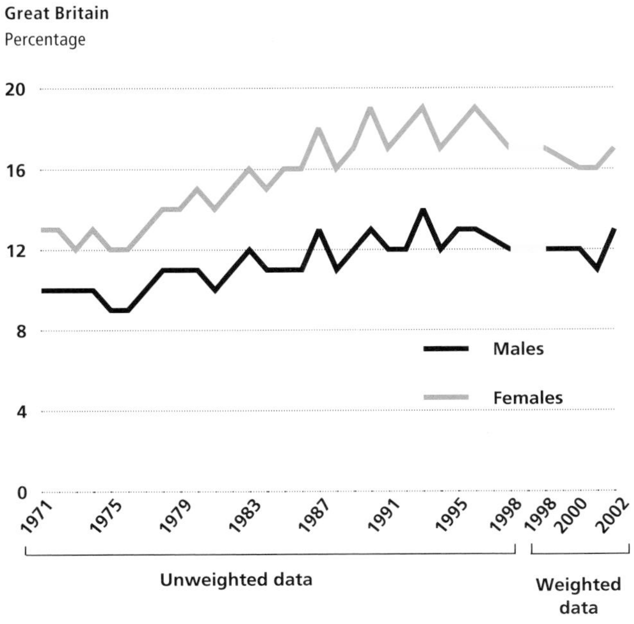

Great Britain
Percentage

Table 7.21 shows the site of consultations for males and females of different ages who consulted a doctor. Older people who consulted a doctor were more likely than younger people to have a home consultation. Between 2% and 3% of people aged under 65 reported having a consultation at home in the two weeks before interview. This compared with 6% of people aged 65 to 74 and 17% of those aged 75 and over. This could be for a variety of reasons. For example: it may be more difficult for older people to get to surgeries and health centres; they may be more likely to suffer from serious debilitating conditions; or they may be more likely to call a surgery out of hours and therefore receive a home consultation.

Table 7.21

NHS GP consultations and economic activity status

Economically inactive men were over twice as likely to consult a NHS doctor in the two weeks before interview than those who were working. The differences were also evident in each age group.

- Men who were economically inactive had an average of seven consultations per year, while those who were working had an average of three.
- Economically inactive women were also more likely to consult a doctor than those who were working (23% compared with 16%). **Table 7.22**

NHS GP consultations and prescriptions

Among people who had consulted a GP in the two weeks before interview, those living in households where the household reference person was in the routine and manual group were more likely to receive a prescription than those in managerial and professional households. Overall, men and women in the oldest age groups (65 and over) were more likely to receive a prescription than those in younger age groups. **Table 7.23**

NHS and private GP consultations

Similar to previous years, only a small percentage of all GP consultations were with private doctors (3% of all consultations). **Table 7.24**

Practice nurse consultations

Questions on consultations with a practice nurse were introduced in 2000. Overall, just over one in 20 people

(6%) reported consulting a practice nurse during the two weeks before interview (broadly unchanged since 2000). Older people were more likely to have visited a practice nurse than younger people (12% of the two oldest age groups visited a practice nurse compared with 1% to 4% of people aged under 45). The proportion of respondents aged 75 and over who had consulted a practice nurse increased from 9% in 2000 to 12% in 2002.

Overall, females were more likely than males to report consulting a practice nurse during the fortnight before interview (7% compared with 5%). These differences were greatest among 16 to 44 year-olds, where women were twice as likely as men to visit a practice nurse (6% compared with 3%). This may be partly due to women visiting practice nurses for reasons associated with family planning and pregnancy. On average there were two consultations with a practice nurse per person, per year. This figure doubled for those aged 65 and over.

Table 7.25

Children's use of other health services

An adult was asked on behalf of each child in the household whether they had used a range of health services (other than a doctor).

It was reported that in the two weeks before interview:
- 2% of children had seen a practice nurse at the GP surgery;
- 3% of children had seen a health visitor at the GP surgery;
- 2% of children had gone to a child health clinic;
- less than 0.5% of children had gone to a child welfare clinic.

Children under five were more likely to have made use of the services. There were no significant differences between boys and girls. **Table 7.26**

Hospital visits

Outpatient visits
The proportion of respondents who reported visiting an outpatient or casualty department at least once in a three-month period before interview has at 14% remained unchanged since 2001. There was a general increase from 10% of all respondents who reported such visits in 1972 to 16% in 1998, but this has declined since.

With the exception of the youngest age group, the likelihood of attending an outpatient or casualty department increased with age. For example, 10% of 5 to 15 year-olds attended an outpatient or casualty department in the three months before interview compared with a quarter of people aged 75 and over. However, the youngest age groups show the largest differences between the proportions of males and females reporting such visits. Among boys, 17% of those aged under five and 11% of those aged 5 to 15 were reported to have attended an outpatient or casualty department, compared with 12% and 9% of girls respectively.   **Table 7.27**

Day patients

The proportion of respondents attending hospitals as day patients has doubled since this question was first asked in 1992 (from 4% to 8% in 2002). Similar to outpatient visits, the proportions of people who reported receiving day-patient treatment generally increased with age. The only significant difference between males and females was in the 16 to 44 age group where 7% of men reported to be in receipt of day-patient treatment compared with 9% of women.   **Table 7.28**

Inpatients

The proportion of respondents who reported an inpatient stay in the 12 months prior to interview has remained unchanged since 2000 (8%). This reflects the

overall trend of little change since 1982. In general, females were slightly more likely to report an inpatient stay than males (8% compared with 7%) but the trend is more complicated when you examine the differences between males and females by age. For example, females were only significantly more likely than males to report an inpatient stay in the 16 to 44 age group (9% compared with 5%). By contrast, males below the age of 16 and in the 65 to 74 age group were significantly more likely than females to report an inpatient stay. At 17%, people aged 75 and over were more likely to report an inpatient stay than any other age group.

Among those who had been an inpatient, the average number of nights spent in hospital during the 12 months before interview was seven. With the exception of children under five, the average number of nights increased with age, from four nights for those aged 16 to 44 to 12 nights for those aged 75 or over.

Table 7.31 shows that in 2002 for every 100 people there was an average of 10 inpatient stays during the 12 months before interview and an average of 113 outpatient attendances per year.

    Tables 7.29–7.31, Figure 7C

Notes and references

1 *Independent Inquiry into Inequalities in Health.* The Stationery Office (London 1998). http://www.archive.official-documents.co.uk/document/doh/ih/ih.htm

2 *Our Healthier Nation.* Department of Health. The Stationery Office (London 1998). http://www.doh.gov.uk/ohn/ohnexec.htm

3 *The NHS Plan.* Department of Health. The Stationery Office (London 2000). http://www.doh.gov.uk/nhsplan/index.htm

4 *Tackling Health Inequalities – consultation on a plan for delivery.* Department of Health. The Stationery Office (London 2001). http://www.doh.gov.uk/healthinequalities/tacklinghealthinequalities.htm
   Tackling Health Inequalities – The results of the consultation exercise. Department of Health. The Stationery Office (London 2002). http://www.doh.gov.uk/healthinequalities/consultationresponse.htm
   Tackling Health Inequalities – A Programme for Action. Department of Health. The Stationery Office (London 2003). http://www.doh.gov.uk/healthinequalities/programmeforaction/index.htm

5 Wilkinson R. and Marmot M. (eds) (2003) *Social Determinants of Health. The Solid Facts.* 2nd Edition WHO. http://www.who.dk/document/e81384.pdf

Figure 7C   **Percentage of males and females reporting an inpatient stay in the 12 months before interview by age, 2002**

Great Britain

Percentage reporting an inpatient stay

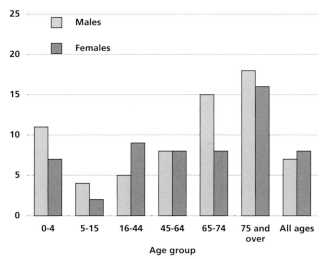

Age group

6   Mossey J. M. and Shapiro, E. (1982) Self-rated health: a predictor of mortality among the elderly. *American Journal of Public Health* 72, pp800–808.
    Idler E.L and Benyamini Y. (1997) Self-Rated Health and Mortality: A Review of Twenty-Seven Community Studies. *Journal of Health and Social Behavior* 38, pp21–37.

7   Respondents are asked 'Do you have any longstanding illness, disability or infirmity? By longstanding, I mean anything that has troubled you over a period of time or that is likely to affect you over a period of time?' It is left to the respondent to define what is meant by longstanding illness, disability or infirmity.

8   Arber S. (1997) Comparing inequalities in women's and men's health: Britain in the 1990s. *Social Science & Medicine* 44, pp773–787.

9   Figures not shown.

10 Traynor A and Walker A. (2003) *People aged 65 and over. Results of a study carried out on behalf of the Department of Health as part of the 2001 General Household Survey.* Office for National Statistics, The Stationery Office : London. http://www.statistics.gov.uk/lib2001/section3730.html

11 This may be due to the small sample sizes in the unemployed group.

12 The interviewers checked whether respondents had more than one complaint. They recorded details of, and coded up to six complaints.

13 Blaxter M. (1987) Self-reported Health, in Cox and Brian D (eds), *The Health and Lifestyle Survey.* Health Promotion Research Trust. London.

14 Bennett N et al. (1995) *Health Survey for England 1993: Appendix D*, OPCS, HMSO: London.

15 Discrepancies do not necessarily indicate that data from self-reported sources are inaccurate. Respondents may not have brought a condition to the attention of a doctor, medical records could be inaccurate, doctors may not have informed patients of their diagnosis, and lay descriptions may be different from those given by a doctor (see Blaxter and Bennett).

16 The GHS does not collect data on people living in institutions because the sample is based on private households. Therefore it is possible that rates of service use are an underestimation of the true population rates, particularly among those aged 75 and over.

Table 7.1   **Self perception of general health during the last 12 months: 1977 to 2002**

Persons aged 16 and over*   **Great Britain**

	Unweighted								Weighted			
	1977	1981	1985	1987	1991	1995	1996	1998	1998	2000	2001	2002
	%	%	%	%	%	%	%	%	%	%	%	%
Percentage who reported their general health was:												
Good	58	62	63	60	63	60	55	59	60	59	59	56
Fairly good	30	26	25	28	25	26	33	27	27	27	27	30
Not good	12	12	12	12	12	14	12	14	14	13	14	14
Weighted base (000's) =100%†									40,884	42,467	41,990	41,876
Unweighted sample†	23125	23242	18575	19477	18174	16724	15684	14410		14113	15385	14815

*   This question was not asked of proxies.

†   Trend tables show unweighted and weighted figures for 1998 to give an indication of the effect of the weighting. For the weighted data (1998 and 2000 to 2002) the weighted base (000's) is the base for percentages. Unweighted data (up to 1998) are based on the unweighted sample.

Table 7.2   **Trends in self-reported sickness by sex and age, 1972 to 2002: percentage of persons who reported**
(a) longstanding illness
(b) limiting longstanding illness
(c) restricted activity in the 14 days before interview

All persons   **Great Britain**

	Unweighted								Weighted				Weighted base 2002 (000's) = 100%*	Unweighted sample* 2002
	1972	1975	1981	1985	1991	1995	1996	1998	1998	2000	2001	2002		
							(a) Longstanding illness							
Percentage who reported:														
Males														
0- 4	5	8	12	11	13	14	14	15	15	14	17	17	1,706	604
5-15†	9	11	17	18	17	20	19	21	21	23	20	21	4,050	1515
16-44†	14	17	22	21	23	23	27	24	24	23	22	23	10,790	3521
45-64	29	35	40	42	42	43	46	44	44	45	44	46	6,692	2463
65-74	48	50	51	55	61	55	61	59	59	61	58	65	2,243	848
75 and over	54	63	60	58	63	56	64	68	68	63	64	71	1,526	591
Total	20	23	28	29	31	31	34	33	33	33	32	34	27,007	9542
Females														
0- 4	3	6	7	9	10	11	13	15	15	13	12	12	1,568	560
5-15†	6	9	13	13	15	17	16	19	19	18	16	19	3,976	1484
16-44†	13	16	21	22	23	22	27	23	23	22	21	25	11,287	3887
45-64	31	33	41	43	41	39	47	43	43	42	42	44	6,916	2654
65-74	48	54	58	56	55	54	58	59	59	54	56	61	2,524	910
75 and over	65	61	70	65	65	66	68	65	65	64	63	72	2,451	838
Total	21	25	30	31	32	31	35	34	34	32	31	35	28,722	10333
All persons														
0- 4	4	7	10	10	12	13	13	15	15	14	14	15	3,274	1164
5-15†	8	10	15	16	16	19	18	20	20	20	18	20	8,027	2999
16-44†	13	16	21	22	23	23	27	24	24	22	22	24	22,077	7408
45-64	30	34	41	43	41	41	47	44	43	44	43	45	13,607	5117
65-74	48	52	55	56	58	55	59	59	59	57	57	63	4,768	1758
75 and over	62	62	67	63	65	63	66	66	66	64	63	72	3,977	1429
Total	21	24	29	30	31	31	35	33	33	32	32	35	55,730	19875

*   Trend tables show unweighted and weighted figures for 1998 to give an indication of the effect of the weighting. For the weighted data (1998 and 2000 to 2002) the weighted base (000's) is the base for percentages. Unweighted data (up to 1998) are based on the unweighted sample. Bases for earlier years are of similar size and can be found in GHS reports for each year.

†   These age-groups were 5-14 and 15-44 in 1972 to 1978.

Table 7.2 - continued

All persons | Great Britain

	Unweighted								Weighted				Weighted base 2002 (000's) = 100%*	Unweighted sample* 2002
	1972	1975	1981	1985	1991	1995	1996	1998	1998	2000	2001	2002		
(b) Limiting longstanding illness														
Percentage who reported:														
Males														
0- 4	..	3	3	4	4	5	4	4	4	4	5	5	1,706	604
5-15†	..	6	8	8	7	8	8	8	8	9	9	8	4,051	1515
16-44†	..	9	10	10	10	12	14	12	12	11	10	12	10,781	3518
45-64	..	24	26	27	25	28	31	28	28	27	28	28	6,692	2463
65-74	..	36	35	38	40	37	42	36	36	38	36	43	2,244	848
75 and over	..	46	44	43	46	41	50	48	48	44	47	52	1,525	591
Total	..	14	16	16	17	18	21	19	19	18	18	20	26,999	9539
Females														
0- 4	..	2	3	3	3	3	4	5	5	4	4	3	1,565	559
5-15†	..	4	6	6	5	8	8	8	8	8	8	9	3,977	1484
16-44†	..	9	11	11	11	13	16	13	13	11	12	14	11,287	3887
45-64	..	22	26	26	25	26	32	29	29	27	26	28	6,915	2654
65-74	..	39	41	38	34	37	40	39	39	35	37	39	2,525	910
75 and over	..	49	56	51	51	52	53	51	51	48	45	53	2,451	838
Total	..	16	19	18	18	20	23	21	21	19	19	22	28,720	10332
All persons														
0- 4	..	2	3	3	4	4	4	4	4	4	4	4	3,271	1163
5-15†	..	5	7	7	6	8	8	8	8	8	8	8	8,026	2999
16-44†	..	9	11	10	10	12	15	13	13	11	11	13	22,068	7405
45-64	..	23	26	26	25	27	32	28	28	27	27	28	13,607	5117
65-74	..	38	38	38	37	37	41	38	37	37	36	41	4,767	1758
75 and over	..	48	52	48	49	48	52	50	50	47	46	53	3,977	1429
Total	..	15	17	17	18	19	22	20	20	19	19	21	55,716	19871
(c) Restricted activity in the 14 days before interview														
Percentage who reported:														
Males														
0- 4	5	10	13	13	11	11	12	10	10	11	9	10	1,706	604
5-15†	6	9	12	11	11	10	10	9	9	10	9	9	4,052	1516
16-44†	7	7	8	9	9	10	13	11	11	10	10	11	10,785	3519
45-64	9	10	12	11	12	15	18	18	18	17	17	17	6,694	2464
65-74	10	8	11	13	14	17	19	18	18	20	16	19	2,244	848
75 and over	10	12	15	17	18	20	23	24	24	23	23	24	1,528	592
Total	7	9	11	11	11	13	15	14	14	13	13	14	27,009	9543
Females														
0- 4	6	8	12	13	10	11	9	8	8	7	8	10	1,571	561
5-15†	5	7	11	12	9	10	9	11	11	11	10	9	3,976	1484
16-44†	8	10	11	13	12	13	15	13	13	12	12	13	11,284	3886
45-64	9	10	13	14	13	17	22	20	20	19	18	19	6,921	2656
65-74	10	12	17	18	16	20	21	23	23	21	21	20	2,531	912
75 and over	14	13	21	23	21	26	25	27	27	27	26	28	2,449	837
Total	8	10	13	14	13	15	17	16	16	15	15	16	28,732	10336
All persons														
0- 4	6	9	13	13	11	11	10	9	9	9	9	10	3,278	1165
5-15†	6	8	12	11	10	10	10	10	10	10	10	9	8,028	3000
16-44†	8	9	10	11	10	12	14	12	12	11	11	12	22,069	7405
45-64	9	10	12	12	13	16	20	19	19	18	17	18	13,615	5120
65-74	10	11	14	16	15	19	20	21	21	21	19	20	4,774	1760
75 and over	13	13	19	21	20	24	24	26	26	25	25	26	3,976	1429
Total	8	9	12	12	12	14	16	15	15	14	14	15	55,740	19879

* Trend tables show unweighted and weighted figures for 1998 to give an indication of the effect of the weighting. For the weighted data (1998 and 2000 to 2002) the weighted base (000's) is the base for percentages. Unweighted data (up to 1998) are based on the unweighted sample. Bases for earlier years are of similar size and can be found in GHS reports for each year.
† These age-groups were 5-14 and 15-44 in 1972 to 1978.

Table 7.3   Acute sickness: average number of restricted activity days per person per year, by sex and age

All persons   Great Britain: 2002

	Number of days			Weighted bases (000's) = 100%			Unweighted sample		
	Males	Females	Total	Males	Females	Total	Males	Females	Total
Age									
0- 4	16	13	15	1,706	1,571	3,277	604	561	1165
5-15	12	12	12	4,052	3,977	8,028	1516	1484	3000
16-44	23	26	24	10,785	11,284	22,069	3519	3886	7405
45-64	43	43	43	6,689	6,918	13,607	2462	2655	5117
65-74	55	52	53	2,244	2,531	4,774	848	912	1760
75 and over	72	82	78	1,528	2,445	3,973	592	836	1428
Total	31	34	33	27,003	28,725	55,729	9541	10334	19875

Table 7.4   Chronic sickness: prevalence of reported longstanding illness by sex, age and socio-economic classification of household reference person

All persons   Great Britain: 2002

Socio-economic classification of household reference person*	Males					Females				
	Age					Age				
	0-15	16-44	45-64	65 and over	Total	0-15	16-44	45-64	65 and over	Total
	Percentage who reported longstanding illness									
Large employers and higher managerial	17	27	45	66	35	11	18	33	61	25
Higher professional	12	18	33	60	25	17	21	37	65	28
Lower managerial and professional	20	19	42	63	30	11	23	41	66	32
Intermediate	18	21	43	70	32	15	27	44	66	38
Small employers and own account	21	22	39	66	33	13	25	32	57	29
Lower supervisory and technical	22	25	52	70	40	19	27	49	67	38
Semi-routine	26	24	58	71	40	23	30	52	66	42
Routine	16	27	52	70	40	21	25	52	74	44
Never worked and long-term unemployed	21	38	[71]	†	37	22	30	[69]	57	37
All persons	20	23	46	68	34	17	25	44	66	35
Weighted bases (000's) = 100%										
Large employers and higher managerial	322	671	511	188	1,692	331	721	431	162	1,645
Higher professional	480	1,128	603	268	2,479	487	977	557	186	2,207
Lower managerial and professional	1,219	2,658	1,569	739	6,185	1,160	2,771	1,740	898	6,569
Intermediate	419	755	421	255	1,850	428	991	584	631	2,634
Small employers and own account	599	1,083	898	364	2,944	488	955	815	321	2,579
Lower supervisory and technical	702	1,320	911	679	3,612	674	1,274	778	635	3,361
Semi-routine	827	1,190	743	530	3,290	875	1,443	893	948	4,159
Routine	717	1,255	882	694	3,548	646	1,191	949	962	3,748
Never worked and long-term unemployed	347	304	111	47	809	309	442	103	211	1,065
All persons	5,756	10,790	6,692	3,769	27,007	5,545	11,287	6,916	4,976	28,724
Unweighted sample										
Large employers and higher managerial	127	233	200	76	636	131	258	175	62	626
Higher professional	186	382	236	107	911	190	350	225	69	834
Lower managerial and professional	455	889	594	295	2233	438	971	678	330	2417
Intermediate	151	240	153	100	644	151	334	222	223	930
Small employers and own account	221	355	328	142	1046	178	334	315	116	943
Lower supervisory and technical	252	425	323	248	1248	241	428	293	220	1182
Semi-routine	305	389	267	196	1157	319	498	334	327	1478
Routine	255	397	312	257	1221	233	402	352	326	1313
Never worked and long-term unemployed	124	92	36	16	268	114	152	36	68	370
All persons	2119	3521	2463	1439	9542	2044	3887	2654	1748	10333

* From April 2001 the National Statistics Socio-economic Classification (NS-SEC) was introduced for all official statistics and surveys. It replaced Social Class based on Occupation and Socio-economic Groups (SEG). Full-time students and persons in inadequately described occupations are not shown as separate categories but are included in the figure for all persons (see Appendix A).

† Base too small for analysis.

Table 7.5 **Chronic sickness: prevalence of reported limiting longstanding illness by sex, age and socio-economic classification of household reference person**

All persons

Great Britain: 2002

Socio-economic classification of household reference person*	Males					Females				
	Age				Total	Age				Total
	0-15	16-44	45-64	65 and over		0-15	16-44	45-64	65 and over	
	Percentage who reported limiting longstanding illness									
Large employers and higher managerial	5	14	20	43	17	5	8	16	43	13
Higher professional	5	9	17	38	13	6	12	22	37	15
Lower managerial and professional	7	9	23	41	16	5	13	25	41	19
Intermediate	6	13	25	49	19	4	16	28	49	24
Small employers and own account	5	11	26	40	18	7	15	21	38	18
Lower supervisory and technical	6	12	37	50	24	9	17	34	45	24
Semi-routine	11	14	36	53	24	9	15	32	46	25
Routine	6	16	37	51	26	9	15	36	56	30
Never worked and long-term unemployed	8	25	[52]	†	23	11	21	[69]	40	27
All persons	7	12	28	47	20	7	14	28	46	22
Weighted bases (000's) = 100%										
Large employers and higher managerial	321	671	511	189	1,692	332	721	431	162	1,646
Higher professional	480	1,128	603	268	2,479	487	977	557	186	2,207
Lower managerial and professional	1,219	2,658	1,569	739	6,185	1,160	2,771	1,740	898	6,569
Intermediate	419	755	421	255	1,850	428	991	585	631	2,635
Small employers and own account	599	1,079	898	363	2,939	488	955	814	321	2,578
Lower supervisory and technical	702	1,317	911	679	3,609	674	1,274	778	635	3,361
Semi-routine	827	1,190	742	529	3,288	871	1,442	892	949	4,154
Routine	718	1,255	882	694	3,549	646	1,191	949	962	3,748
Never worked and long-term unemployed	347	301	111	47	806	310	442	103	211	1,066
All persons	5,756	10,781	6,692	3,769	26,998	5,542	11,287	6,915	4,976	28,720
Unweighted sample										
Large employers and higher managerial	127	233	200	76	636	131	258	175	62	626
Higher professional	186	382	236	107	911	190	350	225	69	834
Lower managerial and professional	455	889	594	295	2233	438	971	678	330	2417
Intermediate	151	240	153	100	644	151	334	222	223	930
Small employers and own account	221	354	328	142	1045	178	334	315	116	943
Lower supervisory and technical	252	424	323	248	1247	241	428	293	220	1182
Semi-routine	305	389	267	196	1157	318	498	334	327	1477
Routine	255	397	312	257	1221	233	402	352	326	1313
Never worked and long-term unemployed	124	91	36	16	267	114	152	36	68	370
All persons	2119	3518	2463	1439	9539	2043	3887	2654	1748	10332

* From April 2001 the National Statistics Socio-economic Classification (NS-SEC) was introduced for all official statistics and surveys. It replaced Social Class based on Occupation and Socio-economic Groups (SEG). Full-time students and persons in inadequately described occupations are not shown as separate categories but are included in the figure for all persons (see Appendix A).

† Base too small for analysis.

Table 7.6   **Acute sickness**

(a) **Prevalence of reported restricted activity in the 14 days before interview, by sex, age, and socio-economic classification of household reference person**

(b) **Average number of restricted activity days per person per year, by sex, age, and socio-economic classification of household reference person**

All persons   Great Britain: 2002

Socio-economic classification of household reference person*	Males					Females				
	Age					Age				
	0-15	16-44	45-64	65 and over	Total	0-15	16-44	45-64	65 and over	Total
(a) Percentage who reported restricted activity in the 14 days before interview										
Large employers and higher managerial	7	11	14	17	12	9	10	19	17	13
Higher professional	10 (9)	10 (11)	11 (14)	17 (20)	11 (12)	6 (8)	12 (13)	17 (17)	17 (21)	12 (14)
Lower managerial and professional	10	11	15	21	13	8	15	16	22	15
Intermediate	10 (9)	10 (10)	17 (17)	16 (19)	13 (13)	7 (7)	9 (11)	20 (17)	26 (23)	15 (14)
Small employers and own account	8	9	17	21	13	7	12	15	19	13
Lower supervisory and technical	7	9	20	21	14	9	15	18	21	16
Semi-routine	9 (9)	14 (13)	21 (21)	18 (22)	15 (16)	12 (10)	14 (14)	20 (21)	22 (25)	17 (17)
Routine	11	15	22	27	18	9	11	24	29	19
Never worked and long-term unemployed	9	9	[27]	†	13	11	21	[54]	34	24
All persons	9	11	17	21	14	9	13	19	24	16
(b) Average number of restricted activity days per person per year										
Large employers and higher managerial	12	23	34	42	26	9	17	29	44	21
Higher professional	11 (12)	20 (20)	23 (32)	45 (51)	22 (25)	7 (8)	21 (23)	34 (33)	49 (54)	23 (27)
Lower managerial and professional	13	20	35	56	27	8	26	34	57	29
Intermediate	17 (11)	24 (21)	43 (42)	49 (62)	30 (30)	9 (10)	17 (21)	44 (40)	73 (68)	35 (32)
Small employers and own account	7	19	42	72	30	11	24	36	57	30
Lower supervisory and technical	13	21	48	64	34	18	31	47	55	37
Semi-routine	18 (15)	26 (27)	59 (53)	53 (68)	36 (38)	18 (17)	30 (28)	44 (52)	64 (69)	38 (41)
Routine	13	34	54	82	44	13	24	64	84	48
Never worked and long-term unemployed	13	20	[78]	†	30	11	49	[128]	111	58
All persons	13	23	43	62	31	12	26	43	67	34
Weighted bases (000's) = 100%										
Large employers and higher managerial	322	667	511	189	1,690	331	721	434	161	1,647
Higher professional	480	1,128	603	268	2,479	490	974	554	186	2,204
Lower managerial and professional	1,218	2,656	1,569	739	6,182	1,160	2,770	1,740	898	6,568
Intermediate	419	758	418	256	1,850	428	991	584	634	2,637
Small employers and own account	599	1,077	898	366	2,940	487	955	815	319	2,576
Lower supervisory and technical	702	1,319	911	679	3,612	674	1,274	778	638	3,365
Semi-routine	833	1,191	743	530	3,296	874	1,442	892	946	4,155
Routine	717	1,255	882	694	3,548	646	1,193	950	962	3,751
Never worked and long-term unemployed	347	306	111	46	811	310	443	102	211	1,066
All persons	5,758	10,785	6,689	3,772	27,003	5,548	11,284	6,918	4,976	28,725
Unweighted sample										
Large employers and higher managerial	127	232	200	76	635	131	258	176	62	627
Higher professional	186	382	236	107	911	191	349	224	69	833
Lower managerial and professional	455	888	594	295	2232	438	971	678	330	2417
Intermediate	151	241	152	100	644	151	334	222	224	931
Small employers and own account	221	353	328	143	1045	178	334	315	115	942
Lower supervisory and technical	252	425	323	248	1248	241	428	293	221	1183
Semi-routine	307	389	267	196	1159	319	498	334	326	1477
Routine	255	397	312	257	1221	233	403	353	326	1315
Never worked and long-term unemployed	124	93	36	16	269	114	152	36	68	370
All persons	2120	3519	2462	1440	9541	2045	3886	2655	1748	10334

* From April 2001 the National Statistics Socio-economic Classification (NS-SEC) was introduced for all official statistics and surveys. It replaced Social Class based on Occupation and Socio-economic Groups (SEG). Full-time students and persons in inadequately described occupations are not shown as separate categories but are included in the figure for all persons (see Appendix A).

† Base too small for analysis.

Table 7.7   Chronic sickness: prevalence of reported longstanding illness by sex, age, and economic activity status

Persons aged 16 and over   Great Britain: 2002

Economic activity status	Men				Women			
	Age				Age			
	16-44	45-64	65 and over	Total	16-44	45-64	65 and over	Total
	Percentage who reported longstanding illness							
Working	20	37	45	27	23	34	46	27
Unemployed	32	55	*	38	31	[52]	*	35
Economically inactive	35	77	70	64	28	60	67	54
Total	23	46	67	38	25	44	66	39
Weighted bases (000's) = 100%								
Working	8,974	5,080	373	14,427	7,891	4,310	170	12,369
Unemployed	487	166	5	658	382	79	6	466
Economically inactive	1,320	1,447	3,388	6,156	3,006	2,528	4,800	10,333
Total	10,781	6,693	3,766	21,241	11,279	6,917	4,976	23,168
Unweighted sample								
Working	2963	1879	146	4988	2715	1641	62	4418
Unemployed	153	57	2	212	131	29	2	162
Economically inactive	403	527	1290	2220	1038	984	1684	3706
Total	3519	2463	1438	7420	3884	2654	1748	8286

* Base too small for analysis.

Table 7.8   Chronic sickness: prevalence of reported limiting longstanding illness by sex, age, and economic activity status

Persons aged 16 and over   Great Britain: 2002

Economic activity status	Men				Women			
	Age				Age			
	16-44	45-64	65 and over	Total	16-44	45-64	65 and over	Total
	Percentage who reported limiting longstanding illness							
Working	9	18	18	12	11	17	17	13
Unemployed	18	26	*	20	18	[34]	*	21
Economically inactive	30	66	50	49	21	46	47	39
Total	12	28	47	23	14	28	46	25
Weighted bases (000's) = 100%								
Working	8,968	5,080	373	14,421	7,891	4,309	169	12,370
Unemployed	488	166	5	658	382	79	6	466
Economically inactive	1,318	1,447	3,389	6,153	3,006	2,527	4,800	10,333
Total	10,774	6,693	3,767	21,232	11,279	6,915	4,975	23,169
Unweighted sample								
Working	2961	1879	146	4986	2715	1641	62	4418
Unemployed	153	57	2	212	131	29	2	162
Economically inactive	402	527	1290	2219	1038	984	1684	3706
Total	3516	2463	1438	7417	3884	2654	1748	8286

* Base too small for analysis.

Table 7.9   **Acute sickness**
            (a)   Prevalence of reported restricted activity in the 14 days before interview, by sex, age and economic activity status
            (b)   Average number of restricted activity days per person per year, by sex, age, and economic activity status

Persons aged 16 and over   Great Britain: 2002

Economic activity status	Men				Women			
	Age				Age			
	16-44	45-64	65 and over	Total	16-44	45-64	65 and over	Total
(a) Percentage who reported restricted activity in the 14 days before interview								
Working	10	12	9	11	12	12	3	12
Unemployed	9	10	*	9	17	[16]	*	18
Economically inactive	18	37	22	25	16	31	24	24
Total	11	17	21	15	13	19	24	17
(b) Average number of restricted activity days per person per year								
Working	19	25	24	21	21	24	3	22
Unemployed	25	26	*	25	37	[31]	*	38
Economically inactive	44	106	66	71	37	75	69	61
Total	22	43	62	36	26	43	67	40
Weighted bases (000's) = 100%								
Working	*8,969*	*5,077*	*373*	*14,419*	*7,888*	*4,307*	*169*	*12,364*
Unemployed	*491*	*165*	*5*	*661*	*382*	*79*	*6*	*467*
Economically inactive	*1,317*	*1,447*	*3,392*	*6,155*	*3,005*	*2,532*	*4,801*	*10,338*
Total	*10,777*	*6,689*	*3,769*	*21,235*	*11,275*	*6,918*	*4,976*	*23,169*
Unweighted sample								
Working	*2961*	*1878*	*146*	*4985*	*2714*	*1640*	*62*	*4416*
Unemployed	*154*	*57*	*2*	*213*	*131*	*29*	*2*	*162*
Economically inactive	*402*	*527*	*1291*	*2220*	*1038*	*986*	*1684*	*3708*
Total	*3517*	*2462*	*1439*	*7418*	*3883*	*2655*	*1748*	*8286*

* Base too small for analysis.

Table 7.10   **Self-reported sickness by sex and Government Office Region: percentage of persons who reported**
(a) longstanding illness
(b) limiting longstanding illness
(c) restricted activity in the 14 days before interview

All persons   Great Britain: 2002

Government Office Region*	(a) Longstanding illness	(b) Limiting longstanding illness	(c) Restricted activity in the 14 days before interview	Weighted base (000's) = 100%	Unweighted sample
Males					
England					
North East	37	22	15	1,193	415
North West	36	22	15	3,100	1094
Yorkshire and the Humber	43	26	19	2,336	823
East Midlands	36	18	12	1,956	712
West Midlands	34	21	12	2,474	882
East of England	35	19	12	2,490	911
London	25	16	12	3,355	1072
South East	33	18	15	3,773	1424
South West	36	20	13	2,504	920
All England	35	20	14	23,181	8253
Wales	38	25	17	1,415	506
Scotland	29	17	11	2,400	780
Great Britain	34	20	14	26,998	9539
Females					
England					
North East	42	29	24	1,384	490
North West	37	23	16	3,459	1245
Yorkshire and the Humber	40	26	22	2,496	891
East Midlands	34	19	15	2,085	775
West Midlands	35	23	15	2,582	925
East of England	36	23	14	2,541	950
London	28	18	14	3,644	1199
South East	35	21	16	4,039	1547
South West	32	19	13	2,551	951
All England	35	22	16	24,783	8973
Wales	38	23	15	1,461	528
Scotland	34	20	14	2,476	831
Great Britain	35	22	16	28,720	10332
All persons					
England					
North East	40	26	20	2,577	905
North West	37	23	15	6,559	2339
Yorkshire and the Humber	42	26	20	4,832	1714
East Midlands	35	19	13	4,041	1487
West Midlands	34	22	13	5,056	1807
East of England	36	21	13	5,031	1861
London	27	17	13	6,999	2271
South East	34	20	16	7,812	2971
South West	34	19	13	5,055	1871
All England	35	21	15	47,964	17226
Wales	38	24	16	2,876	1034
Scotland	31	18	13	4,876	1611
Great Britain	35	21	15	55,718	19871

* The data have not been standardised to take account of age or socio-economic classification.

Table 7.11   **Self-reported sickness by sex and NHS Regional Office area: percentage of persons who reported**
        (a) longstanding illness
        (b) limiting longstanding illness
        (c) restricted activity in the 14 days before interview

All persons   **Great Britain: 2002**

NHS Regional Office area	Males	Females	All persons
	(a) Longstanding illness		
Northern and Yorkshire	41	41	41
Trent	38	36	37
West Midlands	34	35	34
North West	36	37	37
Eastern	35	36	36
London	26	28	27
South East	33	35	34
South West	36	32	34
England	35	35	35
Wales	38	38	38
Scotland	29	34	31
Great Britain	34	35	35
	(b) Limiting longstanding illness		
Northern and Yorkshire	25	27	26
Trent	20	22	21
West Midlands	21	23	22
North West	22	23	23
Eastern	19	23	21
London	16	18	17
South East	18	21	20
South West	20	19	19
England	20	22	21
Wales	25	23	24
Scotland	17	20	18
Great Britain	20	22	21
	(c) Restricted activity in the 14 days before interview		
Northern and Yorkshire	17	22	20
Trent	14	17	15
West Midlands	12	15	13
North West	15	16	15
Eastern	12	14	13
London	12	14	13
South East	15	16	16
South West	13	13	13
England	14	16	15
Wales	17	15	16
Scotland	11	14	13
Great Britain	14	16	15
Weighted bases (000's) =100%			
Northern and Yorkshire	*2,816*	*3,216*	*5,942*
Trent	*2,670*	*2,841*	*5,511*
West Midlands	*2,468*	*2,579*	*5,047*
North West	*3,100*	*3,459*	*6,559*
Eastern	*2,485*	*2,537*	*5,022*
London	*3,360*	*3,649*	*7,009*
South East	*3,765*	*4,033*	*7,798*
South West	*2,517*	*2,559*	*5,076*
England	*23,181*	*24,783*	*47,964*
Wales	*1,415*	*1,461*	*2,876*
Scotland	*2,400*	*2,476*	*4,876*
Great Britain	*26,998*	*28,720*	*55,718*

Table 7.11 - **continued**

All persons

NHS Regional Office area	Males	Females	All persons
Unweighted sample			
Northern and Yorkshire	990	1113	2103
Trent	960	1043	2003
West Midlands	880	924	1804
North West	1094	1245	2339
Eastern	909	948	1857
London	1074	1201	2275
South East	1421	1545	2966
South West	925	954	1879
England	8253	8973	17226
Wales	506	528	1034
Scotland	780	831	1611
Great Britain	9539	10332	19871

Table 7.12   **Chronic sickness: rate per 1000 reporting longstanding condition groups, by sex**

Persons aged 16 and over   Great Britain: 2002

Condition group	Men	Women	Total
XIII Musculoskeletal system	156	179	168
VII Heart and circulatory system	119	119	119
VIII Respiratory system	71	69	70
III Endocrine and metabolic	41	49	45
IX Digestive system	32	39	36
VI Nervous system	29	32	30
V Mental disorders	28	30	29
VI Eye complaints	19	19	19
X Genito-urinary system	18	16	17
VI Ear complaints	17	15	16
II Neoplasms and benign growths	10	14	12
XII Skin complaints	10	10	10
IV Blood and related organs	4	6	5
Other complaints*	2	4	3
I Infectious diseases	2	2	2
Average number of conditions reported by those with a longstanding illness	1.5	1.6	1.5
Weighted bases (000's) = 100%	*21,746*	*23,487*	*45,232*
Unweighted sample	*7579*	*8393*	*15972*

* Including general complaints such as insomnia, fainting, generally run down, old age and general infirmity and non-specific conditions such as war wounds or road accident injuries where no further details were given.

Table 7.13 Chronic sickness: rate per 1000 reporting longstanding condition groups, by age

Persons aged 16 and over   Great Britain: 2002

Condition group	16-44	45-64	65-74	75 and over
XIII Musculoskeletal system	76	213	309	363
VII Heart and circulatory system	18	141	309	386
VIII Respiratory system	60	72	99	88
III Endocrine and metabolic	17	57	102	96
IX Digestive system	22	42	59	66
VI Nervous system	26	35	36	31
V Mental disorders	28	38	16	24
VI Eye complaints	5	16	38	86
X Genito-urinary system	9	18	36	34
VI Ear complaints	7	17	31	49
II Neoplasms and benign growths	4	13	24	42
XII Skin complaints	10	10	9	11
IV Blood and related organs	3	4	5	16
Other complaints*	4	2	2	4
I Infectious diseases	2	2	1	1
Average number of conditions reported by those with a longstanding illness	1.3	1.5	1.7	1.8
Weighted bases (000's) = 100%	*22,690*	*13,757*	*4,792*	*3,993*
Unweighted sample	*7602*	*5169*	*1766*	*1435*

* Including general complaints such as insomnia, fainting, generally run down, old age and general infirmity and non-specific conditions such as war wounds or road accident injuries where no further details were given.

Table 7.14 Chronic sickness: rate per 1000 reporting selected longstanding condition groups, by age and sex

Persons aged 16 and over   Great Britain: 2002

Condition group			16-44	45-64	65-74	75 and over	All ages
XIII	Musculoskeletal system	Men	82	208	285	274	156
		Women	70	218	329	419	179
VII	Heart and circulatory system	Men	17	152	330	398	119
		Women	19	129	291	379	119
VIII	Respiratory system	Men	57	70	114	110	71
		Women	62	73	85	74	69
III	Endocrine and metabolic	Men	13	54	95	102	41
		Women	21	59	109	93	49
IX	Digestive system	Men	22	35	55	60	32
		Women	22	48	63	70	39
VI	Nervous system	Men	22	33	43	34	29
		Women	30	37	30	29	32
Weighted bases (000's) = 100%		*Men*	*11,178*	*6,783*	*2,248*	*1,537*	*21,746*
		Women	*11,512*	*6,974*	*2,544*	*2,457*	*23,487*
Unweighted sample		*Men*	*3641*	*2494*	*849*	*595*	*7579*
		Women	*3961*	*2675*	*917*	*840*	*8393*

Table 7.15   **Chronic sickness: rate per 1000 reporting selected longstanding conditions, by sex and age**

Persons aged 16 and over   Great Britain: 2002

Condition	Men					Women				
	16-44	45-64	65-74	75 and over	All ages	16-44	45-64	65-74	75 and over	All ages
Musculoskeletal (XIII)										
Arthritis and rheumatism	13	86	153	194	63	19	119	219	268	96
Back problems	36	67	59	17	47	33	57	45	36	41
Other bone and joint problems	33	54	73	63	46	18	43	65	115	41
Heart and circulatory (VII)										
Hypertension	8	65	119	84	42	8	74	132	140	55
Heart attack	1	26	70	118	24	0	15	58	90	20
Stroke	1	8	26	39	9	1	8	22	37	9
Other heart complaints	4	42	85	133	33	5	24	56	80	24
Other blood vessel/embolic disorders	2	9	24	23	8	4	7	17	28	9
Respiratory (VIII)										
Asthma	43	39	54	42	43	53	52	48	49	52
Bronchitis and emphysema	1	9	25	34	8	1	10	16	12	6
Hay fever	10	4	4	4	7	5	2	2	0	3
Other respiratory complaints	4	19	31	31	13	4	9	19	13	8
Weighted bases (000's) = 100%	*11,178*	*6,783*	*2,248*	*1,537*	*21,746*	*11,512*	*6,974*	*2,544*	*2,457*	*23,487*
Unweighted sample	*3641*	*2494*	*849*	*595*	*7579*	*3961*	*2675*	*917*	*840*	*8393*

Table 7.16   **Chronic sickness: rate per 1000 reporting selected longstanding condition groups, by socio-economic classification of household reference person**

Persons aged 16 and over   Great Britain: 2002

Condition group		Managerial and professional	Intermediate	Small employers and own account	Lower supervisory and technical	Semi-routine and routine	All persons*
XIII	Musculoskeletal system	130	173	160	188	223	168
VII	Heart and circulatory system	97	120	101	153	152	119
VIII	Respiratory system	57	73	66	81	88	70
III	Endocrine and metabolic	35	49	40	46	61	45
IX	Digestive system	27	36	31	36	50	36
VI	Nervous system	27	35	28	35	35	30
Average number of condition groups reported by those with a longstanding illness		1.42	1.54	1.54	1.54	1.63	1.53
Weighted bases (000's) = 100%		*16,982*	*3,709*	*4,481*	*5,656*	*11,816*	*45,232*
Unweighted sample		*6201*	*1292*	*1606*	*1955*	*4104*	*15972*

* From April 2001 the National Statistics Socio-economic Classification (NS-SEC) was introduced for all official statistics and surveys. It replaced Social Class based on Occupation and Socio-economic Groups (SEG). Full-time students, persons in inadequately described occupations, persons who have never worked and the long-term unemployed are not shown as separate categories, but are included in the figure for all persons (see Appendix A for details).

Table 7.17 **Chronic sickness: rate per 1000 reporting selected longstanding condition groups, by sex and age and socio-economic classification of household reference person**

Persons aged 16 and over   Great Britain: 2002

Condition group	Men				Women				All aged 16 and over			
	16-44	45-64	65 and over	Total	16-44	45-64	65 and over	Total	16-44	45-64	65 and over	Total
XIII Musculoskeletal system												
Managerial and professional	67	157	227	118	65	172	347	141	66	165	289	130
Intermediate	75	182	262	142	89	176	405	187	82	179	348	166
Routine and manual	103	271	321	204	70	286	380	219	86	278	355	212
VII Heart and circulatory system												
Managerial and professional	18	138	345	103	9	97	381	92	14	117	363	97
Intermediate	15	144	309	107	27	122	276	113	21	133	289	110
Routine and manual	20	173	378	149	28	169	336	155	24	171	354	152
VIII Respiratory system												
Managerial and professional	55	57	90	61	52	48	65	53	54	52	77	57
Intermediate	56	61	102	65	76	66	78	73	66	64	87	69
Routine and manual	59	90	129	85	73	101	93	86	66	96	108	86
III Endocrine and metabolic												
Managerial and professional	9	44	106	34	18	51	74	37	13	47	90	35
Intermediate	18	62	90	45	18	43	94	43	18	52	92	44
Routine and manual	16	58	96	47	25	75	115	64	20	66	107	56
IX Digestive system												
Managerial and professional	16	24	47	23	19	45	48	32	18	34	48	27
Intermediate	21	33	65	32	17	35	68	34	19	34	67	33
Routine and manual	29	47	61	42	23	57	77	48	26	52	70	45
VI Nervous system												
Managerial and professional	16	28	42	24	26	35	36	30	21	31	39	27
Intermediate	27	34	35	30	38	29	26	32	32	31	30	31
Routine and manual	26	38	40	33	36	44	29	37	31	41	34	35
Weighted bases (000's) = 100%												
Managerial and professional	4,561	2,697	1,195	8,453	4,525	2,754	1,250	8,529	9,085	5,451	2,446	16,982
Intermediate	1,901	1,325	622	3,848	1,978	1,405	959	4,342	3,880	2,729	1,581	8,190
Routine and manual	3,860	2,558	1,907	8,325	3,966	2,628	2,554	9,148	7,826	5,186	4,461	17,473
Unweighted sample												
Managerial and professional	1538	1035	478	3051	1599	1088	463	3150	3137	2123	941	6201
Intermediate	614	482	243	1339	679	539	341	1559	1293	1021	584	2898

Table 7.18   **Trends in consultations with an NHS GP in the 14 days before interview by sex and age: 1972 to 2002**

All persons   Great Britain

	Unweighted								Weighted				Weighted base 2002 (000's) = 100%*	Unweighted sample* 2002
	1972	1975	1981	1985	1991	1995	1996	1998	1998	2000	2001	2002		
								Percentage consulting GP						
Males														
0- 4	13	13	21	22	23	22	23	18	18	18	18	19	1,706	604
5-15†	7	7	8	9	10	9	9	8	8	8	7	8	4,052	1516
16-44†	8	7	7	7	9	10	10	9	9	8	8	9	10,792	3521
45-64	11	11	12	12	11	14	15	14	14	15	13	15	6,700	2466
65-74	12	12	13	15	17	17	19	17	17	20	18	22	2,243	848
75 and over	19	20	17	19	21	22	21	21	21	20	22	21	1,528	592
Total	10	9	10	11	12	13	13	12	12	12	11	13	27,021	9547
Females														
0- 4	15	13	17	21	21	21	20	18	18	14	18	14	1,571	561
5-15†	6	7	9	11	11	13	9	10	10	9	9	8	3,976	1484
16-44†	15	13	15	17	17	18	20	17	17	16	15	18	11,288	3887
45-64	12	12	13	15	17	17	19	18	18	17	18	17	6,920	2656
65-74	15	16	16	17	19	23	21	19	19	22	18	21	2,531	912
75 and over	20	17	20	20	19	23	23	20	20	22	20	27	2,448	837
Total	13	12	14	16	17	18	19	17	17	16	16	17	28,734	10337
All persons														
0- 4	14	13	19	21	22	21	22	18	18	16	18	17	3,278	1165
5-15†	7	7	9	10	10	11	9	9	9	8	8	8	8,028	3000
16-44†	12	10	11	12	13	14	15	13	13	12	11	14	22,080	7408
45-64	12	11	12	14	14	16	17	16	16	16	16	16	13,620	5122
65-74	14	14	15	16	18	20	20	18	18	21	18	22	4,774	1760
75 and over	20	18	19	20	19	23	22	21	20	21	21	25	3,976	1429
Total	12	11	12	14	14	16	16	14	14	14	13	15	55,756	19884

*   Trend tables show unweighted and weighted figures for 1998 to give an indication of the effect of the weighting. For the weighted data (1998 and 2000 to 2002) the weighted base (000's) is the base for percentages. Unweighted data (up to 1998) are based on the unweighted sample. Bases for earlier years are of similar size and can be found in GHS reports for each year.

†   These age-groups were 5-14 and 15-44 in 1972 to 1978.

Table 7.19 Average number of NHS GP consultations per person per year by sex and age: 1972 to 2002

All persons*   Great Britain

	Unweighted									Weighted			
	1972†	1975	1981	1985	1991	1995	1996	1998		1998	2000	2001	2002
Males													
0- 4	4	4	7	7	7	7	8	6		6	6	6	7
5-15**	2	2	2	3	3	3	3	2		2	2	2	3
16-44**	3	2	2	2	3	3	3	3		3	3	3	3
45-64	4	4	4	4	4	4	5	4		4	5	4	5
65-74	4	4	4	5	5	5	6	5		5	6	5	7
75 and over	7	7	6	6	7	8	7	7		7	6	7	7
Total	3	3	3	3	4	4	4	4		4	4	4	4
Females													
0- 4	5	4	5	7	7	7	6	6		6	4	6	4
5-15**	2	2	3	3	3	4	3	3		3	3	3	2
16-44**	5	4	5	5	5	6	7	5		5	5	5	6
45-64	4	4	4	5	5	5	6	6		6	5	6	5
65-74	5	5	5	5	6	7	7	6		6	7	5	7
75 and over	7	6	6	7	6	7	7	6		6	7	6	9
Total	4	4	4	5	5	6	6	5		5	5	5	6
All persons													
0- 4	4	4	6	7	7	7	7	6		6	5	6	5
5-15**	2	2	3	3	3	3	3	3		3	2	3	2
16-44**	4	3	4	4	4	4	5	4		4	4	4	4
45-64	4	4	4	4	4	5	5	5		5	5	5	5
65-74	4	4	4	5	6	6	6	6		6	6	5	7
75 and over	7	7	6	6	6	7	7	6		6	7	6	8
Total	4	4	4	4	5	5	5	4		4	4	4	5

* Trend tables show unweighted and weighted figures for 1998 to give an indication of the effect of the weighting. Bases for 2002 are shown in Table 7.2. Bases for earlier years can be found in GHS reports for each year.
† 1972 figures relate to England and Wales.
** These age-groups were 5-14 and 15-44 in 1972 to 1978.

Table 7.20 NHS GP consultations: trends in site of consultation: 1971 to 2002

Consultations in the 14 days before interview   Great Britain

Site of consultation	Unweighted									Weighted			
	1971	1975	1981	1985	1991	1995	1996	1998		1998	2000	2001	2002
	%	%	%	%	%	%	%	%		%	%	%	%
Surgery*	73	78	79	79	81	84	84	84		84	86	85	86
Home	22	19	14	14	11	9	8	6		6	5	5	5
Telephone	4	3	7	7	8	7	8	10		10	10	10	9
Weighted base (000's)													
= 100%†										9,658	9,744	9,161	10,284
Unweighted sample†	5031	4455	4704	4123	4228	4385	4341	3504			3294	3418	3656

* Includes consultations with a GP at a health centre and those who had answered 'elsewhere'.
† Trend tables show unweighted and weighted figures for 1998 to give an indication of the effect of the weighting. For the weighted data (1998 and 2000 to 2002) the weighted base (000's) is the base for percentages. Unweighted data (up to 1998) are based on the unweighted sample.

Table 7.21   **Percentage of persons who consulted an NHS GP in the 14 days before interview by sex, and site of consultation, and by age and site of consultation**

Persons who consulted in the 14 days before interview   Great Britain: 2002

Site of consultation*	Total	Males	Females	Age					
				0-4	5-15	16-44	45-64	65-74	75 and over
	%	%	%	%	%	%	%	%	%
Surgery	88	89	88	87	88	90	92	88	75
At home	5	4	5	3	3	2	3	6	17
Telephone	10	9	10	14	11	11	7	8	11
Weighted base (000's) = 100%	*8,451*	*3,453*	*4,998*	*550*	*634*	*3,048*	*2,198*	*1,036*	*985*
Unweighted sample	*3015*	*1229*	*1786*	*194*	*234*	*1036*	*823*	*380*	*348*

* Percentages add to more than 100 because some people consulted at more than one site during the reference period.

Table 7.22   **NHS GP consultations**
         (a) Percentage of persons who consulted a doctor in the 14 days before interview, by sex, age, and economic activity status
         (b) Average number of consultations per person per year, by sex, age, and economic activity status

Persons aged 16 and over   Great Britain: 2002

Economic activity status	Men				Women			
	Age				Age			
	16-44	45-64	65 and over	Total	16-44	45-64	65 and over	Total
(a) Percentage who consulted a GP in the 14 days before interview								
Working	8	12	16	10	18	14	8	16
Unemployed	11	16	*	12	18	[24]	*	19
Economically inactive	14	24	22	21	20	23	25	23
Total	9	15	22	13	18	17	24	19
(b) Average number of consultations per person per year								
Working	3	4	4	3	6	4	2	5
Unemployed	3	5	*	4	6	[8]	*	7
Economically inactive	4	8	7	7	7	7	8	8
Total	3	5	7	4	6	5	8	6
Weighted base (000's) = 100%								
Working	*8,965*	*5,074*	*350*	*14,389*	*7,875*	*4,281*	*161*	*12,317*
Unemployed	*485*	*166*	*5*	*656*	*382*	*79*	*6*	*467*
Economically inactive	*1,318*	*1,447*	*3,391*	*6,156*	*3,009*	*2,532*	*4,803*	*10,344*
Total	*10,768*	*6,687*	*3,746*	*21,201*	*11,266*	*6,892*	*4,970*	*23,128*
Unweighted sample								
Working	*2960*	*1877*	*137*	*4974*	*2709*	*1629*	*59*	*4397*
Unemployed	*152*	*57*	*2*	*211*	*131*	*29*	*2*	*162*
Economically inactive	*402*	*527*	*1291*	*2220*	*1039*	*986*	*1685*	*3710*
Total	*3514*	*2461*	*1430*	*7405*	*3879*	*2644*	*1746*	*8269*

* Base too small for analysis.

Table 7.23 **Percentage of persons consulting an NHS GP in the 14 days before interview who obtained a prescription from the doctor, by sex, age and socio-economic classification of household reference person**

Persons who consulted in the 14 days before interview

Great Britain: 2002

Socio-economic classification of household reference person*	Males					Females				
	Age					Age				
	0-15	16-44	45-64	65 and over	Total	0-15	16-44	45-64	65 and over	Total
	Percentage consulting who obtained a prescription									
Managerial and professional	55	58	65	67	62	58	58	69	69	62
Intermediate	[53]	64	75	[80]	68	[55]	63	64	63	63
Routine and manual	71	71	72	79	74	71	68	73	79	73
All persons consulting	62	64	71	75	68	63	63	70	74	67
Weighted base (000's) = 100%										
Managerial and professional	216	365	336	268	1,185	194	790	424	276	1,684
Intermediate	121	160	219	100	600	85	326	225	230	866
Routine and manual	265	397	406	441	1,509	197	752	494	645	2,088
All persons consulting	662	977	1,000	812	3,451	522	2,071	1,195	1,209	4,997
Unweighted sample										
Managerial and professional	81	126	129	106	442	73	279	167	98	617
Intermediate	43	51	78	39	211	30	114	90	83	317
Routine and manual	93	127	142	161	523	70	257	185	221	733
All persons consulting	238	321	362	307	1228	190	715	460	421	1786

* From April 2001 the National Statistics Socio-economic Classification (NS-SEC) was introduced for all official statistics and surveys. It replaced Social Class based on Occupation and Socio-economic Groups (SEG). Full-time students, persons in inadequately described occupations, persons who have never worked and the long-term unemployed are not shown as separate categories, but are included in the figure for all persons (see Appendix A for details).

Table 7.24   **GP consultations: consultations with a doctor in the 14 days before interview by sex of person consulting and whether consultation was NHS or private**

Consultations in the 14 days before interview   Great Britain: 2002

Type of consultation	Consultations made by males	Consultations made by females	All persons
	%	%	%
NHS	96	98	97
Private	4	2	3
Weighted base (000's) = 100%	*4,934*	*6,931*	*11,865*
Unweighted sample	*1753*	*2462*	*4215*

Table 7.25   **Trends in reported consultations with a practice nurse by sex and age: 2000 to 2002**
         (a)   percentage consulting a practice nurse in the 14 days before interview
         (b)   average number of consultations with a practice nurse per person per year

All persons   Great Britain

	(a) percentage consulting a practice nurse			(b) average number of consultations with a practice nurse per person per year				
	2000	2001	2002	2000	2001	2002	*Weighted base 2002 (000's) = 100%*	*Unweighted sample 2002*
Males								
0- 4	4	3	4	1	1	1	*1,706*	*604*
5-15	2	2	2	0	1	0	*4,052*	*1516*
16-44	2	2	3	1	1	1	*10,798*	*3523*
45-64	5	6	6	1	2	2	*6,700*	*2466*
65-74	10	12	13	3	4	4	*2,243*	*848*
75 and over	8	13	11	3	4	3	*1,528*	*592*
Total	4	5	5	1	1	1	*27,027*	*9549*
Females								
0- 4	5	3	5	1	1	1	*1,571*	*561*
5-15	1	1	1	0	0	0	*3,977*	*1484*
16-44	5	5	6	1	1	2	*11,294*	*3889*
45-64	6	7	7	2	2	2	*6,918*	*2655*
65-74	10	11	11	3	4	3	*2,533*	*913*
75 and over	9	12	12	3	4	4	*2,448*	*837*
Total	5	6	7	2	2	2	*28,741*	*10339*
All persons								
0- 4	4	3	4	1	1	1	*3,277*	*1165*
5-15	1	2	1	0	1	0	*8,028*	*3000*
16-44	3	4	4	1	1	1	*22,092*	*7412*
45-64	6	6	7	2	2	2	*13,618*	*5121*
65-74	10	12	12	3	4	4	*4,777*	*1761*
75 and over	9	12	12	3	4	4	*3,976*	*1429*
Total	5	5	6	1	2	2	*55,768*	*19888*

Table 7.26   **Percentage of children using health services other than a doctor in the 14 days before interview**

All persons aged under 16   Great Britain: 2002

	Male			Female			Total		
	0-4	5-15	Total	0-4	5-15	Total	0-4	5-15	Total
Percentage who reported:*									
Seeing a practice nurse at the GP surgery	4	2	2	5	1	2	4	1	2
Seeing a health visitor at the GP surgery	7	1	3	8	1	3	8	1	3
Going to a child health clinic	6	0	2	5	0	2	5	0	2
Going to a child welfare clinic	0	0	0	0	0	0	0	0	0
None of the above	84	97	93	85	98	94	84	97	93
Weighted base (000's) =100%	*1,706*	*4,052*	*5,758*	*1,571*	*3,977*	*5,548*	*3,277*	*8,028*	*11,306*
Unweighted sample	*604*	*1516*	*2120*	*561*	*1484*	*2045*	*1165*	*3000*	*4165*

*   Percentages may sum to more than 100 as respondents could give more than one answer.

Table 7.27   **Trends in percentages of persons who reported attending an outpatient or casualty department in the 3 months before interview by sex and age: 1972 to 2002**

All persons*   Great Britain

	Unweighted									Weighted			
	1972†	1975	1981	1985	1991	1995	1996	1998		1998	2000	2001	2002
							Percentages						
Males													
0- 4	8	9	12	13	14	12	13	16		16	14	16	17
5-15**	9	8	11	12	11	11	12	12		12	11	10	11
16-44**	11	9	11	12	11	12	13	13		13	12	11	11
45-64	11	10	12	16	15	16	16	17		17	16	16	15
65-74	10	11	14	16	18	21	20	25		25	24	22	24
75 and over	10	12	14	15	22	26	25	29		29	26	31	26
Total	10	10	11	13	13	14	15	16		16	15	14	14
Females													
0- 4	6	8	9	11	11	12	9	13		13	10	11	12
5-15**	6	6	8	9	8	9	10	11		11	8	8	9
16-44**	9	9	11	12	12	12	13	13		13	13	12	12
45-64	11	10	13	15	16	17	18	18		18	16	18	16
65-74	12	12	16	17	18	21	22	21		21	21	21	20
75 and over	13	10	16	17	20	22	24	26		26	24	23	25
Total	10	9	12	13	14	14	15	16		16	15	14	14
All persons													
0- 4	7	9	10	12	13	12	11	14		15	12	13	14
5-15**	8	7	10	10	10	10	11	11		11	10	9	10
16-44**	10	9	11	12	12	12	13	13		13	13	12	12
45-64	11	10	13	15	16	16	17	18		18	16	17	15
65-74	11	11	15	17	18	21	21	23		23	22	21	22
75 and over	12	10	15	16	21	24	24	27		27	25	26	25
Total	10	9	12	13	13	14	15	16		16	15	14	14

*   Trend tables show unweighted and weighted figures for 1998 to give an indication of the effect of the weighting. Bases for 2002 are shown in Table 7.2. Bases for earlier
     years can be found in GHS reports for each year.
†   1972 figures relate to England and Wales.
** These age-groups were 5-14 and 15-44 in 1972 to 1978.

Table 7.28   **Trends in day-patient treatment in the 12 months before interview by sex and age, 1992 to 2002**

All persons   Great Britain

	Unweighted					Weighted				Weighted base 2002 (000's) = 100%*	Unweighted sample* 2002
	1992	1994	1995	1996	1998	1998	2000	2001	2002		
					Percentage receiving day-patient treatment						
Males											
0- 4	4	4	4	5	6	6	6	7	8	1,706	604
5-15	2	3	3	3	4	4	5	4	5	4,052	1516
16-44	4	5	6	5	6	6	6	7	7	10,806	3525
45-64	4	5	7	6	7	7	8	8	8	6,699	2466
65-74	5	6	6	7	6	6	10	8	11	2,244	848
75 and over	4	5	5	6	12	11	7	10	12	1,528	592
Total	4	5	5	5	6	6	7	7	8	27,035	9551
Females											
0- 4	2	3	3	3	5	4	6	4	5	1,571	561
5-15	2	3	2	4	4	4	3	3	4	3,977	1484
16-44	5	7	6	7	8	8	8	8	9	11,294	3889
45-64	5	5	7	8	8	8	9	8	8	6,917	2655
65-74	4	5	5	6	6	6	12	8	10	2,533	913
75 and over	3	5	5	7	8	8	8	8	10	2,442	835
Total	4	5	5	6	7	7	8	7	8	28,734	10337
All persons											
0- 4	3	3	3	4	5	5	6	5	6	3,277	1165
5-15	2	3	3	3	4	4	4	4	4	8,029	3000
16-44	4	6	6	6	7	7	7	7	8	22,099	7414
45-64	5	5	7	7	8	8	8	8	8	13,618	5121
65-74	4	5	6	7	6	6	11	8	10	4,777	1761
75 and over	3	5	5	6	9	9	8	9	11	3,971	1427
Total	4	5	5	6	7	7	7	7	8	55,771	19888

*   Trend tables show unweighted and weighted figures for 1998 to give an indication of the effect of the weighting. For the weighted data (1998 and 2000 to 2002) the weighted base (000's) is the base for percentages. Unweighted data (up to 1998) are based on the unweighted sample. Bases for earlier years are of similar size and can be found in GHS reports for each year.

Table 7.29   Trends in inpatient stays in the 12 months before interview by sex and age, 1982 to 2002

All persons   Great Britain

	Unweighted						Weighted				Weighted base 2002 (000's) = 100%*	Unweighted sample* 2002
	1982	1985	1991	1995	1996	1998	1998	2000	2001	2002		
					Percentage with inpatient stay							
Males												
0- 4	14	12	10	9	9	9	9	8	11	11	1,706	604
5-15	6	8	6	5	5	5	5	5	4	4	4,047	1514
16-44	5	6	6	5	5	5	5	4	5	5	10,808	3526
45-64	8	8	8	9	8	8	8	8	8	8	6,699	2466
65-74	12	13	13	15	13	15	15	13	12	15	2,244	848
75 and over	14	17	20	21	18	21	21	18	19	18	1,528	592
Total	7	8	8	8	7	8	8	7	7	7	27,032	9550
Females												
0- 4	12	8	8	8	7	10	10	6	6	7	1,571	561
5-15	4	5	4	4	4	4	4	3	3	2	3,977	1484
16-44	15	16	15	12	12	11	11	10	9	9	11,294	3889
45-64	8	8	9	8	10	8	9	7	8	8	6,920	2656
65-74	8	18	11	11	12	10	10	13	10	8	2,533	913
75 and over	12	13	16	20	16	15	15	18	15	16	2,442	835
Total	11	11	11	10	10	10	10	9	8	8	28,737	10338
All persons												
0- 4	13	10	9	9	8	9	9	7	9	9	3,277	1165
5-15	5	6	5	4	4	5	5	4	4	3	8,023	2998
16-44	10	11	10	8	9	8	8	7	7	7	22,102	7415
45-64	8	8	8	8	9	8	9	8	8	8	13,620	5122
65-74	10	10	12	13	12	12	12	13	11	11	4,777	1761
75 and over	13	15	18	20	17	17	17	18	17	17	3,970	1427
Total	9	10	10	9	9	9	9	8	8	8	55,769	19888

* Trend tables show unweighted and weighted figures for 1998 to give an indication of the effect of the weighting. For the weighted data (1998 and 2000 to 2002) the weighted base (000's) is the base for percentages. Unweighted data (up to 1998) are based on the unweighted sample. Bases for earlier years are of similar size and can be found in GHS reports for each year.

Table 7.30　　**Average number of nights spent in hospital as an inpatient during the 12 months before interview, by sex and age**

All inpatients　　　　　　　　　　　　　　　　　　　　　　　　　　　　　　　　　　Great Britain: 2002

Age	Male	Female	Total	Male	Female	Total	Male	Female	Total
	Average number of nights			Weighted base (000's) = 100% (all inpatients)			Unweighted sample		
0-4	3	[5]	4	183	116	299	65	41	106
5-15	2	[4]	3	158	98	255	59	36	95
16-44	5	4	4	523	560	1,083	171	193	364
45-64	7	7	7	478	550	1,028	177	206	383
65-74	7	7	7	331	195	526	124	72	196
75 and over	11	14	12	264	355	619	103	123	226
All persons	6	7	7	1,937	1,874	3,811	699	671	1370

Table 7.31　　**Inpatient stays and outpatient attendances**
　　　　　　　(a)　Average number of inpatient stays per 100 persons in a 12 month reference period, by sex and age
　　　　　　　(b)　Average number of outpatient attendances per 100 persons per year, by sex and age

All persons　　　　　　　　　　　　　　　　　　　　　　　　　　　　　　　　　　Great Britain: 2002

Age	(a) Average number of inpatient stays per 100 persons in a 12 month reference period			(b) Average number of outpatient attendances per 100 persons per year			Weighted base (000's) = 100%			Unweighted sample		
	Males	Females	Total	Males	Females	Total	Males	Females	Total	Males	Females	Total
0- 4	14	9	12	108	68	89	1,706	1,571	3,277	604	561	1165
5-15	5	3	4	64	51	57	4,047	3,977	8,023	1514	1484	2998
16-44	6	7	7	82	108	95	10,802	11,288	22,090	3524	3887	7411
45-64	12	11	11	133	122	128	6,697	6,920	13,617	2465	2656	5121
65-74	22	11	16	192	161	176	2,241	2,531	4,772	847	912	1759
75 and over	26	21	23	199	221	212	1,528	2,439	3,967	592	834	1426
Total	10	9	10	109	116	113	27,021	28,726	55,747	9546	10334	19880

Chapter 8 Smoking

Questions about smoking behaviour have been asked of GHS respondents aged 16 and over in alternate years since 1974. Following the review of the GHS, the smoking questions became part of the continuous survey and have been included every year from 2000 onwards.

- The reliability of smoking estimates
- The effect of weighting on the smoking data
- Targets for the reduction of smoking
- Trends in the prevalence of cigarette smoking
- Cigarette smoking and marital status
- Regional variation in cigarette smoking
- Cigarette smoking and socio-economic classification
- Cigarette smoking and economic activity status
- Cigarette consumption
- Cigarette type
- Tar yield
- Cigar and pipe smoking
- Age started smoking
- Dependence on cigarette smoking

This chapter updates information about trends in cigarette smoking presented in previous reports in this series. It also comments on variations according to personal characteristics such as sex, age, socio-economic classification and economic activity status, as well as commenting briefly on the prevalence of cigarette smoking in different parts of Great Britain. Other topics covered include cigarette consumption, type of cigarette smoked and dependence on cigarettes.

The reliability of smoking estimates

As noted in previous GHS reports, it is likely that the GHS underestimates cigarette consumption and (perhaps to a lesser extent) prevalence (the proportion of people who smoke). The evidence suggests that when respondents are asked how many cigarettes they smoke each day, there is a tendency to round the figure down to the nearest multiple of 10. Underestimates of consumption are therefore likely to occur in all age groups.

Under-reporting of prevalence is most likely to occur among young people. To protect their privacy, particularly when they are being interviewed in their parents' home, young people aged 16 and 17 complete the smoking and drinking sections of the questionnaire themselves. Neither the questions nor their responses are heard by anyone else who may be present. This is probably only partially successful in encouraging honest answers[1].

When considering trends in smoking, it is usually assumed that any under-reporting remains constant over time. However, since the prevalence of smoking has fallen, this assumption may not be entirely justified. As smoking has become less acceptable as a social habit, some people may be less inclined to admit how much they smoke – or, indeed, to admit to smoking at all.

The effect of weighting on the smoking data

Weighting for nonresponse was introduced on the GHS in 2000 and was described in detail in the GHS 2000 report[2]. The effect of weighting on the smoking data is slight, increasing the overall prevalence of cigarette smoking by one percentage point. The change occurs because weighting reduces the contribution to the overall figure of those aged 60 and over, among whom prevalence is relatively low.

The weights which should be applied to the 2001 data were re-calculated following revisions to the population estimates to which the data are grossed – see Appendix D for further details. The effect on the data was very slight, and figures published in *Living in Britain 2001* have not been amended in this report.

Targets for the reduction of smoking

In December 1998 *Smoking Kills – a White Paper on Tobacco*[3] was released, which included targets for reducing the prevalence of cigarette smoking among adults in England from 28% in 1996 to 24% by 2010 (with an interim target of 26% by 2005). These targets were based on unweighted GHS data. Since they will now be monitored using weighted data, it is suggested that they should be revised upwards by one percentage point.

Since smoking is estimated to be the cause of about one third of all cancers, reducing smoking is also one of three key commitments at the heart of the NHS *Cancer Plan*[4]. In particular, the *Cancer Plan* focuses on the need to reduce the comparatively high rates of smoking among those in manual socio-economic groups, which result in much higher death rates from cancer among unskilled workers than among professionals. The national target is to reduce the proportion of smokers in manual groups in England from 32% in 1998 to 26% by 2010. Comparisons of weighted and unweighted data suggest that, as with the *Smoking Kills* targets, these should also be increased by one percentage point. These figures may also need further revision in the light of the recent introduction of the new socio-economic classification NS-SEC.

Trends in the prevalence of cigarette smoking

There was a fall of one percentage point in the overall prevalence of cigarette smoking among adults in Great Britain, from 27% in 2001 to 26% in 2002.

The prevalence of cigarette smoking fell substantially in the 1970s and the early 1980s, from 45% in 1974 to 35% in 1982. After 1982, the rate of decline slowed, with prevalence falling by only about one percentage point every two years until 1990, since when it has levelled out. The fall in prevalence between 2001 and 2002 is on the borderline of statistical significance, but it should be noted that even during periods when the prevalence of smoking in the general population is changing little, upward and downward movements in survey estimates are to be expected. This is because of sampling fluctuations that make the detection of trends difficult.

The series of weighted data shown for 1998 and later years suggests that there may have been a recent fall in the prevalence of cigarette smoking among men, though not among women. Prevalence is nevertheless still higher among men, with 27% of men and 25% of women being cigarette smokers in 2002. The fall among men appears to be due to an increase in the proportion who say they have never smoked cigarettes regularly, rather than in the proportion who have given up smoking.

In the 1970s, men were much more likely than women to be smokers. In 1974 for example, 51% of men smoked cigarettes, compared with 41% of women. Since then, the difference in smoking prevalence between men and women has reduced, although it has not disappeared completely.

This change results mainly from a combination of two factors.
- First, there is a cohort effect resulting from the fact that smoking became common among men several decades before it did among women. In the 1970s there was a fall in the proportion of women aged 60 and over who had never smoked regularly.

- Second, men are more likely than women to have given up smoking cigarettes. It should be noted, however, that this difference conceals the fact that a proportion of men who give up smoking cigarettes remain smokers (by continuing to smoke cigars and pipes). This is much less common among women who stop smoking cigarettes.

The effect of weighting on the 1998 and 2000 data suggested that the difference in prevalence between men and women may have been slightly underestimated by the unweighted data shown in previous reports, but this is less clear from the 2001 and 2002 data.

Smoking among different age groups is another key area of interest. Since the early 1990s, the prevalence of cigarette smoking has been higher among those aged 20 to 24 than among those in other age groups. Up to the early twenties, more young people are starting to smoke than are giving up (as shown later in this chapter, almost one in five of those who have smoked at some time in their lives took up the habit after the age of 20).

Since the survey began, there has been considerable fluctuation in prevalence rates among those aged 16 to 19, but this is mainly because of the small sample size in this age group. However, the fall in prevalence among young men aged 16 to 19 from 30% in 2000 to 22% in 2002 is statistically significant. There was no equivalent fall among young women of that age, and in 2002, women aged 16 to 19 were significantly more likely to smoke than were young men of similar age. This is interesting because since the early 1990s surveys of smoking among secondary school children have shown higher prevalence among 15-year-old girls than among boys of the same age, but this has not previously been reflected in the GHS data.

At 15% in 2002, prevalence continues to be lowest among men and women aged 60 and over. Although they are more likely than younger people to have ever been smokers, they are much more likely to have given up.

Tables 8.1–8.3, Figures 8A, 8B

Figure 8A   **Prevalence of cigarette smoking for men and women, 1974 to 2002**

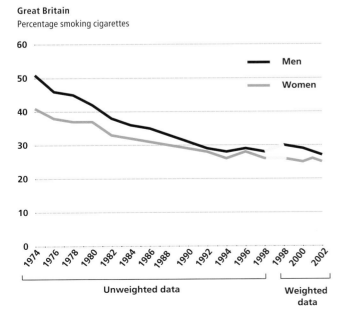

Figure 8B   **Prevalence of cigarette smoking by sex and age, 1980 to 2002**

Great Britain
Percentage smoking cigarettes

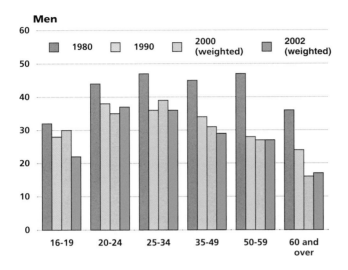

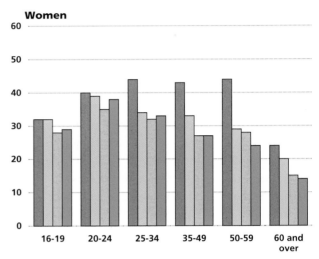

Cigarette smoking and marital status

The prevalence of cigarette smoking varies considerably according to marital status. It is much lower among married people than among those in any of the three other marital status categories (single, cohabiting, and widowed, divorced or separated). This is not explained by the association between age and marital status (for example, married people and those who are widowed, divorced or separated are older, on average, than single people). Table 8.5 shows that in every age group married people were less likely to be smokers than were other respondents (although this is much less marked among those aged 60 and over). For example, about 40% of those aged 25 to 34 who were single or cohabiting were smokers, compared with only 24% of those who were married.   **Tables 8.4–8.5**

Regional variation in cigarette smoking

The data presented so far have been for Great Britain, but the targets included in the White Paper *Smoking Kills* and in the NHS *Cancer Plan* relate to England only. Table 8.6 shows that in 2002, overall prevalence in England was 26%, the same as in Great Britain as a whole.

As in previous years, prevalence in 2002 was significantly higher in Scotland, at 28%, than in the rest of Great Britain. In England, prevalence tended to be higher in the north of the country than in the midlands and the south.   **Tables 8.6–8.7**

Cigarette smoking and socio-economic classification

The National Statistics Socio-economic classification (NS-SEC), which was introduced in 2001, does not allow categories to be collapsed into broad non-manual and manual groupings. Since the *Cancer Plan* targets for England relate particularly to those in the manual socio-economic groups, the old socio-economic groupings have been recreated for this report in Table 8.9. Because of the new occupation coding, the classifications are not exactly the same, and comparisons with previous years should be treated with caution.

The GHS has consistently shown striking differences in the prevalence of cigarette smoking in relation to socio-economic group, with smoking being considerably more prevalent among those in manual groups than among those in non-manual groups. In the 1970s and 1980s, the prevalence of cigarette smoking fell more sharply among those in non-manual than in manual groups, so that differences between the groups became proportionately greater (table not shown). There was little further change in the relative proportions smoking cigarettes during the 1990s.

In England in 2002, 31% of those in manual groups were cigarette smokers, suggesting some progress in relation to the targets set out in the *Cancer Plan*. These are to reduce prevalence among those in the manual group from 33% in 1998 to 27% in 2010, if the targets are increased by one percentage point to allow for the effect of weighting the data. However, since the proportion of those in non-manual groups who are cigarette smokers has fallen by a similar amount (from 22% in 1998 to 20% in 2002) the differential between non-manual and manual has not reduced.

Caution is advisable when making comparisons over this period because the data may have been affected by two matters. These are:

- the change from head of household to household reference person as the basis for assessing socio-economic group;
- the introduction of the revised occupation coding and socio-economic classification.

In England in 2002 32% of men living in households in the manual group smoked cigarettes compared with 21% of those in non-manual households. The corresponding proportions for women were 30% and 20%.

Table 8.10 shows the prevalence of cigarette smoking in 2002 in relation to the eight- and three- category versions of NS-SEC. As was the case with the socio-economic groupings used previously, there were striking differences between the various classes. Prevalence was lowest among those in higher professional and higher managerial households (14% and 16% respectively) and highest (33%) among those whose household reference person was in a semi-routine or routine occupation.

Tables 8.9–8.10

Cigarette smoking and economic activity status

Those who were economically active were more likely to smoke than those who were not, but this is largely explained by the lower prevalence of smoking among those aged 60 and over who form the majority of economically inactive people.

Indeed, among both men and women, prevalence was highest (33%) among economically inactive people aged 16 to 59 compared with 28% of economically active people and only 15% of economically inactive people aged 60 and over. Prevalence was particularly high among economically inactive people aged 16 to 59 whose last job was a routine or manual one, 46% of whom were cigarette smokers.

Table 8.11

Cigarette consumption

Although the prevalence of cigarette smoking changed little during the 1990s, the GHS has shown a continuing fall in the reported number of cigarettes smoked. The fall in consumption has occurred mainly among younger smokers, while the number of cigarettes smoked by those aged 50 and over has changed very little since the mid-1970s.

Most of the decline in consumption in the 1990s is due to a reduction in the proportion of heavy smokers. The proportion of respondents smoking on average 20 or more cigarettes a day fell from 14% of men in 1990 to 11% in 1998, and from 9% to 7% of women over the same period. It has since remained virtually unchanged among both men and women. The proportion of respondents who were light smokers also changed little throughout the 1990s.

In all age groups, respondents are much more likely to be light than heavy smokers, the difference being most pronounced among younger age groups up to the age of 35. For example, 19% of young men and 25% of young women aged 16 to 19 were light smokers in 2002, but only 3% and 4% respectively were heavy smokers.

Tables 8.12–8.13

As in previous years, male smokers smoked more cigarettes a day on average in 2002 than female smokers (15 compared with 13). Cigarette consumption also varied by age. Among both men and women smokers, those aged 35 to 59 smoked the most – men smokers smoked on average 17 or 18 cigarettes a day and women smoked 15 a day.

Earlier GHS reports have shown cigarette consumption levels to be higher among male and female smokers in manual socio-economic groups than among those in non-manual groups. A similar pattern was evident in 2002 in relation to NS-SEC. Smokers in households where the household reference person was in a routine or manual occupation smoked an average of 15 cigarettes a day, compared with 13 a day for those in managerial or professional households.

Tables 8.14–8.15

Cigarette type

Filter cigarettes continue to be the most widely smoked type of cigarette, especially among women. During the 1990s, however, there was a marked increase in the proportion of smokers who said that they smoked mainly hand-rolled tobacco. In 1990, 18% of men smokers and 2% of women smokers said they smoked mainly hand-rolled cigarettes, but by 2002 this had risen to 33% and 13% respectively.

There are likely to be two main reasons for this sharp increase in the use of hand-rolled cigarettes:
- the rise in the real price of packaged cigarettes - hand-rolled ones are cheaper;
- the reduced tar yield of packaged cigarettes (see below) - depending on how they are rolled and smoked, hand-rolled ones may give a higher tar yield.

It is possible that the lessening of the restrictions on the amount of tobacco that can legally be brought into the country and an increase in smuggling have also contributed to the increase in the consumption of hand-rolled tobacco.

The use of hand-rolled tobacco was more common among men aged 35 to 59 than among men of other ages. Among women smokers there was less variation with age, except that only 7% of women smokers aged 60 or over used hand-rolled tobacco.   **Tables 8.16–8.17**

Tar yield[5]

Table 8.18 shows the very marked reduction in the tar yield of cigarettes over the period during which the GHS has been collecting information about brand smoked. In 1986, 40% of those who smoked manufactured cigarettes smoked brands yielding 15mg or more of tar per cigarette. In the following decade, the proportion smoking this type of cigarette fell to zero. Initially, this was partly due to smokers switching to lower tar brands, but the main factor has been the requirement for manufacturers to reduce substantially the tar yields of existing brands. Following legislation in 1992, they were required to reduce the tar yield to no more than 12mg per cigarette by the beginning of 1998. An EU Directive which came into force at the end of 2002 has further reduced the maximum tar yield to 10 mg per cigarette from January 2004[6].

Although there has been a shift towards the cigarettes with the lowest tar yield, the biggest increase has been in what used to be the middle category – cigarettes with a tar yield of 10 but less than 15mg. In 2002 71% of those who smoked manufactured cigarettes smoked brands in this category.

Among smokers aged under 35, differences between men and women in the tar yield of their usual brand were small. Among those aged 35 and over, however, men were much less likely to smoke low tar brands.

There was also a difference in tar yield of cigarettes smoked according to the socio-economic classification of the smoker's household reference person. Cigarettes with the highest tar yield (12 mg or more) were more likely to be smoked by those in routine and manual households than by other smokers: 44% of smokers in routine and manual groups smoked these cigarettes, compared with only 31% of smokers in managerial and professional households.   **Tables 8.18–8.20**

Cigar and pipe smoking

A decline in the prevalence of pipe and cigar smoking among men has been evident since the survey began, with most of the reduction occurring in the 1970s and 1980s. In 2002, only 5% of men smoked at least one cigar a month, compared with 34% in 1974. Only a small number of women smoked cigars in 1974, and since 1978 the percentages have been scarcely measurable on the GHS.

Only 1% of men in 2002 said they smoked a pipe. At 3%, the proportion doing so was higher among men aged 60 and over than among any other age group. Cigar smoking, on the other hand, was not so concentrated among older men, with 2% of men aged 16 to 19 saying they smoked at least one cigar a month.   **Tables 8.21–8.22**

Age started smoking

The White Paper *Smoking Kills*[3] noted that people who start smoking at an early age are more likely than other smokers to smoke for a long period of time and more likely to die prematurely from a smoking-related disease.

Two thirds of respondents who were either current smokers or who had smoked regularly at some time in their lives had started smoking before they were 18, and almost two fifths started before the age of 16. Men were more likely than women to have started smoking before they were 16 (42% of men who had ever smoked regularly, compared with 33% of women). One fifth of women who had ever smoked did not start until they were in their twenties or older, compared with 15% of men.

As the GHS has shown in previous years, there was an association between age started smoking regularly and socio-economic classification based on the current or last job of the household reference person. Of those in managerial and professional households, 30% had started smoking before they were 16, compared with 43% of those in routine and manual households.

Current heavy smokers were much more likely than other current or ex-smokers to have started smoking at an early age. Of those smoking 20 or more cigarettes a day, 50% started smoking regularly before they were 16, compared with only 32% of those currently smoking fewer than 10 cigarettes a day.   **Tables 8.23–8.24**

Dependence on cigarette smoking

For the prevalence of cigarette smoking to reduce, young people have to be discouraged from starting to smoke and existing smokers have to be encouraged to stop. Since 1992, the GHS has asked three questions relevant to the likelihood of a smoker giving up. First, whether they would like to stop smoking, and then two indicators of dependence: whether they think they would find it easy or difficult not to smoke for a whole day; and how soon after waking they smoke their first cigarette.

There has been very little change since 1992 in any of the three dependence measures used. This is perhaps not unexpected, given that there has been little change in the prevalence of cigarette smoking over that period.

For an attempt to stop smoking to be successful, the smoker must want to stop. In 2002, 68% of smokers (68% of men and 69% of women) said they would like to stop smoking altogether. The relationship between wanting to stop smoking and the number of cigarettes smoked is not straightforward. In every survey since the questions were first included in 1992, the proportion wanting to give up has been highest among those smoking on average 10-19 cigarettes a day.

It is interesting that it is not the heaviest smokers who are most likely to want to stop. This may be because they feel it would be too difficult or because they have been discouraged from wanting to stop by previous unsuccessful attempts. Furthermore, some previously heavy smokers who would like to give up may have cut down their consumption prior to an attempt to do so.

In 2002, 57% of smokers felt that it would be either very or fairly difficult to go without smoking for a whole day. Not surprisingly, heavier smokers were more likely to say they would find it difficult – 81% of those smoking 20 or more cigarettes a day did so, compared with only 22% of those smoking fewer than 10 cigarettes a day.

Since women are less likely to be heavy smokers than men, it might be expected that women would be less likely to say they would find it hard to stop smoking for a day. As in previous years, however, this was not the case. Overall, 59% of women, compared with 56% of men, said they would find it difficult not to smoke for a day. The difference was particularly marked among those smoking 20 or more cigarettes a day: 86% of women, compared with 77% of men, said they would find it difficult.

In 2002, 15% of smokers had their first cigarette within five minutes of waking up. Heavy smokers were more likely than light smokers to smoke immediately on waking up, with 31% of those smoking 20 or more cigarettes a day doing so, compared with only 3% of those smoking fewer than 10 a day. The proportions of men and women smokers who said they had their first cigarette within five minutes of waking were similar.

Women smokers are therefore more likely to perceive themselves as dependent despite the fact that on average they smoke fewer cigarettes a day than men. There is no statistically significant difference between men and women smokers in the proportions wanting to give up, nor in the more objective of the two indicators of dependence (how soon they smoke after waking up).

Smokers in managerial and professional households were more likely than smokers in routine and manual households to say they would like to give up smoking altogether (72% compared with 66%). In general, this association was evident at each consumption level.

Overall, smokers in routine and manual households were more likely than others to say they would find it difficult to go without smoking for a whole day (62% compared with 50% of those in managerial and professional households). However, once amount smoked was taken into account (smokers in that group smoke more on average than smokers in other social classes) the pattern of association was less clear.

Similarly, smokers in routine and manual households were more likely than others to have their first cigarette of the day within five minutes of waking up. The difference was particularly marked among those smoking 20 or more cigarettes a day. Of heavy smokers in routine and manual households, 32% smoked within five minutes of waking, compared with only 25% of those in managerial and professional households.

Tables 8.25–8.30

Notes and references

1 See *Chapter 4, General Household Survey 1992*, HMSO 1994. This includes a discussion of the differences found when smoking prevalence reported by young adults on the GHS was compared with prevalence reported on surveys of smoking among secondary school children.

2 See *Appendix D, Living in Britain: results from the 2000 General Household Survey.* The Stationery Office (London 2001).

3 *Smoking Kills – a White Paper on Tobacco.* The Stationery Office (London 1998).

4 *The NHS Cancer Plan.*, Department of Health, 2000 (www.doh.gov.uk/cancer/cancerplan.htm)

5 An error was found in the automated procedure for coding the brand of cigarette smoked which was introduced when the GHS moved to computerised interviewing in April 1994. The net effect of this was that from 1994 to 2000, some brands were wrongly assigned to a low tar category. The coding procedure was revised for the 2001 survey. Corrected data for 1998 and 2000 are given in Table 8.18.

6 In practice, some brands will continue to have a higher measured tar yield: the Directive relates to the tar yield as declared by the manufacturer, but this is permitted to vary by up to 15% from the yield as measured for the Laboratory of the Government Chemist.

Table 8.1　　**Prevalence of cigarette smoking by sex and age: 1974 to 2002**

Persons aged 16 and over　　　　　　　　　　　　　　　　　　　　　　　　　　　　　　　　　　Great Britain

Age	Unweighted									Weighted				Weighted base 2002 (000's) =100%*	Unweighted sample* 2002
	1974	1978	1982	1986	1990	1992	1994	1996	1998	1998	2000	2001	2002		
							Percentage smoking cigarettes								
Men															
16-19	42	35	31	30	28	29	28	26	30	30	30	25	22	1,089	349
20-24	52	45	41	41	38	39	40	43	42	41	35	40	37	1,384	420
25-34	56	48	40	37	36	34	34	38	37	38	39	38	36	3,292	1063
35-49	55	48	40	37	34	32	31	30	32	33	31	31	29	5,491	1884
50-59	53	48	42	35	28	28	27	28	27	28	27	26	27	3,374	1241
60 and over	44	38	33	29	24	21	18	18	16	16	16	16	17	4,931	1880
All aged 16 and over	51	45	38	35	31	29	28	29	28	30	29	28	27	19,561	6837
Women															
16-19	38	33	30	30	32	25	27	32	31	32	28	31	29	1,122	388
20-24	44	43	40	38	39	37	38	36	39	39	35	35	38	1,586	510
25-34	46	42	37	35	34	34	30	34	33	33	32	31	33	3,704	1282
35-49	49	43	38	34	33	30	28	30	28	29	27	28	27	6,175	2207
50-59	48	42	40	35	29	29	26	26	27	27	28	25	24	3,516	1333
60 and over	26	24	23	22	20	19	17	19	16	16	15	17	14	6,134	2231
All aged 16 and over	41	37	33	31	29	28	26	28	26	26	25	26	25	22,236	7951
Total															
16-19	40	34	30	30	30	27	27	29	31	31	29	28	25	2,211	737
20-24	48	44	40	39	38	38	39	39	40	40	35	37	38	2,971	930
25-34	51	45	38	36	35	34	32	36	35	35	35	34	34	6,996	2345
35-49	52	45	39	36	34	31	30	30	30	31	29	29	28	11,666	4091
50-59	51	45	41	35	29	29	27	27	27	28	27	26	26	6,889	2574
60 and over	34	30	27	25	21	20	17	18	16	16	16	17	15	11,065	4111
All aged 16 and over	45	40	35	33	30	28	27	28	27	28	27	27	26	41,798	14788

* Trend tables show unweighted and weighted figures for 1998 to give an indication of the effect of the weighting. For the weighted data (1998 and 2000 to 2002) the weighted base (000's) is the base for percentages. Unweighted data (up to 1998) are based on the unweighted sample. Bases for earlier years are of similar size and can be found in GHS reports for each year.

Table 8.2　Ex-regular cigarette smokers by sex and age: 1974 to 2002

Persons aged 16 and over　　　　　　　　　　　　　　　　　　　　　　　　　　　　　　　　Great Britain

Age	Unweighted									Weighted				Weighted base 2002 (000's) =100%*	Unweighted sample* 2002
	1974	1978	1982	1986	1990	1992	1994	1996	1998	1998	2000	2001	2002		
						Percentage of ex-regular cigarette smokers									
Men															
16-19	3	4	4	5	4	5	5	5	5	5	3	4	3	1,089	349
20-24	9	9	9	11	8	8	7	10	8	9	7	9	7	1,384	420
25-34	18	18	20	20	16	16	16	13	13	13	12	15	13	3,292	1063
35-49	21	26	32	33	32	29	27	27	22	21	20	20	20	5,491	1884
50-59	30	35	38	38	42	41	40	41	41	40	36	36	35	3,374	1241
60 and over	37	43	47	52	52	55	55	55	54	54	52	47	51	4,931	1880
All aged 16 and over	23	27	30	32	32	32	31	32	31	29	27	27	28	19,561	6837
Women															
16-19	4	5	6	7	6	5	6	5	7	8	6	6	5	1,122	388
20-24	9	8	9	9	8	9	10	11	8	8	11	12	10	1,586	510
25-34	12	14	15	16	14	15	14	13	14	14	13	16	16	3,704	1282
35-49	10	13	15	20	20	22	21	18	19	19	19	19	17	6,175	2207
50-59	13	18	19	18	20	22	22	25	25	25	24	24	26	3,516	1333
60 and over	11	16	20	23	27	29	29	28	29	29	29	29	30	6,134	2231
All aged 16 and over	11	14	16	18	19	21	21	20	21	20	20	21	21	22,236	7951

* Trend tables show unweighted and weighted figures for 1998 to give an indication of the effect of the weighting. For the weighted data (1998 and 2000 to 2002) the weighted base (000's) is the base for percentages. Unweighted data (up to 1998) are based on the unweighted sample. Bases for earlier years are of similar size and can be found in GHS reports for each year.

Table 8.3　Percentage who have never smoked cigarettes regularly by sex and age: 1974 to 2002

Persons aged 16 and over　　　　　　　　　　　　　　　　　　　　　　　　　　　　　　　　Great Britain

Age	Unweighted									Weighted				Weighted base 2002 (000's) =100%*	Unweighted sample* 2002
	1974	1978	1982	1986	1990	1992	1994	1996	1998	1998	2000	2001	2002		
						Percentage who have never smoked regularly									
Men															
16-19	56	61	65	65	68	67	67	69	64	65	67	71	75	1,089	349
20-24	38	46	50	47	54	52	53	47	49	50	58	51	55	1,384	420
25-34	26	33	39	43	48	50	50	49	50	49	49	47	51	3,292	1063
35-49	24	26	28	30	34	39	42	43	46	45	49	49	51	5,491	1884
50-59	16	17	20	26	31	31	33	31	32	32	37	38	38	3,374	1241
60 and over	18	18	20	19	24	24	27	28	30	30	32	36	32	4,931	1880
All aged 16 and over	25	29	32	34	37	38	40	40	41	42	44	45	46	19,561	6837
Women															
16-19	58	62	64	62	62	70	67	63	62	61	66	63	66	1,122	388
20-24	47	49	51	54	53	54	52	54	53	53	54	53	52	1,586	510
25-34	42	44	48	48	52	51	55	53	53	53	54	53	51	3,704	1282
35-49	41	44	47	46	48	49	51	52	52	52	54	53	55	6,175	2207
50-59	38	39	41	47	51	49	52	49	48	48	48	51	50	3,516	1333
60 and over	63	60	57	55	54	52	54	53	55	56	56	54	55	6,134	2231
All aged 16 and over	49	49	51	51	52	52	54	53	53	53	54	53	54	22,236	7951

* Trend tables show unweighted and weighted figures for 1998 to give an indication of the effect of the weighting. For the weighted data (1998 and 2000 to 2002) the weighted base (000's) is the base for percentages. Unweighted data (up to 1998) are based on the unweighted sample. Bases for earlier years are of similar size and can be found in GHS reports for each year.

Table 8.4   **Cigarette-smoking status by sex and marital status**

Persons aged 16 and over   Great Britain: 2002

Marital status		Current cigarette smokers			Current non-smokers of cigarettes		Weighted base (000's) = 100%	Unweighted sample
		Light (under 20 per day)	Heavy (20 or more per day)	Total	Ex-regular cigarette smokers	Never or only occasionally smoked cigarettes		
Men								
Single	%	24	8	32	11	57	4,608	1418
Married/cohabiting	%	14	10	24	32	44	12,907	4719
Married couple	%	12	9	21	34	45	10,936	4048
Cohabiting couple	%	27	15	42	20	39	1,971	671
Widowed/divorced/separated	%	19	14	34	35	32	2,046	700
All aged 16 and over	%	17	10	27	28	46	19,561	6837
Women								
Single	%	29	6	35	12	54	4,168	1420
Married/cohabiting	%	16	6	22	22	56	13,417	4909
Married couple	%	13	6	19	22	58	11,373	4209
Cohabiting couple	%	28	9	36	20	43	2,044	700
Widowed/divorced/separated	%	17	8	26	25	49	4,652	1622
All aged 16 and over	%	18	7	25	21	54	22,236	7951
Total								
Single	%	26	7	33	12	55	8,776	2838
Married/cohabiting	%	15	8	23	27	50	26,324	9628
Married couple	%	13	7	20	28	52	22,309	8257
Cohabiting couple	%	27	12	39	20	41	4,015	1371
Widowed/divorced/separated	%	18	10	28	28	44	6,698	2322
All aged 16 and over	%	18	8	26	24	50	41,798	14788

Table 8.5   Cigarette-smoking status by age and marital status

Persons aged 16 and over   Great Britain: 2002

Marital status	Age					Total
	16-24	25-34	35-49	50-59	60 and over	
	Percentage smoking cigarettes					
Single	31	41	34	28	18	33
Married couple	27	24	22	22	14	20
Cohabiting couple	42	39	41	32	26	39
Widowed/divorced/separated	*	52	42	38	17	28
All aged 16 and over	32	34	28	26	15	26
Weighted base (000's) = 100%						
Single	*4,103*	*2,306*	*1,417*	*440*	*510*	*8,776*
Married couple	*255*	*2,881*	*7,562*	*5,018*	*6,591*	*22,309*
Cohabiting couple	*789*	*1,494*	*1,235*	*326*	*171*	*4,015*
Widowed/divorced/separated	*34*	*314*	*1,452*	*1,106*	*3,792*	*6,698*
All aged 16 and over	*5,182*	*6,996*	*11,666*	*6,889*	*11,065*	*41,798*
Unweighted sample						
Single	*1333*	*734*	*444*	*145*	*182*	*2838*
Married couple	*81*	*1003*	*2732*	*1928*	*2513*	*8257*
Cohabiting couple	*242*	*499*	*431*	*129*	*70*	*1371*
Widowed/divorced/separated	*11*	*109*	*484*	*372*	*1346*	*2322*
All aged 16 and over	*1667*	*2345*	*4091*	*2574*	*4111*	*14788*

* Base too small for analysis.

Table 8.6   Prevalence of cigarette smoking by sex and country of Great Britain: 1978 to 2002

Persons aged 16 and over   Great Britain

Country	Unweighted								Weighted				Weighted base 2002 (000's) =100%*	Unweighted sample* 2002
	1978	1982	1986	1990	1992	1994	1996	1998	1998	2000	2001	2002		
	Percentage smoking cigarettes													
Men														
England	44	37	34	31	29	28	28	28	29	29	28	27	*16,806*	*5916*
Wales	44	36	33	30	32	28	28	28	29	25	27	27	*972*	*349*
Scotland	48	45	37	33	34	31	33	33	35	30	32	29	*1,783*	*572*
Great Britain	45	38	35	31	29	28	29	28	30	29	28	27	*19,561*	*6837*
Women														
England	36	32	31	28	27	25	27	26	26	25	25	25	*19,176*	*6896*
Wales	37	34	30	31	33	27	27	26	27	24	26	27	*1,112*	*407*
Scotland	42	39	35	35	34	29	31	29	29	30	30	28	*1,948*	*648*
Great Britain	37	33	31	29	28	26	28	26	26	25	26	25	*22,236*	*7951*
Total														
England	40	35	32	29	28	26	28	27	28	27	27	26	*35,983*	*12812*
Wales	40	35	31	31	32	27	27	27	28	25	27	27	*2,083*	*756*
Scotland	45	42	36	34	34	30	32	30	31	30	31	28	*3,731*	*1220*
Great Britain	40	35	33	30	28	27	28	27	28	27	27	26	*41,798*	*14788*

* Trend tables show unweighted and weighted figures for 1998 to give an indication of the effect of the weighting. For the weighted data (1998 and 2000 to 2002) the weighted base (000's) is the base for percentages. Unweighted data (up to 1998) are based on the unweighted sample. Bases for earlier years are of similar size and can be found in GHS reports for each year.

Table 8.7   **Prevalence of cigarette smoking by sex and Government Office Region: 1998 to 2002**

Persons aged 16 and over   Great Britain

Government Office Region	Weighted				Weighted base 2002 (000's) = 100%*	Unweighted sample* 2002
	1998	2000	2001	2002		
			Percentage smoking cigarettes			
Men						
England						
North East	28	27	33	24	857	294
North West	29	29	28	28	2,251	797
Yorkshire and the Humber	30	29	30	27	1,722	595
East Midlands	27	27	28	24	1,398	511
West Midlands	32	27	27	25	1,739	609
East of England	26	27	27	25	1,812	663
London	34	31	29	29	2,428	750
South East	28	28	26	27	2,849	1054
South West	26	30	27	27	1,751	643
All England	29	29	28	27	16,806	5916
Wales	29	25	27	27	972	349
Scotland	35	30	32	29	1,783	572
Great Britain	30	29	28	27	19,561	6837
Women						
England						
North East	30	28	26	29	1,145	400
North West	32	30	29	28	2,694	971
Yorkshire and the Humber	28	26	28	27	1,966	695
East Midlands	26	24	27	24	1,584	589
West Midlands	26	24	22	21	1,992	706
East of England	24	23	25	25	1,941	729
London	27	24	26	21	2,747	888
South East	21	23	23	25	3,164	1191
South West	25	24	22	24	1,945	727
All England	26	25	25	25	19,176	6896
Wales	27	24	26	27	1,112	407
Scotland	29	30	30	28	1,948	648
Great Britain	26	25	26	25	22,236	7951
All persons						
England						
North East	29	27	29	27	2,001	694
North West	31	30	29	28	4,945	1768
Yorkshire and the Humber	29	28	29	27	3,688	1290
East Midlands	27	25	28	24	2,982	1100
West Midlands	29	26	24	23	3,731	1315
East of England	25	25	26	25	3,753	1392
London	31	27	27	24	5,175	1638
South East	24	25	24	26	6,013	2245
South West	25	27	24	25	3,695	1370
All England	28	27	27	26	35,983	12812
Wales	28	25	27	27	2,083	756
Scotland	31	30	31	28	3,731	1220
Great Britain	28	27	27	26	41,798	14788

* Bases for earlier years are of similar size and can be found in GHS reports for each year.

Table 8.8   Cigarette-smoking status by sex and age: England

Persons aged 16 and over

England: 2002

Cigarette-smoking status and number of cigarettes smoked per day	Age							Total
	16-24	25-34	35-44	45-54	55-64	65-74	75 and over	
	%	%	%	%	%	%	%	%
Men								
Current smokers:								
Less than 10	12	12	6	5	5	4	2	7
10, less than 20	14	15	10	10	8	6	3	10
20 or more	5	9	12	14	12	6	3	10
Total current cigarette smokers*	31	36	28	29	24	16	8	27
Ex-regular smokers	5	13	19	28	42	51	60	28
Never or only occasionally smoked cigarettes	64	51	53	43	33	33	32	45
Weighted base (000's) = 100%	*2,097*	*2,882*	*3,325*	*2,766*	*2,582*	*1,884*	*1,270*	*16,806*
Unweighted sample	*661*	*933*	*1129*	*1010*	*970*	*718*	*495*	*5916*
Women								
Current smokers:								
Less than 10	12	12	7	5	5	4	4	7
10, less than 20	18	15	12	11	9	6	4	11
20 or more	3	6	8	10	8	5	2	6
Total current cigarette smokers*	34	33	27	25	23	15	9	25
Ex-regular smokers	8	17	16	22	27	29	34	21
Never or only occasionally smoked cigarettes	59	51	57	53	50	56	57	54
Weighted base (000's) = 100%	*2,324*	*3,235*	*3,731*	*3,153*	*2,584*	*2,107*	*2,043*	*19,176*
Unweighted sample	*777*	*1122*	*1316*	*1195*	*1022*	*764*	*700*	*6896*
Total								
Current smokers:								
Less than 10	12	12	7	5	5	4	3	7
10, less than 20	16	15	11	10	9	6	3	11
20 or more	4	7	10	12	10	5	2	8
Total current cigarette smokers*	32	34	28	27	24	16	9	26
Ex-regular smokers	7	15	17	25	35	39	44	24
Never or only occasionally smoked cigarettes	61	51	55	48	42	45	48	50
Weighted base (000's) = 100%	*4,421*	*6,116*	*7,056*	*5,920*	*5,166*	*3,991*	*3,313*	*35,983*
Unweighted sample	*1438*	*2055*	*2445*	*2205*	*1992*	*1482*	*1195*	*12812*

* Includes those for whom number of cigarettes was not known.

Table 8.9   **Prevalence of cigarette smoking by sex and whether household reference person is in a non-manual or manual socio-economic group: England, 1992 to 2002***

Persons aged 16 and over   England

Socio-economic group of household reference person†	Unweighted				Weighted				Weighted base 2002 (000's) = 100%**	Unweighted sample** 2002
	1992	**1994**	**1996**	**1998**	**1998**	**2000**	**2001#**	**2002**		
					Percentage smoking cigarettes					
Men										
Non-manual	22	21	21	21	22	24	22	21	8,505	3095
Manual	35	34	35	34	35	34	34	32	7,592	2607
Total††	29	28	28	28	29	29	28	27	16,806	5916
Women										
Non-manual	23	21	22	21	22	22	20	20	10,025	3706
Manual	30	30	33	31	31	29	31	30	8,025	2820
Total††	27	25	27	26	26	25	25	25	19,176	6896
All persons										
Non-manual	23	21	22	21	22	23	21	20	18,530	6801
Manual	33	32	34	32	33	31	32	31	15,617	5427
Total††	28	26	28	27	28	27	27	26	35,983	12812

*   Figures for 1992 to 1996 are taken from the Department of Health bulletin *Statistics on Smoking: England, 1978 onwards.* Figures for 2001 and 2002 are based on the new NS-SEC classification recoded to produce SEG and should therefore be treated with caution.

†   Head of household in years before 2000.

** Trend tables show unweighted and weighted figures for 1998 to give an indication of the effect of the weighting. For the weighted data (1998 and 2000 to 2002) the weighted base (000's) is the base for percentages. Unweighted data (up to 1998) are based on the unweighted sample. Bases for earlier years are of similar size and can be found in GHS reports for each year.

†† Persons whose head of household/household reference person was a full time student, in the Armed forces, had an inadequately described occupation or had never worked are not shown as separate categories but are included in the total. The total also includes some missing cases.

In 2001 there was an error in the variable used to create this table. This has been corrected and the smoking figures for both manual men and women have increased by one percentage point on the figures given in this table in the 2001 report. All other figures are unchanged.

Table 8.10 **Prevalence of cigarette smoking by sex and socio-economic classification based on the current or last job of the household reference person**

Persons aged 16 and over

Great Britain: 2002

Socio-economic classification of household reference person	Men	Women	Total
	Percentage smoking cigarettes		
Managerial and professional			
Large employers and higher managerial	17	16	16
Higher professional	14 / 20	14 / 18	14 / 19
Lower managerial and professional	24	19	21
Intermediate			
Intermediate	25 / 27	24 / 25	25 / 26
Small employers and own account	29	25	27
Routine and manual			
Lower supervisory and technical	30	29	29
Semi routine	32 / 32	33 / 31	33 / 32
Routine	35	32	33
Total*	27	25	26
Weighted base (000's) = 100%			
Large employers and higher managerial	1,264	1,272	2,536
Higher professional	1,874	1,633	3,507
Lower managerial and professional	4,633	5,179	9,812
Intermediate	1,288	2,153	3,441
Small employers and own account	2,073	1,960	4,033
Lower supervisory and technical	2,734	2,617	5,350
Semi routine	2,253	3,165	5,418
Routine	2,571	2,972	5,544
Total*	19,492	22,090	41,582
Unweighted sample			
Large employers and higher managerial	470	480	950
Higher professional	680	611	1291
Lower managerial and professional	1659	1895	3554
Intermediate	444	760	1204
Small employers and own account	731	717	1448
Lower supervisory and technical	935	916	1851
Semi routine	780	1117	1897
Routine	879	1035	1914
Total*	6817	7902	14719

* From April 2001 the National Statistics Socio-economic classification (NS-SEC) was introduced for all official statistics and surveys. It replaced Social Class based on Occupation and Socio-economic Groups (SEG). Persons whose household reference person was a full-time student, had an inadequately described occupation, had never worked or was long term unemployed are not shown as separate categories but are included in the figure for all persons (see Appendix A).

Table 8.11   **Prevalence of cigarette smoking by sex and socio-economic classification based on own current or last job, whether economically active or inactive, and, for economically inactive persons, age**

Persons aged 16 and over

Great Britain: 2002

Socio-economic classification*	Men					Women					All persons				
	Active	Inactive 16-59	Inactive 60 and over	Total inactive	Total	Active	Inactive 16-59	Inactive 60 and over	Total inactive	Total	Active	Inactive 16-59	Inactive 60 and over	Total inactive	Total
Percentage smoking cigarettes															
Managerial and professional	22	18	8	9	19	21	23	9	14	19	21	21	8	12	19
Intermediate	30	41	17	23	28	26	27	11	16	22	27	30	13	18	24
Routine and manual	36	55	20	28	33	36	42	17	27	31	36	46	18	27	32
Total*	29	35	16	22	27	28	32	14	22	25	28	33	15	22	26
Weighted bases (000's) = 100%															
Managerial and professional	5,383	247	1,240	1,487	6,871	4,171	604	1,041	1,645	5,816	9,554	851	2,281	3,132	12,686
Intermediate	2,389	197	582	779	3,168	3,123	626	1,268	1,895	5,018	5,512	824	1,850	2,674	8,186
Routine and manual	5,236	651	2,076	2,727	7,970	4,175	1,853	2,908	4,762	8,946	9,411	2,504	4,984	7,489	16,916
Total*	13,748	1,875	3,928	5,803	19,561	12,339	4,299	5,590	9,889	22,236	26,087	6,173	9,518	15,691	41,798
Unweighted sample															
Managerial and professional	1908	89	494	583	2491	1497	227	391	618	2115	3405	316	885	1201	4606
Intermediate	831	64	226	290	1121	1125	232	467	699	1824	1956	296	693	989	2945
Routine and manual	1758	218	764	982	2742	1478	664	1030	1694	3175	3236	882	1794	2676	5917
Total*	4735	605	1494	2099	6837	4400	1531	2017	3548	7951	9135	2136	3511	5647	14788

* From April 2001 the National Statistics Socio-economic classification (NS-SEC) was introduced for all official statistics and surveys. It replaced Social Class based on Occupation and Socio-economic Groups (SEG). Full-time students, persons in inadequately described occupations, those who had never worked and the long term unemployed are not shown as separate categories but are included in the figure for all persons (see Appendix A).

Table 8.12   Cigarette-smoking status by sex: 1974 to 2002

Persons aged 16 and over   Great Britain

	Unweighted									Weighted			
	1974	1978	1982	1986	1990	1992	1994	1996	1998	1998	2000	2001	2002
						Percentages							
Men													
Current cigarette smokers													
Light (under 20 per day)	25	22	20	20	17	17	17	17	18	19	18	19	17
Heavy (20 or more per day)	26	23	18	15	14	12	12	11	10	11	10	10	10
Total current cigarette smokers	51	45	38	35	31	29	28	29	28	30	29	28	27
Ex-regular cigarette smokers	23	27	30	32	32	32	31	32	31	29	27	27	28
Never or only occasionally smoked cigarettes	25	29	32	34	37	38	40	40	41	42	44	45	46
Weighted base (000's) = 100%*										19,229	20,350	19,913	19,561
Unweighted sample*	9852	10480	9199	8874	8106	8417	7642	7172	6579		6593	7055	6837
Women													
Current cigarette smokers													
Light (under 20 per day)	28	23	22	21	20	19	18	19	19	19	19	19	18
Heavy (20 or more per day)	13	13	11	10	9	9	8	8	7	7	6	7	7
Total current cigarette smokers	41	37	33	31	29	28	26	28	26	26	25	26	25
Ex-regular cigarette smokers	11	14	16	18	19	21	21	20	21	20	20	21	21
Never or only occasionally smoked cigarettes	49	49	51	51	52	52	54	53	53	53	54	53	54
Weighted base (000's) = 100%										21,654	22,044	21,987	22,236
Unweighted sample*	11480	12156	10641	10304	9445	9764	9108	8501	7830		7496	8299	7951

* Trend tables show unweighted and weighted figures for 1998 to give an indication of the effect of the weighting. For the weighted data (1998 and 2000 to 2002) the weighted base (000's) is the base for percentages. Unweighted data (up to 1998) are based on the unweighted sample.

Table 8.13   **Cigarette-smoking status by sex and age**

Persons aged 16 and over

Great Britain: 2002

Age		Current cigarette smokers			Current non-smokers of cigarettes		Weighted base (000's) = 100%	Unweighted sample
		Light (under 20 per day)	Heavy (20 or more per day)	All current smokers	Ex-regular cigarette smokers	Never or only occasionally smoked cigarettes		
Men								
16-19	%	19	3	22	3	75	1,089	349
20-24	%	31	6	37	7	55	1,384	420
25-34	%	26	9	36	13	51	3,292	1063
35-49	%	16	13	29	20	51	5,491	1884
50-59	%	13	14	27	35	38	3,374	1241
60 and over	%	10	7	17	51	32	4,931	1880
All aged 16 and over	%	17	10	27	28	46	19,561	6837
Women								
16-19	%	25	4	29	5	66	1,122	388
20-24	%	34	4	38	10	52	1,586	510
25-34	%	27	6	33	16	51	3,704	1282
35-49	%	18	9	27	17	55	6,175	2207
50-59	%	16	9	24	26	50	3,516	1333
60 and over	%	10	5	14	30	55	6,134	2231
All aged 16 and over	%	18	7	25	21	54	22,236	7951
Total								
16-19	%	22	3	25	4	71	2,211	737
20-24	%	32	5	38	9	54	2,971	930
25-34	%	27	8	34	15	51	6,996	2345
35-49	%	17	11	28	18	53	11,666	4091
50-59	%	14	11	26	30	44	6,889	2574
60 and over	%	10	5	15	40	45	11,065	4111
All aged 16 and over	%	18	8	26	24	50	41,798	14788

Table 8.14   **Average daily cigarette consumption per smoker by sex and age: 1974 to 2002**

Current cigarette smokers aged 16 and over

Great Britain

Age	Unweighted									Weighted				Weighted base 2002 (000's) =100%*	Unweighted sample* 2002
	1974	1978	1982	1986	1990	1992	1994	1996	1998	1998	2000	2001	2002		
	Mean number of cigarettes per day														
Men															
16-19	16	14	12	12	13	12	10	12	10	10	12	11	11	237	75
20-24	19	17	16	15	16	13	13	14	14	13	12	12	12	518	157
25-34	19	19	17	16	16	14	15	15	13	13	13	13	13	1,178	380
35-49	20	20	20	19	19	19	18	18	17	18	17	17	17	1,585	528
50-59	18	20	18	17	17	18	20	17	18	18	17	18	18	901	319
60 and over	14	15	16	15	15	15	14	15	16	16	15	15	16	823	305
All aged 16 and over	18	18	17	16	17	16	16	16	16	15	15	15	15	5,242	1764
Women															
16-19	12	13	11	11	11	10	10	10	10	10	10	12	12	326	110
20-24	14	14	14	12	13	13	13	11	12	11	10	11	10	601	194
25-34	15	16	16	14	15	14	14	13	12	12	12	12	12	1,209	422
35-49	15	16	15	16	15	16	15	16	15	15	14	15	15	1,692	593
50-59	13	14	14	14	15	15	15	16	15	15	15	15	15	853	317
60 and over	10	11	11	12	12	12	13	13	12	12	12	12	13	874	319
All aged 16 and over	13	14	14	14	14	14	14	14	13	13	13	13	13	5,554	1955

* Trend tables show unweighted and weighted figures for 1998 to give an indication of the effect of the weighting. For the weighted data (1998 and 2000 to 2002) the weighted base (000's) is the base for percentages. Unweighted data (up to 1998) are based on the unweighted sample. Bases for earlier years are of similar size and can be found in GHS reports for each year.

Table 8.15   **Average daily cigarette consumption per smoker by sex, and socio-economic classification based on the current or last job of the household reference person**

Current cigarette smokers aged 16 and over   Great Britain: 2002

Socio-economic classification of household reference person*	Men	Women	Total
	Mean number of cigarettes a day		
Managerial and professional			
Large employers and higher managerial	14	11	12
Higher professional	12 ⎤ 13	11 ⎤ 12	11 ⎤ 13
Lower managerial and professional	14	12	13
Intermediate			
Intermediate	14 ⎤ 16	13 ⎤ 13	13 ⎤ 15
Small employers and own account	18	14	16
Routine and manual			
Lower supervisory and technical	16	15	15
Semi routine	15 ⎤ 16	13 ⎤ 14	14 ⎤ 15
Routine	17	14	15
Total*	15	13	14
Weighted base (000's) = 100%			
Large employers and higher managerial	215	198	413
Higher professional	253	232	485
Lower managerial and professional	1,108	990	2,098
Intermediate	325	520	844
Small employers and own account	596	496	1,092
Lower supervisory and technical	813	750	1,563
Semi routine	731	1,046	1,777
Routine	886	945	1,831
*Total**	5,218	5,511	10,729
Unweighted sample			
Large employers and higher managerial	75	71	146
Higher professional	89	85	174
Lower managerial and professional	382	354	736
Intermediate	111	182	293
Small employers and own account	199	178	377
Lower supervisory and technical	270	260	530
Semi routine	245	368	613
Routine	297	330	627
*Total**	1757	1941	3698

* From April 2001 the National Statistics Socio-economic classification (NS-SEC) was introduced for all official statistics and surveys. It replaced Social Class based on Occupation and Socio-economic Groups (SEG). Persons whose household reference person was a full-time student, had an inadequately described occupation, had never worked or was long term unemployed are not shown as separate categories but are included in the figure for all persons (see Appendix A).

Table 8.16   Type of cigarette smoked by sex: 1974 to 2002

Current cigarette smokers aged 16 and over   Great Britain

Type of cigarette smoked	Unweighted									Weighted			
	1974	1978	1982	1986	1990	1992	1994	1996	1998	1998	2000	2001	2002
	%	%	%	%	%	%	%	%	%	%	%	%	%
Men													
Mainly filter	69	75	72	78	80	80	78	75	74	74	69	68	66
Mainly plain	18	11	7	4	2	2	2	1	1	1	1	1	1
Mainly hand-rolled	13	14	21	18	18	18	21	23	25	25	31	31	33
Weighted base (000's) = 100%*										5,687	5,802	5,643	5,246
Unweighted sample*	4993	4646	3469	3072	2510	2473	2150	2052	1857		1796	1911	1765
Women													
Mainly filter	91	95	94	96	97	97	96	93	92	92	89	87	86
Mainly plain	8	4	3	1	1	1	1	1	1	1	1	1	1
Mainly hand-rolled	1	1	3	2	2	2	4	6	7	8	10	12	13
Weighted base (000's) = 100%*										5,735	5,619	5,635	5,560
Unweighted sample*	4600	4421	3522	3192	2748	2698	2336	2341	2044		1900	2101	1957

* Trend tables show unweighted and weighted figures for 1998 to give an indication of the effect of the weighting. For the weighted data (1998 and 2000 to 2002) the weighted base (000's) is the base for percentages. Unweighted data (up to 1998) are based on the unweighted sample.

Table 8.17   Type of cigarette smoked by sex and age

Current cigarette smokers aged 16 and over   Great Britain: 2002

Type of cigarette smoked	Age					All aged 16 and over
	16-24	25-34	35-49	50-59	60 and over	
	%	%	%	%	%	%
Men						
Mainly filter	72	72	62	62	63	66
Mainly plain	3	0	0	0	2	1
Mainly hand-rolled	25	28	38	38	35	33
Weighted base (000's) = 100%	755	1,180	1,586	901	823	5,246
Unweighted sample	232	381	528	319	305	1765
Women						
Mainly filter	84	86	83	87	92	86
Mainly plain	3	1	0	0	0	1
Mainly hand-rolled	13	14	17	12	7	13
Weighted base (000's) = 100%	927	1,209	1,694	856	874	5,560
Unweighted sample	304	422	594	318	319	1957
Total						
Mainly filter	79	79	73	74	78	76
Mainly plain	3	0	0	0	1	1
Mainly hand-rolled	19	21	27	25	21	23
Weighted base (000's) = 100%	1,682	2,389	3,281	1,757	1,697	10,806
Unweighted sample	536	803	1122	637	624	3722

Table 8.18   Tar yield per cigarette: 1986 to 2002

Current smokers of manufactured cigarettes

Great Britain

Tar yield	Unweighted					Weighted			
	1986	1988	1990	1992	1998	1998	2000	2001	2002
	%	%	%	%	%	%	%	%	%
<10mg	19	21	24	25	28	28	27	26	27
10<15mg	32	58	54	68	70	69	71	71	71
15+mg	40	17	19	4	0	0	0	0	0
No regular brand/new brand/ don't know	10	4	4	3	2	2	2	2	2
*Weighted base (000's) =100%**						9,568	9,104	8,850	8,317
*Unweighted sample**	5620	5363	4739	4662	3288		2955	3174	2870

* Trend tables show unweighted and weighted figures for 1998 to give an indication of the effect of the weighting. For the weighted data (1998 and 2000 to 2002) the weighted base (000's) is the base for percentages. Unweighted data (up to 1998) are based on the unweighted sample.

Table 8.19   Tar yields by sex and age

Current smokers of manufactured cigarettes* aged 16 and over

Great Britain: 2002

		Tar yield						Weighted base (000's) = 100%	Unweighted sample
		Less than 4mg	4<8mg	8<10mg	10<12mg	12<15mg	No regular brand/ don't know tar yield		
Men									
16-19	%	0	9	1	48	40	1	188	59
20-24	%	0	29	1	31	35	3	374	113
25-34	%	2	26	3	34	33	3	855	274
35-49	%	1	14	4	43	35	3	983	330
50-59	%	1	11	5	39	42	2	560	199
60 and over	%	2	7	10	37	40	4	531	198
Total	%	1	17	4	38	37	3	3,492	1173
Women									
16-19	%	0	12	4	29	55	0	292	98
20-24	%	0	24	2	33	41	1	516	166
25-34	%	2	23	8	26	38	2	1,043	363
35-49	%	3	17	10	31	36	2	1,415	497
50-59	%	3	14	14	35	34	0	750	278
60 and over	%	4	11	17	28	38	3	810	295
Total	%	3	17	10	30	38	2	4,825	1697
Total									
16-19	%	0	11	3	36	49	1	480	157
20-24	%	0	26	2	32	38	2	890	279
25-34	%	2	24	6	30	36	3	1,898	637
35-49	%	2	16	8	36	36	3	2,398	827
50-59	%	2	13	10	37	37	1	1,310	477
60 and over	%	3	10	14	31	39	3	1,342	493
Total	%	2	17	8	34	37	2	8,317	2870

* Thirty-three per cent of male smokers and 13 per cent of female smokers said they mainly smoked hand-rolled cigarettes and have been excluded from this analysis.

Table 8.20	Tar yields by sex and socio-economic classification based on the current or last job of the household reference person

Current smokers of manufactured cigarettes* aged 16 and over					Great Britain: 2002

Socio-economic classification of household reference person†		Tar yields						Weighted base (000's) = 100%	Unweighted sample
		Less than 4mg	4<8mg	8<10mg	10<12mg	12<15mg	No regular brand don't know tar yield		
Men									
Managerial and professional	%	3	24	4	36	31	2	1,238	428
Intermediate	%	0	14	5	41	34	5	686	230
Routine and manual	%	1	10	4	39	43	2	1,396	464
Total	%	1	17	4	38	36	3	3,472	1167
Women									
Managerial and professional	%	4	28	10	25	31	2	1,289	463
Intermediate	%	3	21	11	32	32	1	919	325
Routine and manual	%	2	10	11	32	44	1	2,309	806
Total	%	3	17	10	30	38	2	4,783	1683
All persons									
Managerial and professional	%	3	26	7	30	31	2	2,527	891
Intermediate	%	2	18	8	36	33	3	1,606	555
Routine and manual	%	1	10	8	35	44	2	3,705	1270
Total	%	2	17	8	34	37	2	8,255	2850

* Thirty-three per cent of male smokers and 13 per cent of female smokers said they mainly smoked hand-rolled cigarettes and have been excluded from this analysis.

† From April 2001 the National Statistics Socio-economic classification (NS-SEC) was introduced for all official statistics and surveys. It replaced Social Class based on Occupation and Socio-economic Groups (SEG). Persons whose household reference person was a full-time student, had an inadequately described occupation, had never worked or was long term unemployed are not shown as separate categories but are included in the figure for all persons (see Appendix A).

Table 8.21   Prevalence of smoking by sex and type of product smoked: 1974 to 2002

Persons aged 16 and over   **Great Britain**

	Unweighted									Weighted			
	1974	1978	1982	1986	1990	1992	1994	1996	1998	1998	2000	2001	2002
					Percentage smoking								
Men													
Cigarettes*	51	45	38	35	31	29	28	29	28	30	29	28	27
Pipe	12	10	..	6	4	4	3	2	2	2	2	2	1
Cigars†	34	16	12	10	8	7	6	6	6	6	5	5	5
All smokers**	64	55	45††	44	38	36	33	33	33	34	32	32	30
Weighted base (000's) = 100%§										19,225	20,350	19,972	19,561
Unweighted sample§	9862	10439	9171	8884	8119	8427	7662	7186	6579		6593	7074	6835
Women													
Cigarettes*	41	37	33	31	29	28	26	28	26	26	25	26	25
Cigars†	3	1	0	1	0	0	0	0	0	0	0	0	0
All smokers**	41	37	34	31	29	28	26	28	26	27	26	26	25
Weighted base (000's) = 100%§										21,653	22,044	22,032	22,236
Unweighted sample§	11419	12079	10559	10312	9455	9772	9137	8512	7830		7496	8317	7951

*   Figures for cigarettes include all smokers of manufactured and hand-rolled cigarettes.
†   For 1974 the figures include occasional cigar smokers, that is, those who smoked less than one cigar a month.
** The percentages for cigarettes, pipes and cigars add to more than the percentage for all smokers because some people smoked more than one type of product.
†† In 1982 men were not asked about pipe smoking, and therefore the figures for all smokers exclude those who smoked only a pipe.
§   Trend tables show unweighted and weighted figures for 1998 to give an indication of the effect of the weighting. For the weighted data (1998 and 2000 to 2002) the weighted base (000's) is the base for percentages. Unweighted data (up to 1998) are based on the unweighted sample.

Table 8.22   Prevalence of smoking by sex and age and type of product smoked

Persons aged 16 and over   **Great Britain: 2002**

Age	Men						Women				
	Cigarettes*	Pipe†	Cigars†	All smokers**	Weighted base (000's) = 100%	Unweighted sample	Cigarettes*	Cigars†	All smokers**	Weighted base (000's) = 100%	Unweighted sample
	Percentage smoking						**Percentage smoking**				
16-19	22	0	2	23	1,089	347	29	1	29	1,122	388
20-24	37	1	3	38	1,384	420	38	1	38	1,586	510
25-29	36	0	5	37	1,431	452	32	0	32	1,631	555
30-34	36	0	5	38	1,861	611	33	0	33	2,072	727
35-49	29	1	6	32	5,491	1884	27	0	27	6,175	2207
50-59	27	2	6	32	3,374	1241	24	0	24	3,516	1333
60 and over	17	3	4	22	4,931	1880	14	0	14	6,134	2231
All aged 16 and over	27	1	5	30	19,561	6835	25	0	25	22,236	7951

*   Figures for cigarettes include all smokers of both manufactured and hand-rolled cigarettes.
†   Young people aged 16-17 were not asked about cigar or pipe-smoking.
** The percentages for cigarettes, pipes and cigars add to more than the percentage for all smokers because some people smoked more than one type of product.

Table 8.23   **Age started smoking regularly by sex and socio-economic classification based on the current or last job of the household reference person**

Pesons aged 16 and over who had ever smoked regularly   Great Britain: 2002

Age started smoking regularly	Socio-economic classification of household reference person*			
	Managerial and professional	Intermediate	Routine and manual	Total
	%	%	%	%
Men				
Under 16	34	45	47	42
16–17	29	27	28	28
18–19	20	15	13	16
20–24	13	9	10	11
25 and over	4	4	3	4
Weighted base (000's) = 100%	*3,656*	*1,905*	*4,530*	*10,469*
Unweighted sample	*1335*	*673*	*1570*	*3696*
	%	%	%	%
Women				
Under 16	25	30	39	33
16–17	27	29	28	28
18–19	23	19	15	18
20–24	18	14	10	13
25 and over	7	8	8	7
Weighted base (000's) = 100%	*3,203*	*1,895*	*4,488*	*10,067*
Unweighted sample	*1175*	*679*	*1573*	*3589*
	%	%	%	%
All persons				
Under 16	30	38	43	38
16–17	28	28	28	28
18–19	21	17	14	17
20–24	15	12	10	12
25 and over	6	6	5	5
Weighted base (000's) = 100%	*6,859*	*3,800*	*9,018*	*20,537*
Unweighted sample	*2510*	*1352*	*3143*	*7285*

* From April 2001 the National Statistics Socio-economic classification (NS-SEC) was introduced for all official statistics and surveys. It replaced Social Class based on Occupation and
  Socio-economic Groups (SEG). Persons whose household reference person was a full-time student, had an inadequately described occupation, had never worked or was long term
  unemployed are not shown as separate categories but are included in the figure for all persons (see Appendix A).

Table 8.24   **Age started smoking regularly by sex, whether current smoker and if so, cigarettes smoked a day**

Persons aged 16 and over who had ever smoked regularly   **Great Britain: 2002**

Age started smoking regularly	Current smoker				Ex-regular smoker	All who have ever smoked regularly
	20 or more a day	10–19 a day	0–9 a day	All current smokers*		
	%	%	%	%	%	%
Men						
Under 16	55	45	31	45	39	42
16–17	25	27	30	27	29	28
18–19	9	15	21	14	17	16
20–24	9	10	14	11	11	11
25 and over	2	3	5	3	4	4
Weighted base (000's) = 100%	*1,885*	*1,987*	*1,267*	*5,143*	*5,354*	*10,497*
Unweighted sample	*640*	*672*	*418*	*1731*	*1973*	*3704*
	%	%	%	%	%	%
Women						
Under 16	44	37	33	38	28	33
16–17	29	30	28	29	27	28
18–19	14	15	17	16	21	18
20–24	8	12	13	11	15	13
25 and over	5	6	9	6	9	7
Weighted base (000's) = 100%	*1,465*	*2,481*	*1,564*	*5,515*	*4,614*	*10,129*
Unweighted sample	*527*	*868*	*545*	*1942*	*1668*	*3610*
	%	%	%	%	%	%
All persons						
Under 16	50	41	32	41	34	38
16–17	26	29	29	28	28	28
18–19	11	15	19	15	19	17
20–24	9	11	13	11	13	12
25 and over	3	4	7	5	6	5
Weighted base (000's) = 100%	*3,350*	*4,469*	*2,831*	*10,658*	*9,968*	*20,627*
Unweighted sample	*1167*	*1540*	*963*	*3673*	*3641*	*7314*

*   Includes a few smokers who did not say how many cigarettes a day they smoked.

Table 8.25   **Proportion of smokers who would like to give up smoking altogether, by sex and number of cigarettes smoked per day: 1992 to 2002**

Current cigarette smokers aged 16 and over   Great Britain

Number of cigarettes smoked a day	Unweighted				Weighted				Weighted base 2002 (000's) = 100%*	Unweighted sample* 2002
	1992	1994	1996	1998	1998	2000	2001	2002		
	Percentage who would like to stop altogether									
Men										
20 or more	68	70	66	69	69	74	70	68	*1,912*	*649*
10-19	70	72	69	73	73	76	71	71	*2,007*	*678*
0-9	58	61	62	62	62	64	62	62	*1,324*	*437*
All smokers†	66	69	66	69	69	72	68	68	*5,248*	*1766*
Women										
20 or more	70	69	69	68	68	73	66	67	*1,471*	*529*
10-19	72	71	70	75	75	76	67	71	*2,484*	*869*
0-9	58	62	59	65	65	63	60	67	*1,599*	*557*
All smokers†	68	68	67	70	70	71	65	69	*5,560*	*1957*
Total										
20 or more	69	70	68	69	69	74	68	68	*3,382*	*1178*
10-19	71	71	70	74	74	76	69	71	*4,491*	*1547*
0-9	58	61	60	64	64	63	61	65	*2,923*	*994*
All smokers†	67	68	67	69	69	72	66	68	*10,808*	*3723*

*   Trend tables show unweighted and weighted figures for 1998 to give an indication of the effect of the weighting. For the weighted data (1998 and 2000 to 2002) the weighted base (000's) is the base for percentages. Unweighted data (up to 1998) are based on the unweighted sample. Bases for earlier years are of similar size and can be found in GHS reports for each year.
†   Includes a few smokers who did not say how many cigarettes a day they smoked.

Table 8.26   **Proportion of smokers who would find it difficult to go without smoking for a day, by sex and number of cigarettes smoked per day: 1992 to 2002**

Current cigarette smokers aged 16 and over   Great Britain

Number of cigarettes smoked a day	Unweighted				Weighted				Weighted base 2002 (000's) = 100%*	Unweighted sample* 2002
	1992	1994	1996	1998	1998	2000	2001	2002		
	Percentage who would find it difficult not to smoke for a day									
Men										
20 or more	76	78	78	78	78	78	74	77	*1,902*	*646*
10-19	54	57	54	54	54	56	55	57	*1,986*	*671*
0-9	20	17	20	25	23	14	21	23	*1,320*	*436*
All smokers†	55	56	56	56	56	53	52	56	*5,215*	*1755*
Women										
20 or more	86	86	87	87	86	88	87	86	*1,466*	*527*
10-19	68	68	66	66	65	67	65	66	*2,466*	*862*
0-9	23	20	24	24	25	22	24	21	*1,599*	*557*
All smokers†	61	60	61	59	59	58	58	59	*5,536*	*1948*
Total										
20 or more	80	82	83	82	82	82	80	81	*3,368*	*1173*
10-19	61	63	60	61	60	62	61	62	*4,452*	*1533*
0-9	21	19	23	24	24	18	22	22	*2,919*	*993*
All smokers†	58	59	58	58	57	56	55	57	*10,751*	*3703*

*   Trend tables show unweighted and weighted figures for 1998 to give an indication of the effect of the weighting. For the weighted data (1998 and 2000 to 2002) the weighted base (000's) is the base for percentages. Unweighted data (up to 1998) are based on the unweighted sample. Bases for earlier years are of similar size and can be found in GHS reports for each year.
†   Includes a few smokers who did not say how many cigarettes a day they smoked.

Table 8.27 **Proportion of smokers who have their first cigarette within five minutes of waking, by sex and number of cigarettes smoked per day: 1992 to 2002**

Current cigarette smokers aged 16 and over

Great Britain

Number of cigarettes smoked a day	Unweighted				Weighted				Weighted base 2002 (000's) = 100%*	Unweighted sample* 2002
	1992	1994	1996	1998	1998	2000	2001	2002		
	Percentage smoking within 5 minutes of waking									
Men										
20 or more	29	31	29	31	32	30	30	31	1,912	649
10-19	10	13	9	11	11	13	11	11	2,007	678
0-9	2	2	3	2	2	2	3	3	1,315	434
All smokers†	16	18	16	16	17	16	15	16	5,240	1763
Women										
20 or more	29	34	32	31	31	32	35	31	1,468	528
10-19	10	9	11	12	12	12	12	12	2,477	867
0-9	1	0	1	1	1	2	2	2	1,593	555
All smokers†	14	14	15	14	14	14	15	14	5,544	1952
Total										
20 or more	29	33	30	31	31	31	32	31	3,379	1177
10-19	10	11	10	12	12	13	11	11	4,484	1545
0-9	2	1	2	2	2	2	2	3	2,908	989
All smokers†	15	16	15	15	15	15	15	15	10,784	3715

*   Trend tables show unweighted and weighted figures for 1998 to give an indication of the effect of the weighting. For the weighted data (1998 and 2000 to 2002) the weighted base (000's) is the base for percentages. Unweighted data (up to 1998) are based on the unweighted sample. Bases for earlier years are of similar size and can be found in GHS reports for each year.

†   Includes a few smokers who did not say how many cigarettes a day they smoked.

Table 8.28   **Proportion of smokers who would like to give up smoking altogether, by sex, socio-economic classification of household reference person, and number of cigarettes smoked a day**

Current cigarette smokers aged 16 and over   Great Britain: 2002

Number of cigarettes smoked a day	Socio-economic classification*			
	Managerial and professional	Intermediate	Routine and manual	Total
	Percentage who would like to stop altogether			
Men				
20 or more	69	75	65	68
10–19	76	69	70	71
0–9	67	49	61	62
All smokers†	71	67	66	68
Women				
20 or more	75	65	64	67
10–19	78	67	69	71
0–9	66	73	62	67
All smokers†	73	68	66	69
Total				
20 or more	71	71	64	68
10–19	77	67	69	71
0–9	66	63	61	65
All smokers†	72	67	66	68
Weighted base (000's) = 100%				
Men				
20 or more	477	376	975	1,912
10–19	541	331	982	2,007
0–9	558	214	472	1,324
All smokers	1,576	923	2,433	5,248
Women				
20 or more	280	277	820	1,471
10–19	612	437	1,282	2,484
0–9	528	301	639	1,599
All smokers	1,419	1,016	2,747	5,560
Total				
20 or more	757	653	1,796	3,382
10–19	1,153	468	2,265	4,491
0–9	1,085	515	1,111	2,923
All smokers	2,996	1,939	5,181	10,808
Unweighted sample				
Men				
20 or more	168	128	326	649
10–19	190	111	330	678
0–9	188	71	156	437
All smokers	546	311	813	1766
Women				
20 or more	102	101	293	529
10–19	219	153	445	869
0–9	189	106	220	557
All smokers	510	360	960	1957
Total				
20 or more	270	229	619	1178
10–19	409	264	775	1547
0–9	377	177	376	994
All smokers	1056	671	1773	3723

* From April 2001 the National Statistics Socio-economic classification (NS-SEC) was introduced for all official statistics and surveys. It replaced Social Class based on Occupation and Socio-economic Groups (SEG). Persons whose household reference person was a full-time student, had an inadequately described occupation, had never worked or was long term unemployed are not shown as separate categories but are included in the figure for all persons (see Appendix A).
† Includes a few smokers who did not say how many cigarettes a day they smoked.

Table 8.29   Proportion of smokers who would find it difficult to go without smoking for a day, by sex, socio-economic classification of household reference person, and number of cigarettes smoked a day

Current cigarette smokers aged 16 and over

Great Britain: 2002

Number of cigarettes smoked a day	Socio-economic classification*			Total
	Managerial and professional	Intermediate	Routine and manual	
	Percentage who would find it difficult to stop for a day			
Men				
20 or more	74	79	78	77
10–19	52	52	60	57
0–9	21	24	23	23
All smokers†	48	57	60	56
Women				
20 or more	85	89	85	86
10–19	68	61	67	66
0–9	16	19	26	21
All smokers†	52	56	63	59
Total				
20 or more	78	83	81	81
10–19	61	57	64	62
0–9	18	21	25	22
All smokers†	50	56	62	57
Weighted base (000's) = 100%				
Men				
20 or more	*477*	*373*	*970*	*1,902*
10–19	*541*	*327*	*968*	*1,986*
0–9	*558*	*214*	*469*	*1,320*
All smokers	*1,576*	*917*	*2,409*	*5,215*
Women				
20 or more	*280*	*274*	*818*	*1,466*
10–19	*607*	*437*	*1,269*	*2,466*
0–9	*528*	*301*	*638*	*1,599*
All smokers	*1,414*	*1,013*	*2,731*	*5,536*
Total				
20 or more	*757*	*647*	*1,787*	*3,368*
10–19	*1,148*	*764*	*2,237*	*4,452*
0–9	*1,085*	*515*	*1,107*	*2,919*
All smokers	*2,990*	*1,930*	*5,141*	*10,751*
Unweighted sample				
Men				
20 or more	*168*	*127*	*324*	*646*
10–19	*190*	*110*	*325*	*671*
0–9	*188*	*71*	*155*	*436*
All smokers	*546*	*309*	*805*	*1755*
Women				
20 or more	*102*	*100*	*292*	*527*
10–19	*217*	*153*	*440*	*862*
0–9	*189*	*106*	*220*	*557*
All smokers	*508*	*359*	*954*	*1948*
Total				
20 or more	*270*	*227*	*616*	*1173*
10–19	*407*	*263*	*765*	*1533*
0–9	*377*	*177*	*375*	*993*
All smokers	*1054*	*668*	*1759*	*3703*

* From April 2001 the National Statistics Socio-economic classification (NS-SEC) was introduced for all official statistics and surveys. It replaced Social Class based on Occupation and Socio-economic Groups (SEG). Persons whose household reference person was a full-time student, had an inadequately described occupation, had never worked or was long term unemployed are not shown as separate categories but are included in the figure for all persons (see Appendix A).

† Includes a few smokers who did not say how many cigarettes a day they smoked.

Table 8.30   **Proportion of smokers who have their first cigarette within five minutes of waking, by sex, socio-economic classification of household reference person, and number of cigarettes smoked a day**

Current cigarette smokers aged 16 and over

Great Britain: 2002

Number of cigarettes smoked a day	Socio-economic classification*			
	Managerial and professional	Intermediate	Routine and manual	Total
	Percentage who smoke within 5 minutes of waking			
Men				
20 or more	23	30	33	31
10–19	6	11	12	11
0–9	1	0	4	3
All smokers†	9	16	19	16
Women				
20 or more	30	29	31	31
10–19	8	11	13	12
0–9	0	3	4	2
All smokers†	10	13	17	14
Total				
20 or more	25	29	32	31
10–19	7	11	13	11
0–9	1	2	4	3
All smokers†	9	15	18	15
Weighted base (000's) = 100%				
Men				
20 or more	*477*	*376*	*975*	*1,912*
10–19	*541*	*331*	*982*	*2,007*
0–9	*556*	*211*	*469*	*1,315*
All smokers	*1,574*	*920*	*2,430*	*5,240*
Women				
20 or more	*280*	*274*	*820*	*1,468*
10–19	*609*	*437*	*1,282*	*2,477*
0–9	*525*	*301*	*639*	*1,593*
All smokers	*1,413*	*1,013*	*2,747*	*5,544*
Total				
20 or more	*757*	*651*	*1,796*	*3,379*
10–19	*1,150*	*768*	*2,265*	*4,484*
0–9	*1,080*	*512*	*1,107*	*2,908*
All smokers	*2,987*	*1,933*	*5,177*	*10,784*
Unweighted sample				
Men				
20 or more	*168*	*128*	*326*	*649*
10–19	*190*	*111*	*330*	*678*
0–9	*187*	*70*	*155*	*434*
All smokers	*545*	*310*	*812*	*1763*
Women				
20 or more	*102*	*100*	*293*	*528*
10–19	*218*	*153*	*445*	*867*
0–9	*188*	*106*	*220*	*555*
All smokers	*508*	*359*	*960*	*1952*
Total				
20 or more	*270*	*228*	*619*	*1177*
10–19	*408*	*264*	*775*	*1545*
0–9	*375*	*176*	*375*	*989*
All smokers	*1053*	*669*	*1772*	*3715*

* From April 2001 the National Statistics Socio-economic classification (NS-SEC) was introduced for all official statistics and surveys. It replaced Social Class based on Occupation and Socio-economic Groups (SEG). Persons whose household reference person was a full-time student, had an inadequately described occupation, had never worked or was long term unemployed are not shown as separate categories but are included in the figure for all persons (see Appendix A).
† Includes a few smokers who did not say how many cigarettes a day they smoked.

Chapter 9 Drinking

Questions about drinking alcohol were included in the General Household Survey every two years from 1978 to 1998. Following the review of the GHS, the questions about drinking in the last seven days form part of the continuous survey, and have been included every year from 2000 onwards.

Before 1988 questions about drinking were asked only of those aged 18 and over, but since then respondents aged 16 and 17 have answered the questions using a self-completion questionnaire.

Measuring alcohol consumption

Obtaining reliable information about drinking behaviour is difficult, so social surveys consistently record lower levels of consumption than would be expected from data on alcohol sales. This is partly because people may consciously or unconsciously under-estimate how much alcohol they consume. Drinking at home is particularly likely to be under-estimated because the quantities consumed are not measured and are likely to be larger than those dispensed in licensed premises.

There are different methods for obtaining survey information on drinking behaviour. One approach is to ask people to recall all episodes of drinking during a set period[1]. However, this is time-consuming and is not suitable for the GHS, where drinking is only one of a number of subjects covered.

Since 1998 the GHS has used two measures of alcohol consumption:
- maximum daily amount drunk last week;
- average weekly alcohol consumption.

Maximum daily amount drunk last week

These questions have been included in the GHS since 1998, following the publication in 1995 of an inter-departmental review of the effects of drinking[2]. This concluded that it was more appropriate to set benchmarks for daily than for weekly consumption of alcohol, partly because of concern about the health and social risks associated with single episodes of intoxication. The report considered that regular consumption of between three and four units[3] a day for men and two to three units a day for women does not carry a significant health risk, but that consistently drinking above these levels is not advised.

The government's advice on sensible drinking is now based on these daily benchmarks, and GHS data are used to monitor the extent to which people are following the advice given. Respondents are asked on how many days they drank alcohol during the previous week. They are then asked how much of each of six different types of

drink (normal strength beer, strong beer [6% or greater ABV[4]], wine, spirits, fortified wines and alcopops) they drank on their heaviest drinking day during the previous week. These amounts are added to give an estimate of the maximum number of units the respondent had drunk on that day.

Average weekly alcohol consumption

The current measure of average weekly alcohol consumption has been used on the GHS since 1986, and was developed in response to earlier medical guidelines on drinking relating to maximum recommended weekly amounts of alcohol. Its use continues to provide a consistent measure of alcohol consumption through which trends can be monitored. Respondents are asked how often over the last year they have drunk normal strength beer, strong beer, wine, spirits, fortified wines and alcopops, and how much they have usually drunk on any one day. This information is combined to give an estimate of the respondent's weekly alcohol consumption (averaged over a year) in units of alcohol[5].

Since 1998, the estimates of alcohol consumption (both the maximum drunk on any one day in the last week and the weekly average) have been under reported due to an error in the conversion of bottled beers from pints to units. Bottled beers account for only about 2% of average weekly alcohol consumption, and the effect of correcting the error was to increase total average weekly consumption from 11.9 to 12.1 units in 2001. Alcohol consumption in the last seven days is presented in bands, and most figures are unchanged by the correction. The tables show amended figures for 2001. Data for 1998 and 2000 are also affected but due to the small difference demonstrated by the 2001 correction these figures have not been changed.

Frequency of drinking during the last week

Patterns of drinking behaviour in 2002 were broadly the same as those described in earlier GHS reports.

Men were more likely than women to have had an alcoholic drink in the previous week (74% of men and 59% of women had a drink on at least one day during the previous week). The proportions drinking last week also varied between age groups. On the whole, among both men and women, those aged 16 to 24 and those aged 65 and over were least likely to report drinking alcohol during the previous week.

Men also drank on more days of the week than women. More than one in five men (22%) compared with one in eight women (13%) had drunk on at least five of the preceding seven days. Meanwhile, 13% of men, but only 8% of women, had drunk alcohol every day during the previous week.

Among both men and women, older people drank more frequently than younger people. For example, 21% of men and 11% of women aged 65 and over had drunk every day during the previous week, compared with only 4% of men and 2% of women aged 16 to 24.

Table 9.1, Figure 9A

Maximum daily amount drunk last week

Two measures of daily consumption are shown in the tables. The first is the proportion exceeding the recommended daily benchmarks (men drinking more than four units and women drinking more than three units[2] in one day). The second measure is intended to indicate heavy drinking that would be likely to lead to intoxication. Although people vary in their susceptibility to the effect of alcohol, this level is taken as a rough guide to be more than eight units on one day for men and more than six units for women.

Trends in daily drinking

The GHS has only included questions about the maximum daily amount drunk last week since 1998, so these data provide no evidence on long-term trends. However, Table 9.2 shows that there were no significant changes between 1998 and 2002 in the proportions of men and women and of different age groups who had an alcoholic drink in the previous week, nor in the proportion who had drunk on five or more days in the week.

There was little change overall in the proportions of men drinking more than the daily benchmark of four units or drinking heavily (more than eight units) on at least one day in the previous week. The proportion of the youngest group of men drinking more than four units fell, however, from 52% in 1998 to 49% in 2002, and the proportion drinking more than eight units fell from 39% to 35% over the same period. Although these changes are not statistically significant, the fact that the downward trend has been fairly consistent from survey to survey suggests that the decline may be a real one.

On the other hand, there appears to be an increase between 1998 and 2002 in heavy drinking among women. The proportion of women who had drunk more than six units on at least one day in the previous week rose from 8% in 1998 to 10% in 2000 and remained at that level in 2001 and 2002. The increase over this period occurred only among women under the age of 45: among those aged 16 to 24, for example, the proportion who had drunk more than six units on at least one day in the previous week rose from 24% to 28%. **Table 9.2**

Daily drinking and sex, age and marital status

Men were much more likely than women to have exceeded the daily benchmarks on at least one day during the previous week - 38% of men compared with 23% of women - and they were twice as likely as women to have drunk heavily (21% compared with 10%).

It was noted earlier that young people drink less frequently than older people. However, among both men and women, those aged 16 to 24 were significantly more likely than respondents in other age groups to have exceeded the recommended number of daily units on at least one day. Almost half of young men (49%) aged 16 to 24 had exceeded four units on at least one day during the previous week compared with 16% of men aged 65 and over. Among women, 42% of those in the youngest

Figure 9A   **Percentage of men and women who had drunk alcohol on 5 days or more in the week prior to interview by age, 2002**

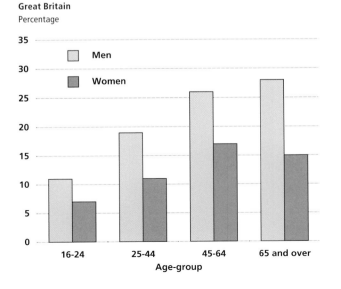

Great Britain
Percentage

age group had exceeded three units on at least one day compared with only 5% of those aged 65 and over.

Similar patterns were evident for heavy drinking: 35% of men aged 16 to 24 (but only 5% of those aged 65 and over) had drunk more than eight units on at least one day during the previous week. Among young women aged 16 to 24, 28% had drunk heavily on at least one day during the preceding week compared with only 1% of women in the oldest age group.

Analysis of alcohol consumption by marital status is complicated by the strong association between marital status and age. This is only partly controlled for in Table 9.4, where two age groups are shown. For example, among those aged 16 to 44, single men are more likely than married men to drink heavily. This may, however, be due to the fact that within this age group single men are, on average, younger than married men, so the difference may be due to their age rather than their marital status.

However, it is clear that alcohol consumption is higher than average among divorced and separated men aged 45 and over (20% of this group had drunk more than eight units on at least one day in the last week, compared with only 12% of married or cohabiting men in the same age group). Divorced and separated women aged 45 and over, however, were no more likely than women in general in that age group to have drunk heavily in the previous week.   **Tables 9.3 – 9.4, Figures 9B and 9C**

Daily drinking and socio-economic characteristics

The link between alcohol consumption and socio-economic characteristics is an important focus of analysis for the GHS. A review of information on inequalities in health, undertaken by the Department of Health, noted that both mortality and morbidity show a clear association with socio-economic position, with death rates much higher among unskilled men than among those in professional households[6]. Some 20,000 deaths a year are thought to be attributable to alcohol misuse[7]. However, the GHS has shown over many years that there is little difference in usual weekly alcohol consumption between those in non-manual and manual households. Where differences do exist, it has been those in the non-manual categories who tended to have the higher weekly consumption.

The National Statistics Socio-economic Classification (NS-SEC) was introduced in April 2001 for all official statistics and surveys. NS-SEC classifies occupations according to different criteria from those used by the Social Class and Socio-economic Group classifications which it has replaced. In addition, the occupational classification underpinning the groupings also changed in 2001. The new NS-SEC is not designed to be collapsed into broad non-manual and manual groupings.

Tables 9.5 and 9.6 show four indicators of last week's drinking in relation to the eight- and three-category versions of NS-SEC that are based on the current or last

Figure 9B   **Percentage exceeding daily benchmarks by age: men, 2002**

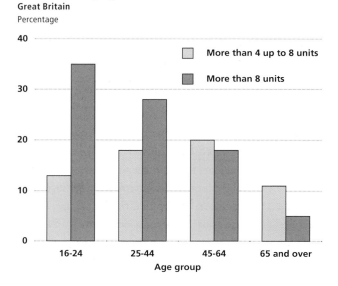

Figure 9C   **Percentage exceeding daily benchmarks by age: women, 2002**

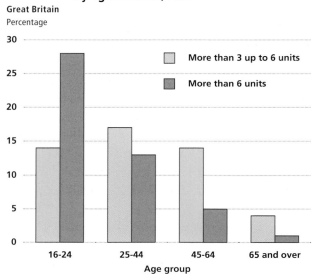

job of the household reference person. These are the proportions of adults who:

- drank alcohol last week;
- drank on five or more days last week;
- drank more than four/three units on at least one day last week;
- drank more than eight/six units on at least one day last week.

Among both men and women, those in large employer/ higher managerial households were the most likely to have drunk alcohol in the previous week, (and those in households where the household reference person was in a routine occupation the least likely). Similar differences were apparent in the proportions drinking on five or more days in the previous week. For example 77% of women in large employer/higher managerial households had a drink in the last week and 22% had done so on five or more days. This compared with only 45% and 7% respectively of women in households where the household reference person was in a routine occupation.

Variations in amounts drunk, however, were much less marked. Apart from men and women in large employer/ higher managerial households (who were more likely than those in other groups to have drunk more than four units [men] and three units [women]) there were no clear patterns of difference according to household socio-economic class. In all other groups two fifths of men or fewer had drunk more than four units, and just over one fifth of women had drunk more than three units.

There were no significant differences between the socio-economic classes in the proportion who had drunk heavily on at least one day in the previous week. Those in large employer/higher managerial households were not significantly more likely to have done so than those in other types of household. In all socio-economic classes, about one in five men had drunk more than eight units, and about one in ten women more than six units, on at least one day in the previous week.

Tables 9.5–9.6

Daily drinking and household income
In general, the higher the level of gross weekly household income, the more likely both men and women were to have drunk alcohol in the previous week and to have exceeded the recommended number of units for daily consumption. Among men in households with a gross

weekly income of over £1,000, 85% had had a drink in the previous week, and 49% had drunk more than four units on at least one day. Only 59% of men in households with an income of £200 or less, however, had had a drink and only 28% had drunk more than four units on any one day.

Similar differences according to household income were evident among women, although the proportions drinking and the proportions exceeding recommended levels were lower than for men.

Men and women in households with a gross weekly income of over £1,000 were twice as likely as those in households with a gross weekly household income of £200 or less to have drunk more than eight and six units respectively on at least one day in the previous week.

Table 9.7

Daily drinking, economic activity status and earnings from employment
Among men aged 16 to 64, those in employment were more likely than those who were economically inactive to have drunk heavily during the previous week (27% of working men, 26% of unemployed men and 18% of economically inactive men had done so). Lower levels of drinking among economically inactive men are partly due to the large proportion of this group who are aged 60 to 64.

One in seven working women (14%) aged 16 to 64 and a similar proportion of those who were unemployed had drunk more than six units on at least one day during the previous week, compared with only 8% of women who were economically inactive. **Table 9.8**

Among those working full time, there was some variation in drinking behaviour in relation to earnings from employment, but the pattern of association was different to that described above in relation to household income.

There was a strong association between earnings and the frequency of drinking in the previous week - men and women who were high earners were more likely both to have drunk alcohol at all and to have drunk on five or more days. Furthermore, differences in drinking frequency between men and women were much smaller among high earners than among those at the other end of the earnings scale. Among those with gross earnings of more than £800 a week, 32% of men and 31% of women had drunk alcohol on five or more days in the previous week. Among those earning £200 or less, however, 18% of men and only 12% of women had done so.

Variation in maximum amounts drunk on any one day in the previous week was, on the whole, less marked and the pattern of association less clear. Among men, high earners were no more likely than other men in full-time work to have drunk more than four units on any one day in the previous week. They were also less likely to have drunk more than eight units (22% of men earning more than £800 a week had drunk more than eight units on at least one day in the last week, compared with 28% of all men in full time employment).

Among women in full-time work, high earners were more likely than other women to have drunk more than three units on at least one day in the previous week. Of women earning more than £800 a week, 43% had drunk more than three units on at least one day in the last week, compared with 32% of women earning £200 or less.

However, high earning women were no more likely than other women to have drunk more than six units on one day. **Table 9.9**

Regional variation in daily drinking

The extent to which drinking habits and alcohol consumption vary in different parts of Great Britain is always of interest, but care should be taken in interpreting the results for any one year. This is because sample sizes in some regions are small, making them subject to relatively high levels of sampling error.

In 2002 men and women living in Wales and Scotland were more likely than those living in England to have drunk heavily on at least one day in the previous week. Of men in Scotland and Wales, 26% had drunk more than eight units of alcohol on at least one day during the previous week, compared with 21% in England. There was a similar pattern of differences among women: 12% of women in Scotland and 15% in Wales had drunk more than six units on at least one day in the previous week, compared with 10% of women in England.

Earlier GHS reports showed that among the regions of England average weekly alcohol consumption tended to be higher in the north than in the south. They also

showed that this was mainly the case for men - differences for women were much less marked.

Maximum recommended daily amounts, however, appear to be much more likely to be exceeded by both men and women in the north (North East, North West, and Yorkshire and the Humber) than in the rest of England.

In 2002 the North East was the English region which had the highest proportions drinking more than the daily benchmark amounts (49% of men and 32% of women). The proportions doing so were also higher than those in Scotland (44% and 26% respectively) and Wales (42% and 29%). The equivalent figures for the South East were 33% of men and 22% of women.

The same broad pattern of regional variation in daily drinking has been evident since these questions were first included in 1998. As noted above, however, sample sizes in some regions are small and some fluctuation in results from year to year is to be expected. This can affect whether a particular region or country appears to have a high or low consumption level relative to other areas.

For example, the proportions of men and women drinking more than the recommended daily maximum amounts in 2002 were relatively high in Wales. The sample size is small though, and the increases compared with 2001 were not statistically significant. Similarly, the proportions of women exceeding three and six units in the East Midlands were relatively high in 2001, but in 2002 they had fallen back to be more in line with the lower levels found in the south of England. Neither of these changes is likely to reflect real changes in behaviour in the population from which the sample is drawn. **Tables 9.10–9.11**

Average weekly alcohol consumption

As noted in the introduction, the main GHS measure of drinking behaviour until 1998 was average weekly alcohol consumption. This has been retained, primarily to give a continuing indication of trends in drinking behaviour[8].

The average weekly alcohol consumption measure should not necessarily be expected to show the same patterns of variation in relation to respondents' characteristics as the measure of daily consumption.

Trends in weekly alcohol consumption

Consideration of trends is complicated by the introduction of weighting. This increased the proportion of men drinking more than 21 units a week in 1998 by about one percentage point. The comparison of weighted and unweighted figures for later years, although not shown in the tables, is similar.

Since the late 1980s the GHS has shown a slight increase in overall weekly alcohol consumption among men and a much more marked one among women. Since 2000, however, there has been some indication of a slight decline. The proportion of men drinking more than 21 units a week on average fell from 29% to 28% in 2001 and 27% in 2002. This appears to be due to a fall in consumption among younger men (in particular those aged 16 to 24). Because of the relatively small sample size, however, the decline in that age group is not large enough to be statistically significant. There was also a fall in the proportion of women drinking more than 14 units a week (from 17% in 2000 to 15% in 2001, rising back to 17% in 2002).

There has been no significant change in the proportion of men drinking more than 50 units a week on average, nor in the proportion of women drinking more than 35 units. **Table 9.12**

In 2002, men drank an average of 17.2 units a week (equivalent to just over 8½ pints of beer), the same as the figure for the previous year. Women drank an average of 7.6 units a week, less than half as much as men. The increase of 0.1 units for women compared with 2001 is not statistically significant, but continues the steady rise in women's consumption over the past decade.

The average consumption of young men aged 16 to 24 was lower in 2002 than in 2001, as would be expected from the fall in the proportion drinking more than 21 units a week. Among young women of the same age, however, average consumption continues to rise, increasing from 12.6 units in 2000 to 14.1 in 2001 and 2002. It has almost doubled in the ten years since 1992, when they drank on average 7.3 units a week.

Among both men and women, alcohol consumption was highest among those aged 16 to 24, and then declined with increasing age. Overall, in 2002 men's consumption was more than twice that of women but the difference was less marked among younger than older people. This

again reflects the trend that has occurred in recent years for women's consumption to increase relative to that of men, particularly among younger age groups.

Table 9.13

Weekly alcohol consumption and household socio-economic classification

The relationship between weekly alcohol consumption and socio-economic classification was similar to that shown earlier in relation to daily amounts. Average weekly consumption was highest among men and women in large employer/higher managerial households, at 19.9 and 9.4 units respectively.

Apart from the high consumption in that particular group, there was no clear socio-economic gradient in relation to alcohol consumption among men. Using the three-category classification, average consumption was 17.3 units a week among men in managerial and professional households, 17.9 units among men in intermediate households and 16.8 units among those in routine or manual households.

The pattern among women was slightly clearer. Average weekly consumption was highest, at 8.3 units, in the managerial and professional group, and lowest (at 6.5 units) among those in routine and manual worker households. **Table 9.14**

Weekly alcohol consumption, income and economic activity status

Average weekly alcohol consumption was higher among men and women in high income households than among other men and women. Among those living in households with a gross income of more than £1,000 a week, men drank on average 20.7 units as week, and women 10.3 units. These levels compared with 14.4 units and 5.3 units respectively among those in households with an income of £200 or less.

Among those in full-time employment, however, there was no significant variation in average weekly alcohol consumption according to earnings. **Tables 9.15–9.17**

Weekly alcohol consumption and Government Office Region

The pattern of regional differences in England in relation to average weekly alcohol consumption was similar to that seen for maximum amounts consumed on any day in the previous week. Overall, weekly consumption was highest in two of the same three regions (North East, and Yorkshire and the Humber). The third highest was the South West, rather than the North West.

Similarly, average weekly consumption was higher in Wales than in either England or Scotland. The higher proportion exceeding the daily benchmark amounts in Scotland is not, however, reflected in average weekly consumption, which was slightly higher in Scotland than in England for men, but considerably lower for women. Overall weekly consumption was 11.9 units in Scotland, 12.0 units in England and 13.8 in Wales.

As noted above, sample sizes in some regions are small so some fluctuation in results from year to year is to be expected. This can affect whether a particular region or country appears to have a high or low consumption level relative to that of other areas, and may not be due to real differences in the population from which the sample is drawn.

Table 9.18

Notes and references

1 Goddard E. *Obtaining information about drinking through surveys of the general population.* National Statistics Methodology Series NSM24 (ONS 2001).

2 *Sensible drinking: the report of an inter-departmental group*, Department of Health 1995.

3 One unit of alcohol is obtained from half a pint of normal strength beer, lager or cider, a single measure of spirits, one glass of wine, or one small glass of port, sherry or other fortified wine.

4 Alcohol by volume.

5 The method used to calculate an individual's average weekly alcohol consumption is to multiply the number of units of each type drunk on a usual drinking day by the frequency with which it was drunk, using the factors shown below, and totalling across all drinks.

Drinking frequency	Multiplying factor
Almost every day	7.0
5 or 6 days a week	5.5
3 or 4 days a week	3.5
Once or twice a week	1.5
Once or twice a month	0.375 (1.5/4)
Once every couple of months	0.115 (6/52)
Once or twice a year	0.029 (1.5/52)

6 Drever F, Bunting J, Harding D. *Male mortality from major causes of death* (in Drever F, Whitehead M, Eds. Health inequalities: decennial supplement: DS Series no.15. London: The Stationery Office, 1997) quoted in *Independent Inquiry into Inequalities in Health Report.* London: The Stationery Office 1998.

7 Prime Minister's Strategy Unit Alcohol Harm Reduction Project: Interim Analytical Report. Available on the web at www.number10.gov.uk/files/pdf/SU%20interim_report2.pdf

8 The earliest year shown in these trend tables is 1988, the first year in which data were collected from 16 and 17 year olds.

Table 9.1　　**Whether drank last week and number of drinking days by sex and age**

Persons aged 16 and over　　　　　　　　　　　　　　　　　　　　　　Great Britain: 2002

Drinking days last week	Age				
	16-24	25-44	45-64	65 and over	Total
	%	%	%	%	%
Men					
0	31	23	24	33	26
1	20	17	15	16	17
2	18	18	15	12	16
3	13	14	11	7	12
4	7	8	9	4	7
5	4	6	5	4	5
6	2 ⎱ 11	4 ⎱ 19	4 ⎱ 26	3 ⎱ 28	4 ⎱ 22
7	4	9	17	21	13
% who drank last week	69	77	76	67	74
Weighted base (000's) =100%	*2,479*	*7,156*	*6,243*	*3,660*	*19,538*
Unweighted sample	*770*	*2364*	*2296*	*1398*	*6828*
	%	%	%	%	%
Women					
0	39	35	37	54	41
1	23	22	17	16	19
2	17	14	14	8	13
3	9	10	9	5	9
4	6	7	6	3	5
5	3	4	4	2	3
6	2 ⎱ 7	2 ⎱ 11	3 ⎱ 17	2 ⎱ 15	2 ⎱ 13
7	2	6	10	11	8
% who drank last week	61	65	63	46	59
Weighted base (000's) =100%	*2,707*	*7,988*	*6,700*	*4,819*	*22,214*
Unweighted sample	*897*	*2782*	*2571*	*1692*	*7942*
	%	%	%	%	%
All persons					
0	36	30	31	45	34
1	21	20	16	16	18
2	17	16	14	10	14
3	11	12	10	6	10
4	6	7	7	3	6
5	4	5	5	3	4
6	2 ⎱ 9	3 ⎱ 15	3 ⎱ 21	2 ⎱ 21	3 ⎱ 17
7	3	7	13	16	10
% who drank last week	64	70	69	55	66
Weighted base (000's) =100%	*5,187*	*15,144*	*12,943*	*8,479*	*41,752*
Unweighted sample	*1667*	*5146*	*4867*	*3090*	*14770*

Table 9.2   Drinking last week by sex and age: 1998 to 2002

Persons aged 16 and over

Great Britain

Drinking last week	Men					Women			
	1998	2000	2001	2002		1998	2000	2001	2002
					Percentages				
Drank last week									
16-24	70	70	70	69		62	64	59	61
25-44	79	78	78	77		65	67	66	65
45-64	77	77	76	76		61	61	61	63
65 and over	65	67	68	67		45	43	45	46
Total	75	75	75	74		59	60	59	59
Drank on 5 or more days									
16-24	13	11	14	11		8	7	8	7
25-44	21	19	19	19		12	11	11	11
45-64	29	26	25	26		15	15	17	17
65 and over	25	28	27	28		14	14	15	15
Total	23	22	22	22		13	13	13	13
Drank more than 4/3 units* **on at least one day**									
16-24	52	50	50	49		42	42	40	42
25-44	48	45	49	46		28	31	31	31
45-64	37	38	37	38		17	19	19	19
65 and over	16	16	18	16		4	4	5	5
Total	39	39	40	38		21	23	23	23
Drank more than 8/6 units* **on at least one day**									
16-24	39	37	37	35		24	27	27	28
25-44	29	27	30	28		11	13	14	13
45-64	17	17	17	18		5	5	5	5
65 and over	4	5	5	5		1	1	1	1
Total	22	21	22	21		8	10	10	10
Weighted base *(000's) = 100%*									
16-24	*2,366*	*2,687*	*2,485*	*2,485*		*2,580*	*2,633*	*2,549*	*2,704*
25-44	*7,528*	*7,936*	*7,799*	*7,154*		*7,995*	*8,091*	*8,104*	*7,977*
45-64	*5,868*	*6,212*	*6,139*	*6,236*		*6,306*	*6,588*	*6,566*	*6,700*
65 and over	*3,412*	*3,534*	*3,488*	*3,659*		*4,744*	*4,742*	*4,765*	*4,822*
Total	*19,174*	*20,369*	*19,911*	*19,534*		*21,625*	*22,054*	*21,985*	*22,202*
Unweighted base									
16-24	*699*	*791*	*774*	*770*		*809*	*814*	*911*	*897*
25-44	*2400*	*2311*	*2589*	*2364*		*2910*	*2732*	*3044*	*2782*
45-64	*2132*	*2186*	*2288*	*2296*		*2364*	*2357*	*2536*	*2571*
65 and over	*1330*	*1310*	*1403*	*1398*		*1738*	*1588*	*1808*	*1692*
Total	*6561*	*6598*	*7054*	*6828*		*7821*	*7491*	*8299*	*7942*

* The first of each pair of figures relates to men, and the second, to women.

Table 9.3   **Maximum daily amount drunk last week by sex and age**

Persons aged 16 and over

Great Britain: 2002

Maximum daily amount	Age				
	16-24	25-44	45-64	65 and over	Total
	%	%	%	%	%
Men					
Drank nothing last week	32	24	24	33	27
Up to 4 units	20	30	38	50	35
More than 4, up to 8 units	13 ⎤ 49	18 ⎤ 46	20 ⎤ 38	11 ⎤ 16	17 ⎤ 38
More than 8 units	35 ⎦	28 ⎦	18 ⎦	5 ⎦	21 ⎦
Weighted base (000's) =100%	*2,485*	*7,154*	*6,236*	*3,659*	*19,534*
Unweighted sample	*772*	*2363*	*2294*	*1398*	*6827*
	%	%	%	%	%
Women					
Drank nothing last week	40	35	38	54	41
Up to 3 units	18	34	43	41	36
More than 3, up to 6 units	14 ⎤ 42	17 ⎤ 31	14 ⎤ 19	4 ⎤ 5	13 ⎤ 23
More than 6 units	28 ⎦	13 ⎦	5 ⎦	1 ⎦	10 ⎦
Weighted base (000's) =100%	*2,704*	*7,977*	*6,700*	*4,822*	*22,202*
Unweighted sample	*896*	*2778*	*2571*	*1693*	*7938*
	%	%	%	%	%
All persons*					
Drank nothing last week	36	30	31	45	34
Up to 4/3 units	19	32	41	45	36
More than 4/3, up to 8/6 units	14 ⎤ 45	17 ⎤ 38	17 ⎤ 28	7 ⎤ 10	15 ⎤ 30
More than 8/6 units	32 ⎦	20 ⎦	11 ⎦	2 ⎦	15 ⎦
Weighted base (000's) =100%	*5,189*	*15130*	*12,936*	*8,481*	*41,736*
Unweighted sample	*1668*	*5141*	*4865*	*3091*	*14765*

* The first of each pair of figures shown relates to men, and the second, to women.

Table 9.4   Drinking last week, by sex, age and marital status

Persons aged 16 and over

Great Britain: 2002

Marital status	Men			Women			All persons		
	16-44	45 and over	Total	16-44	45 and over	Total	16-44	45 and over	Total
Percentage who drank last week									
Single	70	66	70	63	49	61	67	59	66
Married/cohabiting	78	75	76	64	61	62	70	68	69
Divorced/separated	77	67	70	62	55	58	67	60	62
Widowed	*	61	60	*	41	42	*	46	46
Total	75	73	74	64	56	59	69	63	66
Percentage who drank on five or more days last week									
Single	11	20	13	8	10	9	10	15	11
Married/cohabiting	21	28	25	12	19	15	16	23	20
Divorced/separated	17	24	22	8	12	11	11	17	15
Widowed	*	29	28	*	13	13	*	16	16
Total	17	27	22	10	16	13	13	21	17
Percentage who drank more than 4/3 units on at least one day last week†									
Single	50	38	48	43	13	39	47	27	44
Married/cohabiting	44	29	36	28	16	21	35	23	28
Divorced/separated	51	37	41	33	15	22	39	24	29
Widowed	*	17	17	*	5	6	*	8	8
Total	47	30	38	34	13	23	40	21	30
Percentage who drank more than 8/6 units on at least one day last week†									
Single	34	19	32	26	4	23	31	12	28
Married/cohabiting	27	12	18	12	4	8	19	8	13
Divorced/separated	34	20	24	16	4	9	21	11	15
Widowed	*	6	6	*	1	1	*	2	2
Total	30	13	21	17	3	10	23	8	15
Weighted base (000's) =100%									
Single	*3,914*	*695*	*4,608*	*3,605*	*552*	*4,157*	*7,519*	*1,247*	*8,766*
Married/cohabiting	*5,336*	*7,543*	*12,879*	*6,228*	*7,170*	*13,399*	*11,564*	*14,714*	*26,278*
Divorced/separated	*368*	*917*	*1,285*	*830*	*1,276*	*2,106*	*1,197*	*2,193*	*3,390*
Widowed	*21*	*741*	*761*	*18*	*2,523*	*2,541*	*38*	*3,264*	*3,302*
Total	*9,638*	*9,895*	*19,534*	*10,681*	*11,522*	*22,202*	*20,319*	*21,417*	*41,736*
Unweighted sample									
Single	*1189*	*229*	*1418*	*1219*	*197*	*1416*	*2408*	*426*	*2834*
Married/cohabiting	*1830*	*2879*	*4709*	*2156*	*2746*	*4902*	*3986*	*5625*	*9611*
Divorced/separated	*110*	*301*	*411*	*293*	*453*	*746*	*403*	*754*	*1157*
Widowed	*6*	*283*	*289*	*6*	*868*	*874*	*12*	*1151*	*1163*
Total	*3135*	*3692*	*6827*	*3674*	*4264*	*7938*	*6809*	*7956*	*14765*

* Base too small for analysis.

† The first of each pair of figures shown relates to men, and the second, to women.

Table 9.5　　Drinking days last week, by sex, and socio-economic classification based on the current or last job of the household reference person

Persons aged 16 and over　　　　　　　　　　　　　　　　　　　　　　　　　　　　　　　　Great Britain: 2002

Socio-economic classification of household reference person*	Men	Women	All persons
Percentage who drank last week			
Managerial and professional			
Large employers and higher managerial	89	77	83
Higher professional	79 81	72 70	76 76
Lower managerial and professional	80	68	74
Intermediate			
Intermediate	72 71	59 60	64 65
Small employers/own account	70	61	66
Routine and manual			
Lower supervisory and technical	73	57	65
Semi-routine	69 69	51 51	58 59
Routine	64	45	54
Total*	74	59	66
Percentage who drank on 5 or more days last week			
Managerial and professional			
Large employers and higher managerial	31	22	27
Higher professional	28 28	20 18	25 23
Lower managerial and professional	27	17	22
Intermediate			
Intermediate	19 21	11 13	14 16
Small employers/own account	22	15	19
Routine and manual			
Lower supervisory and technical	21	12	16
Semi-routine	15 17	9 9	12 13
Routine	15	7	11
Total*	22	13	17
Weighted bases (000's) =100%			
Large employers and higher managerial	1,261	1,270	2,531
Higher professional	1,868	1,630	3,498
Lower managerial and professional	4,625	5,168	9,794
Intermediate	1,285	2,148	3,433
Small employers/own account	2,080	1,957	4,037
Lower supervisory and technical	2,738	2,623	5,361
Semi-routine	2,245	3,160	5,405
Routine	2,564	2,969	5,533
Total*	19,468	22,061	41,529
Unweighted sample			
Large employers and higher managerial	469	479	948
Higher professional	678	610	1288
Lower managerial and professional	1656	1891	3547
Intermediate	443	758	1201
Small employers/own account	733	716	1449
Lower supervisory and technical	936	918	1854
Semi-routine	777	1115	1892
Routine	877	1034	1911
Total*	6808	7891	14699

* From April 2001 the National Statistics Socio-economic Classification (NS-SEC) was introduced for all official statistics and surveys. It replaced Social Class based on Occupation and Socio-economic Groups (SEG). Persons whose household reference person was a full-time student, had an inadequately described occupation, had never worked or was long-term unemployed are not shown as separate categories, but are included in the figure for all persons (see AppendixA).

Table 9.6   **Maximum number of units drunk on at least one day last week, by sex, and socio-economic classification based on the current or last job of the household reference person**

Persons aged 16 and over   Great Britain: 2002

Socio-economic classification of household reference person*	Men	Women	All persons
	Percentage who drank more than 4/3 units on at least one day last week†		
Managerial and professional			
Large employers and higher managerial	43	32	38
Higher professional	38 | 40	22 | 24	31 | 32
Lower managerial and professional	40	23	31
Intermediate			
Intermediate	37 | 37	22 | 22	28 | 29
Small employers/own account	37	22	29
Routine and manual			
Lower supervisory and technical	39	24	32
Semi-routine	38 | 37	22 | 22	29 | 29
Routine	35	19	26
Total*	38	23	30
	Percentage who drank more than 8/6 units on at least one day last week†		
Managerial and professional			
Large employers and higher managerial	23	10	17
Higher professional	20 | 22	8 | 10	15 | 16
Lower managerial and professional	22	11	16
Intermediate			
Intermediate	20 | 21	9 | 8	13 | 14
Small employers/own account	21	8	15
Routine and manual			
Lower supervisory and technical	23	9	16
Semi-routine	22 | 21	10 | 9	15 | 15
Routine	18	8	13
Total*	21	10	15
Weighted bases (000's) =100%			
Large employers and higher managerial	*1,261*	*1,270*	*2,532*
Higher professional	*1,872*	*1,625*	*3,496*
Lower managerial and professional	*4,622*	*5,162*	*9,784*
Intermediate	*1,285*	*2,148*	*3,433*
Small employers/own account	*2,080*	*1,954*	*4,034*
Lower supervisory and technical	*2,734*	*2,623*	*5,357*
Semi-routine	*2,245*	*3,157*	*5,401*
Routine	*2,564*	*2,975*	*5,538*
Total	*19,464*	*22,049*	*41,513*
Unweighted sample			
Large employers and higher managerial	*469*	*479*	*948*
Higher professional	*679*	*608*	*1287*
Lower managerial and professional	*1655*	*1889*	*3544*
Intermediate	*443*	*758*	*1201*
Small employers/own account	*733*	*715*	*1448*
Lower supervisory and technical	*935*	*918*	*1853*
Semi-routine	*777*	*1114*	*1891*
Routine	*877*	*1036*	*1913*
Total	*6807*	*7887*	*14694*

*   From April 2001 the National Statistics Socio-economic classification (NS-SEC) was introduced for all official statistics and surveys. It replaced Social Class based on Occupation and Socio-economic Group (SEG). Persons whose household reference person was a full-time student, had an inadequately described occupation, had never worked or was long-term unemployed are not shown as separate categories, but are included in the figure for all persons.

†   The first of each pair of figures shown relates to men, and the second, to women.

Table 9.7   **Drinking last week by sex and usual gross weekly household income**

Persons aged 16 and over   **Great Britain: 2002**

Drinking last week	Usual gross weekly household income (£)						
	Up to 200.00	200.01 -400.00	400.01 -600.00	600.01 -800.00	800.01 -1000.00	1000.01 or more	Total*
				Percentages			
Drank last week							
Men	59	68	77	79	81	85	74
Women	45	57	65	64	70	76	59
All persons	50	62	71	72	76	81	66
Drank on 5 or more days							
Men	17	20	21	23	23	30	22
Women	10	11	15	13	14	21	13
All persons	13	15	18	18	19	26	17
Drank more than 4/3 units on at least one day†							
Men	28	32	40	44	44	49	38
Women	15	21	28	28	29	32	23
All persons	20	26	34	36	36	41	30
Drank more than 8/6 units on at least one day†							
Men	14	17	22	27	26	28	21
Women	6	10	12	13	12	13	10
All persons	9	13	17	20	19	21	15
Weighted base (000's) =100%							
Men	*3,119*	*3,566*	*3,617*	*2,834*	*1,590*	*2,860*	*19,538*
Women	*5,000*	*4,118*	*3,665*	*2,804*	*1,608*	*2,657*	*22,214*
All persons	*8,119*	*7,684*	*7,282*	*5,638*	*3,198*	*5,516*	*41,752*
Unweighted sample							
Men	*1075*	*1239*	*1261*	*984*	*561*	*1024*	*6828*
Women	*1747*	*1479*	*1311*	*997*	*586*	*981*	*7942*
All persons	*2822*	*2718*	*2572*	*1981*	*1147*	*2005*	*14770*

*   Includes people for whom income data was not known.
†   The first of each pair of figures shown relates to men, and the second, to women.

Table 9.8   **Drinking last week by sex and economic activity status**

Persons aged 16-64

Great Britain: 2002

Drinking last week	Economic activity status			
	Working	Unemployed	Economically inactive	Total
			Percentages	
Drank last week				
Men	79	65	60	75
Women	70	59	49	63
All persons	74	63	52	69
Drank on 5 or more days				
Men	22	14	16	20
Women	14	9	10	13
All persons	18	12	12	16
Drank more than 4/3 units on at least one day*				
Men	46	39	32	43
Women	32	36	19	28
All persons	39	38	23	35
Drank more than 8/6 units on at least one day*				
Men	27	26	18	25
Women	14	15	8	12
All persons	21	21	11	19
Weighted base (000's) =100%				
Men	12,747	610	2,509	15,867
Women	11,702	446	5,227	17,375
All persons	24,449	1,056	7,736	33,241
Unweighted sample				
Men	4386	196	845	5427
Women	4175	155	1913	6243
All persons	8561	351	2758	11670

* The first of each pair of figures relates to men, and the second, to women.

Table 9.9   **Drinking last week by sex and usual gross weekly earnings**

Persons aged 16-64 in full-time employment   Great Britain: 2002

Drinking last week	Usual gross weekly earnings (£)						Total
	Up to 200.00	200.01 -300.00	300.01 -400.00	400.01 -600.00	600.01 -800.00	800.01 or more	
				Percentages			
Drank last week							
Men	68	75	80	83	87	86	80
Women	71	69	74	77	79	88	74
All persons	69	72	78	81	85	86	78
Drank on 5 or more days							
Men	18	19	17	21	32	32	22
Women	12	11	14	19	20	31	15
All persons	15	15	16	20	29	32	19
Drank more than 4/3 units on at least one day*							
Men	45	46	49	46	49	45	47
Women	32	35	35	36	32	43	35
All persons	38	41	44	43	45	44	43
Drank more than 8/6 units on at least one day*							
Men	28	29	32	27	26	22	28
Women	18	16	16	16	11	16	16
All persons	23	23	27	24	23	21	24
Weighted base (000's) =100%							
Men	*912*	*2,042*	*2,272*	*2,879*	*1,199*	*1,187*	*10,492*
Women	*1,045*	*1,826*	*1,177*	*1,338*	*376*	*276*	*6,039*
All persons	*1,958*	*3,868*	*3,450*	*4,217*	*1,575*	*1,463*	*16,530*
Unweighted sample							
Men	*303*	*683*	*765*	*1002*	*422*	*428*	*3603*
Women	*364*	*640*	*412*	*476*	*134*	*98*	*2124*
All persons	*667*	*1323*	*1177*	*1478*	*556*	*526*	*5727*

* The first of each pair of figures shown relates to men, and the second, to women.

Table 9.10 Drinking last week, by sex and Government Office Region

Persons aged 16 and over

Great Britain: 2002

Government Office Region	Drinking last week					
	Drank last week	Drank on 5 or more days last week	Drank more than 4/3 units on at least one day*	Drank more than 8/6 units on at least one day*	Weighted base (000's) = 100%	Unweighted sample
			Percentages			
Men						
North East	74	19	49	29	857	294
North West	71	20	42	24	2,242	794
Yorkshire and the Humber	76	22	44	26	1,717	593
East Midlands	76	25	41	22	1,395	510
West Midlands	73	22	36	18	1,736	608
East of England	74	24	31	15	1,820	665
London	68	19	32	19	2,427	750
South East	74	27	33	17	2,841	1051
South West	78	25	37	21	1,748	642
England	73	23	37	21	16,782	5907
Wales	74	19	42	26	969	348
Scotland	75	17	44	26	1,783	572
Great Britain	74	22	38	21	19,534	6827
Women						
North East	63	13	32	17	1,145	400
North West	59	11	27	11	2,689	969
Yorkshire and the Humber	64	12	26	14	1,959	693
East Midlands	56	14	18	8	1,584	589
West Midlands	56	14	19	8	1,989	705
East of England	60	18	19	6	1,946	731
London	52	11	18	8	2,738	885
South East	62	14	22	8	3,160	1189
South West	66	18	23	9	1,938	724
England	60	14	22	10	19,148	6885
Wales	63	13	29	15	1,109	406
Scotland	56	9	26	12	1,946	647
Great Britain	59	13	23	10	22,202	7938
All persons						
North East	68	15	39	22	2,002	694
North West	65	15	34	17	4,931	1763
Yorkshire and the Humber	70	16	34	20	3,675	1286
East Midlands	66	19	29	14	2,979	1099
West Midlands	64	18	27	13	3,725	1313
East of England	67	21	25	11	3,766	1396
London	59	15	25	13	5,165	1635
South East	68	20	27	12	6,000	2240
South West	72	21	30	15	3,686	1366
England	66	18	29	15	35,930	12792
Wales	68	16	35	20	2,078	754
Scotland	65	13	35	19	3,729	1219
Great Britain	66	17	30	15	41,736	14765

* The first of each pair of figures shown relates to men, and the second, to women.

Table 9.11 Percentages who drank more than 4 units and 8 units (men) and 3 units and 6 units (women) on at least one day by sex and Government Office Region: 1998 to 2002

Persons aged 16 and over   Great Britain

Government Office Region	1998	2000	2001	2002	1998	2000	2001	2002	Weighted base 2002 (000's) = 100%*	Unweighted sample* 2002
	Percentage who drank more than 4 units on at least one day last week				Percentage who drank more than 8 units on at least one day last week					
Men										
North East	46	44	47	49	24	25	29	29	857	294
North West	46	45	49	42	28	24	29	24	2,242	794
Yorkshire and the Humber	41	42	44	44	25	23	27	26	1,717	593
East Midlands	42	43	43	41	21	22	22	22	1,395	510
West Midlands	42	35	34	36	26	17	18	18	1,736	608
East of England	35	31	34	31	15	18	20	15	1,820	665
London	33	31	36	32	19	17	20	19	2,427	750
South East	37	39	34	33	20	22	18	17	2,841	1051
South West	37	35	38	37	20	20	20	21	1,748	642
England	39	38	39	37	22	21	22	21	16,782	5907
Wales	40	41	37	42	23	23	21	26	969	348
Scotland	40	45	48	44	24	29	30	26	1,783	572
Great Britain	39	39	40	38	22	21	22	21	19,534	6827
	Percentage who drank more than 3 units on at least one day last week				Percentage who drank more than 6 units on at least one day last week					
Women										
North East	23	25	29	32	8	13	15	17	1,145	400
North West	26	28	28	27	11	13	13	11	2,689	969
Yorkshire and the Humber	19	23	25	26	7	10	12	14	1,959	693
East Midlands	21	23	27	18	8	8	12	8	1,584	589
West Midlands	21	19	17	19	9	7	7	8	1,989	705
East of England	18	20	20	19	7	8	8	6	1,946	731
London	17	19	18	18	7	7	8	8	2,738	885
South East	20	22	21	22	7	10	7	8	3,160	1189
South West	21	21	23	23	8	9	9	9	1,938	724
England	21	22	22	22	8	9	10	10	19,148	6885
Wales	22	24	22	29	10	11	11	15	1,109	406
Scotland	28	29	26	26	12	12	13	12	1,946	647
Great Britain	21	23	23	23	8	10	10	10	22,202	7938

* Trend tables show unweighted and weighted figures for 1998 to give an indication of the effect of the weighting. For the weighted data (1998 and 2000 to 2002) the weighted base (000's) is the base for percentages. Unweighted data (up to 1998) are based on the unweighted sample. Bases for earlier years are of similar size and can be found in GHS reports for each year.

Table 9.12   **Weekly alcohol consumption level: percentage exceeding specified amounts by sex and age: 1988 to 2002**

Persons aged 16 and over   **Great Britain**

Age	Unweighted data						Weighted data				*Weighted base 2002 (000's) = 100%**	*Unweighted sample* 2002
	1988	1992	1994	1996	1998		1998	2000	2001	2002		
					Percentage of men who drank more than 21 units							
Men												
16-24	31	32	29	35	36		38	41	40	37	*2,471*	*767*
25-44	34	31	30	30	27		28	30	30	29	*7,160*	*2365*
45-64	24	25	27	26	30		30	28	26	28	*6,243*	*2296*
65 and over	13	15	17	18	16		16	17	15	15	*3,662*	*1399*
Total	26	26	27	27	27		28	29	28	27	*19,536*	*6827*
					Percentage of men who drank more than 50 units							
Men												
16-24	10	9	9	10	13		14	14	15	12	*2,471*	*767*
25-44	9	8	7	6	6		6	7	7	8	*7,160*	*2365*
45-64	6	6	6	5	6		7	6	5	6	*6,243*	*2296*
65 and over	2	2	3	3	3		3	3	2	3	*3,662*	*1399*
Total	7	6	6	6	6		7	7	7	7	*19,536*	*6827*
					Percentage of women who drank more than 14 units							
Women												
16-24	15	17	19	22	25		25	33	32	33	*2,708*	*897*
25-44	14	14	15	16	16		16	19	17	19	*7,988*	*2782*
45-64	9	11	12	13	16		15	14	14	14	*6,700*	*2571*
65 and over	4	5	7	7	6		6	7	6	7	*4,824*	*1694*
Total	10	11	13	14	15		15	17	15	17	*22,220*	*7944*
					Percentage of women who drank more than 35 units							
Women												
16-24	3	4	4	5	6		7	9	10	10	*2,708*	*897*
25-44	2	2	2	2	2		2	3	3	3	*7,988*	*2782*
45-64	1	1	2	2	2		2	2	2	2	*6,700*	*2571*
65 and over	0	0	1	1	1		1	1	1	1	*4,824*	*1694*
Total	2	2	2	2	2		2	3	3	3	*22,220*	*7944*

* Trend tables show unweighted and weighted figures for 1998 to give an indication of the effect of the weighting. For the weighted data (1998 and 2000 to 2002) the weighted base (000's) is the base for percentages. Unweighted data (up to 1998) are based on the unweighted sample. Bases for earlier years are of similar size and can be found in GHS reports for each year.

Table 9.13 Average weekly alcohol consumption by sex and age: 1992 to 2002

Persons aged 16 and over

Great Britain

Age	Unweighted				Weighted				Weighted base 2002 (000's) = 100%*	Unweighted sample* 2002
	1992	1994	1996	1998	1998	2000	2001	2002		
				Mean number of units per week						
Men										
16-24	19.1	17.4	20.3	23.6	25.5	25.9	24.8	21.5	2,471	767
25-44	18.2	17.5	17.6	16.5	17.1	17.7	18.4	18.7	7,160	2365
45-64	15.6	15.5	15.6	17.3	17.4	16.8	16.1	17.5	6,243	2296
65 and over	9.7	10.0	11.0	10.7	10.6	11.0	10.8	10.7	3,662	1399
Total	15.9	15.4	16.0	16.4	17.1	17.4	17.2	17.2	19,536	6827
Women										
16-24	7.3	7.7	9.5	10.6	11.0	12.6	14.1	14.1	2,708	897
25-44	6.3	6.2	7.2	7.1	7.1	8.1	8.3	8.4	7,988	2782
45-64	5.3	5.3	5.9	6.4	6.4	6.2	6.8	6.7	6,700	2571
65 and over	2.7	3.2	3.5	3.3	3.2	3.5	3.6	3.8	4,824	1694
Total	5.4	5.4	6.3	6.4	6.5	7.1	7.5	7.6	22,220	7944
All persons										
16-24	12.9	12.3	14.7	16.6	18.0	19.3	19.4	17.6	5,179	1664
25-44	11.8	11.4	11.9	11.4	12.0	12.9	13.3	13.3	15,148	5147
45-64	10.2	10.2	10.5	11.6	11.7	11.4	11.3	11.9	12,943	4867
65 and over	5.6	6.0	6.8	6.5	6.3	6.7	6.6	6.8	8,487	3093
Total	10.2	10.0	10.7	11.0	11.5	12.0	12.1	12.1	41,756	14771

* Trend tables show unweighted and weighted figures for 1998 to give an indication of the effect of the weighting. For the weighted data (1998 and 2000 to 2002) the weighted base (000's) is the base for percentages. Unweighted data (up to 1998) are based on the unweighted sample. Bases for earlier years are of similar size and can be found in GHS reports for each year.

Table 9.14 **Average weekly alcohol consumption, by sex and socio-economic classification based on the current or last job of the household reference person**

Persons aged 16 and over Great Britain: 2002

Socio-economic classification of household reference person*	Men	Women	All persons
	Mean number of units per week		
Managerial and professional			
Large employers and higher managerial	19.9	9.4	14.7
Higher professional	15.7 17.3	8.3 8.3	12.3 12.7
Lower managerial and professional	17.3	8.1	12.4
Intermediate			
Intermediate	17.3 17.9	7.1 7.5	10.9 12.2
Small employers/own account	18.3	8.0	13.3
Routine and manual			
Lower supervisory and technical	18.5	7.6	13.1
Semi-routine	16.2 16.8	6.9 6.5	10.8 11.3
Routine	15.5	5.2	10.0
Total*	17.2	7.6	12.1
Weighted bases (000's) =100%			
Large employers and higher managerial	1,259	1,270	2,529
Higher professional	1,872	1,630	3,502
Lower managerial and professional	4,625	5,168	9,794
Intermediate	1,288	2,151	3,438
Small employers/own account	2,080	1,957	4,037
Lower supervisory and technical	2,735	2,623	5,358
Semi-routine	2,245	3,160	5,405
Routine	2,561	2,972	5,534
Total	19,467	22,066	41,533
Unweighted sample			
Large employers and higher managerial	468	479	947
Higher professional	679	610	1289
Lower managerial and professional	1656	1891	3547
Intermediate	444	759	1203
Small employers/own account	733	716	1449
Lower supervisory and technical	935	918	1853
Semi-routine	777	1115	1892
Routine	876	1035	1911
Total	6807	7893	14700

* From April 2001 the National Statistics Socio-economic classification (NS-SEC) was introduced for all official statistics and surveys. It replaced Social Class based on Occupation and Socio-economic Group (SEG). Persons whose household reference person was a full-time student, had an inadequately described occupation, had never worked or was long-term unemployed are not shown as separate categories, but are included in the figure for all persons.

Table 9.15 **Average weekly alcohol consumption, by sex and usual gross weekly household income**

Persons aged 16 and over

Great Britain: 2002

Usual gross weekly household income (£)	Men	Women	All persons	Weighted base (000's) =100%			Unweighted sample		
				Men	Women	All persons	Men	Women	All persons
	Mean number of units per week								
Up to 200.00	14.4	5.3	8.8	3,123	5,005	8,128	1076	1749	2825
200.01-400.00	15.6	7.1	11.0	3,561	4,120	7,681	1237	1480	2717
400.01-600.00	16.9	8.6	12.7	3,621	3,671	7,292	1262	1313	2575
600.01-800.00	19.6	8.8	14.2	2,834	2,807	5,641	984	998	1982
800.01-1000.00	18.2	8.8	13.5	1,590	1,611	3,201	561	587	1148
1000.01 or more	20.7	10.3	15.6	2,854	2,657	5,511	1022	981	2003
Total*	17.2	7.6	12.1	19,536	22,220	41,756	6827	7944	14771

* Includes people for whom income data was not known.

Table 9.16 **Average weekly alcohol consumption, by sex and economic activity status**

Persons aged 16 to 64

Great Britain: 2002

Economic activity status	Men	Women	All persons	Weighted base (000's) =100%			Unweighted sample		
				Men	Women	All persons	Men	Women	All persons
	Mean number of units per week								
Working	19.1	9.7	14.6	12,746	11,711	24,457	4385	4178	8563
Unemployed	19.9	12.0	16.6	617	446	1,063	198	155	353
Economically inactive	16.2	6.0	9.3	2,504	5,230	7,733	843	1914	2757
Total	18.7	8.7	13.4	15,866	17,386	33,253	5426	6247	11673

Table 9.17 **Average weekly alcohol consumption, by sex and usual gross weekly earnings**

Persons aged 16-64 in full time employment

Great Britain: 2002

Usual gross weekly earnings (£)	Men	Women	All persons	Weighted base (000's) =100%			Unweighted sample		
				Men	Women	All persons	Men	Women	All persons
	Mean number of units per week								
Up to 200.00	17.9	12.5	15.0	910	1,049	1,959	302	365	667
200.01-300.00	20.9	10.4	16.0	2,049	1,826	3,875	685	640	1325
300.01-400.00	20.5	9.8	16.9	2,272	1,180	3,452	765	413	1178
400.01-600.00	18.3	10.5	15.8	2,879	1,338	4,217	1002	476	1478
600.01-800.00	20.1	9.7	17.6	1,199	379	1,578	422	135	557
800.01 or more	17.8	10.8	16.5	1,187	276	1,463	428	98	526
Total	19.4	10.7	16.2	10,496	6,048	16,543	3604	2127	5731

Table 9.18   Average weekly alcohol consumption, by sex and Government Office Region

Persons aged 16 and over

Great Britain: 2002

Government Office Region	Men	Women	All persons	Weighted base (000's) =100%			Unweighted sample		
				Men	Women	All persons	Men	Women	All persons
	Mean number of units per week								
North East	20.8	9.4	14.3	857	1,145	2,002	294	400	694
North West	17.0	8.1	12.2	2,245	2,692	4,936	795	970	1765
Yorkshire and the Humber	20.0	8.6	13.9	1,720	1,966	3,685	594	695	1289
East Midlands	18.4	7.6	12.7	1,395	1,584	2,979	510	589	1099
West Midlands	15.9	7.2	11.3	1,740	1,989	3,729	609	705	1314
East of England	15.6	7.6	11.5	1,820	1,944	3,764	665	730	1395
London	14.4	5.5	9.7	2,427	2,738	5,165	750	885	1635
South East	16.8	7.4	11.9	2,836	3,160	5,996	1049	1189	2238
South West	17.6	8.3	12.7	1,743	1,943	3,686	640	726	1366
England	17.0	7.6	12.0	16,781	19,160	35,941	5906	6889	12795
Wales	19.2	9.0	13.8	972	1,112	2,083	349	407	756
Scotland	17.7	6.7	11.9	1,783	1,948	3,731	572	648	1220
Great Britain	17.2	7.6	12.1	19,536	22,220	41,756	6827	7944	14771

Chapter 10 **Contraception**

The General Household Survey (GHS) is one of the two main sources of information on contraceptive use in Great Britain, the other being the National Statistics Omnibus Survey.[1] Where the two sources cover precisely the same topic for the same year and geographical coverage, the GHS is the better source. Questions on contraception were first included in the GHS in 1983 and have been repeated regularly since then. Before 2002, the most recent questions appeared in the 1998 GHS.

- Current use of contraception

- Current use of contraception and age

- Trends in contraceptive use

    Contraceptive pill

    Sterilisation

    Condoms

- Current use of contraception and marital status

- Change to and from the condom

- Women not 'at risk' of pregnancy

- Use of the pill and the condom as usual methods of contraception by women 'at risk' of pregnancy

- Use of emergency contraception

- Obtaining emergency contraception

Current use of contraception

In 2002 nearly three quarters of women aged 16 to 49 (72%) used at least one form of contraception. This figure is similar to that found in 1998, 1995 and 1993. The most common methods for avoiding pregnancy used by women in this age range included:
- the contraceptive pill (used by 26% of women);
- surgical sterilisation of either the woman or her partner (21% of women);
- the male condom (used by partners of 19% of women).

Of women questioned in 2002, 28% were *not* using any form of contraception. Of these:
- half were not in a sexual relationship (14% of all women aged 16 to 49);
- almost a quarter were either pregnant or wanting to conceive (6% of all women in the age group).

Most of the analyses presented in this chapter focus on the three most popular methods of contraception used by women.   **Table 10.1**

Current use of contraception and age

As in previous years, whether or not women were using contraception and the type of method used varied significantly with age.
- Over half (52%) of women aged 16 to 17 did not have a current sexual partner and were not using any method of contraception. The percentage of women in this age group using at least one method of contraception was 40% - a statistically significant rise from the 29% recorded in 1998. Among contraceptive users in this age group, almost equal proportions used the pill and the condom.
- Use of the contraceptive pill was particularly common among women aged between 18 and 29. For example, more than half of 20 to 24 year olds (53%) used this form of contraception. Among women aged 30 and over, however, the pill was a significantly less popular method of contraception (28% of women aged 30 to 34 used this form of contraception).
- The proportions of women using surgical sterilisation as a method of contraception increased in each age group beyond the age of 30, reflecting the fact that older women are more likely to have completed their families.   **Table 10.2, Figure 10A**

Trends in contraceptive use

Trends in the use of contraceptive methods by women aged 16 to 49 have been monitored since 1986 when questions about contraception were first addressed to all women in this age group. Previously only women aged 18 to 49 were asked these questions. Since 1986, the proportion of women using at least one method of contraception has varied little. The contraceptive pill, surgical sterilisation (both male and female) and the male condom have remained the three most commonly used methods of contraception. Trends in the use of each of these three methods are described below.

Table 10.1

Contraceptive pill

There has been a small increase in the proportion of women using the contraceptive pill, from 23% in 1986 to 26% in 2002. Trends in use of the pill varied among women of different ages. However, it should be noted that some trends should be interpreted with caution because of the small sample sizes for the younger age groups.

Figure 10A   **Proportion of women who were currently using the pill, male condom, or were sterilised or their partner was sterilised, or were not currently in a heterosexual relationship by age, 2002**

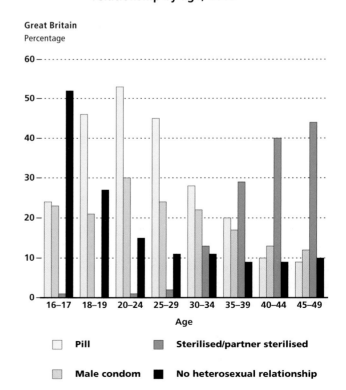

Great Britain
Percentage

Pill | Sterilised/partner sterilised
Male condom | No heterosexual relationship

- Since 1998, there has been a significant increase in the number of women using the pill in three age groups – 16-17 year olds (from 17% to 24%), 30-34 year olds (from 24% to 28%) and 45-49 year olds (from 3% to 9%). In all other age groups, the level of use remained similar between 1998 and 2002.

- Although there was a decrease between 1995 and 1998 in the percentage of women aged 30 to 34 who used the pill as their usual form of contraception, there was an increase (to 28%) in 2002.   **Table 10.3**

Sterilisation

Trends in the use of sterilisation as a method of contraception also varied according to the age of women. Since 1986, the proportion of women and partners sterilised for contraceptive reasons has declined significantly among people aged 25 to 39. Among women aged 45 to 49 and their partners, however, use of sterilisation increased significantly, from 35% in 1986 to 44% in 2002. This is in part due to the fact that women are now completing their families later in life.

                                       Table 10.4

Condoms

There has been a steady increase in the proportion of women whose partners use condoms, from 13% in 1986 to 19% in 1998 and 2002. This increase was not observed across all age groups.

- Among 35 to 44 year old women and their partners, condom use has remained fairly constant since 1986.

- There has been a decline in condom use among women aged 45 to 49 and their partners.

- Condom use has increased significantly among women under the age of 35 since 1986.   **Table 10.5**

Current use of contraception and marital status

Table 10.6 shows current use of contraception by marital status. One third of single women (33%) and almost one third (30%) of women who were widowed, divorced or separated said that they had no current sexual partner. A further 7% of single women and 11% of previously married women said they did have a partner, but did not use a method of contraception for various reasons. Among married/cohabiting women, one fifth said they did not use contraception.

- Single women were more likely to use the pill than any other method of contraception (37%, compared with 23% using the condom and 4% surgical

sterilisation). This compares with 23% of married or cohabiting women using the pill and 17% of divorced, widowed or separated women.

- 29% of women who were married or cohabiting were sterilised (or had partners who were sterilised) compared with 24% of women who were widowed, divorced or separated, and 4% of single women.

- Condom use was most popular among single women - just under one quarter (23%) of single women used this method of contraception with their partners, compared with around one fifth (19%) of married/ cohabiting women, and just one in ten women who had previously been married (10%).

                      Table 10.6, Figure 10B

Table 10.7 looks at contraception use by age and marital status.

- Whereas, overall, single women were more likely to use the pill than women who were married/ cohabiting and women who were widowed, divorced or separated, this was not the case for younger single women. Over half of women aged 16 to 24 (55%) who were married/cohabiting used the pill, compared with just over two fifths of single women in this age group (42%).

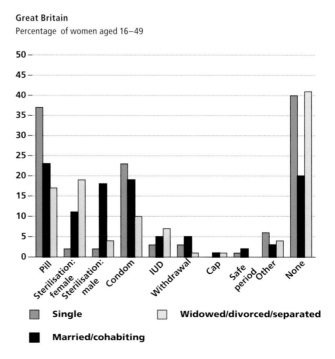

Figure 10B   **Current use of contraception by marital status, 2002**

Great Britain
Percentage of women aged 16–49

Legend: ▨ Single   ▢ Widowed/divorced/separated   ■ Married/cohabiting

- Rates of male and female sterilisation were also related to age, but were consistently highest amongst women aged 35-49 who were either married or cohabiting.
- While single women were more likely to use condoms, rates of condom use among single women aged 16 to 34 were similar to those of married/cohabiting women. **Table 10.7**

Change to and from the condom

Women were asked about their use of condoms over the two-year period prior to interview. Among current users of a contraceptive method:

- 15% had used the condom as their main method throughout the two-year period, 9% had changed to the condom and were currently using it and 5% had switched from the condom to some other method.
- women aged 16 to 17 were much more likely than older women to have used the condom at some stage over the past two years (58% compared, for example, with 40% of 18 to 19 year olds);
- single women were more likely to have used the condom throughout the two years prior to interview than married/cohabiting women (38% compared with 26%);
- single women were more than twice as likely as their married or cohabiting counterparts to have changed to the condom from some other method (15% compared with 7%). **Table 10.8**

Women not 'at risk' of pregnancy

Absence of a sexual relationship and sterilisation of either a woman or her partner carry no 'risk' (or very low 'risk') of pregnancy. Table 10.9 shows the proportion of women in different groups who had no sexual partner or who depended on sterilisation (of self or partner) to avoid pregnancy. A high proportion of women classified as being at 'low risk' of becoming pregnant was seen among:

- young women aged 16 to 19, of whom 41% claimed to have no sexual partner;
- women aged 45 to 49, of whom 44% were sterilised or who had partners who were sterilised, and women aged 40 to 44, of whom 40% were sterilised or who had partners who were sterilised;
- women with two or more children (36% of women with two children were sterilised, or had partners who were sterilised, compared with just 12% of

women with one child and 5% with no children);
- women who said they were not likely to have any (more) children (33% of whom were either sterilised themselves or had a partner who was sterilised, compared with 1% of women who were planning to have (more) children).

Those women who were employed in routine manual occupations were more likely than other women to be at low risk of pregnancy because either they or their partner had been sterilised. Likewise, women who had no formal educational qualifications were more likely to be at low risk of becoming pregnant than women with such qualifications. **Table 10.9**

Use of the pill and the condom as usual methods of contraception by women 'at risk' of pregnancy

In contrast to those women described as not 'at risk' of becoming pregnant, women aged 16 to 49 who were in a heterosexual relationship, who were not pregnant or sterilised and whose partners were not sterilised, are described here as being 'at risk' of pregnancy. Note that being 'at risk' of pregnancy is not the direct inverse of being 'not at risk' (or at low risk), and that the groups are not mutually exclusive. Table 10.10 shows women 'at risk' of pregnancy and their use of the pill and/or condom as their usual method(s) of contraception. In total, 34% of women 'at risk' of becoming pregnant used the pill as their usual method of contraception, 23% used condoms with their partners, whilst 8% used both the pill and condom as their usual methods of contraception. Of women 'at risk' of pregnancy, the remaining 35% used other methods of contraception, or no method at all.

In terms of age, younger women considered to be 'at risk' of becoming pregnant were more likely to use the pill as their usual method of contraception compared with older women. Nearly two fifths (39%) of 16-19 year olds used the pill, 46% of 20-24 year olds, and almost half (47%) of 25-29 year olds. This compared with just over one third (34%) of 30-34 year old women at risk of pregnancy who used the pill and smaller proportions in the older age groups.

By contrast, use of the male condom was more common among women aged 30 and over who were 'at risk' of pregnancy, compared with younger women. Around a quarter of 'at risk' women aged 30 and over, who did not use the pill, used condoms with their partners, compared with 18% of 'at risk' 16-19 year olds and 16% of 20-24 year olds in this group.

Younger women at risk of pregnancy aged 16 to 24 were also more likely to use the pill in combination with condoms as their usual methods of contraception, compared with older women at risk of pregnancy. This age difference was also reflected in differences between 'at risk' women of different marital statuses. Nearly one fifth (19%) of single women at risk of pregnancy used both the pill and condoms, compared with 7% of 'at risk' women who were either widowed, divorced or separated (i.e. those not cohabiting with their husband or partner).

There were also marked differences between those women 'at risk' of pregnancy who had formal educational qualifications and those with no formal qualifications. Women with formal qualifications who were 'at risk' were twice as likely to use condoms with their partner than women with no qualifications in this group, 24% compared with 12% respectively. By contrast, over half (53%) of 'at risk' women with no formal qualifications used neither pill nor condom, compared with just one third (33%) of those with formal qualifications. **Table 10.10**

Use of emergency contraception

Questions on emergency contraception were first included on the GHS in 1993. These related to the use of emergency contraception during the two years prior to interview. In 2002 the reference period was changed from two years to one year, and some additional questions were included. Two forms of emergency contraception are available to women to use after intercourse: hormonal emergency contraception (the 'morning after pill'); and the emergency IUD (intrauterine device). Women aged 16 to 49, who were not sterilised (and whose partners were not sterilised) were asked about their awareness and experiences of these methods of contraception.

In 2002 more than nine in ten (93%) women aged 16 to 49 had heard of the 'morning after pill' compared with just over half (51%) who were aware of the emergency IUD. The likelihood of a woman knowing about

hormonal emergency contraception (i.e. the 'morning after pill') was not associated with age (although women in the sample aged 16 to 17 were slightly less likely to have heard of the 'morning after pill' than those aged 25 and over). However, older women were more likely to have heard of the emergency IUD. For example, only about 40% of the two youngest age groups had heard of the emergency IUD, compared with six in ten women aged 45 to 49 (60%). **Table 10.11**

Seven percent of women aged 16 to 49, who were not sterilised and whose partners were not sterilised, had used a method of emergency contraception in the 12 months prior to the interview. Out of the two methods of emergency contraception available (pill and IUD) very few women had had an IUD fitted. Just 3% of those women who had used emergency contraception and less than 0.5% of all women in this group had had an IUD fitted in the 12 months prior to interview (table not shown), the majority using the emergency contraceptive pill. Because the proportion of the sample that had had an IUD fitted for the purposes of emergency contraception was so small, Table 10.12 only includes information about women who had used the 'morning after pill' in the 12 months prior to interview. Six per cent of women had used the emergency contraceptive pill once in the past year, 1% had used it twice and less than 0.5% had used it more than twice.

Hormonal emergency contraception was most likely to have been used by:
- women under 25 (who were twice as likely as women aged 30 and over to have used it once in the past year);
- single women (who were twice as likely as married or cohabiting women to have used it in the past year);
- women with no children (7% had used it once compared with 4% of women with one or two children);
- women with formal educational qualifications as opposed to those with none;
- women who said they definitely or probably would have children or more children;
- women who said they were currently using a method of contraception.

There were no significant differences with respect to the use of emergency contraception on the basis of their socio-economic classification. **Table 10.12**

Obtaining emergency contraception

Emergency contraception is available to all women from GPs, family planning clinics, NHS walk-in centres, genito-urinary medicine/sexual health clinics and some hospital accident and emergency departments. In 2001 the hormonal emergency contraceptive pill became available over the counter to women aged 16 and over. In the 2002 GHS, women who had used the 'morning after pill' at least once during the year prior to interview were asked for the first time about where they had obtained their emergency contraception. Of all women who had used the 'morning after pill' in the year prior to interview, 47% had obtained it from their own GP or practice nurse, 23% from a chemist/pharmacy and 17% from a family planning clinic. There were some differences on the basis of age, marital status and socio-economic classification in where women had obtained the 'morning after pill', but these differences did not reach statistical significance on account of the small sample sizes.   **Table 10.13**

References

1   The latest Omnibus figures on contraceptive use are: Dawe, F. & Meltzer, H. *Contraception and Sexual Health, 2002.* Office for National Statistics (London 2003).

Table 10.1   **Women aged 16-49: trends in contraceptive use: 1986 to 2002**

Women aged 16-49   **Great Britain**

Current usual method of contraception	Unweighted						Weighted	
	1986	**1989**	**1991**	**1993**	**1995**	**1998**	**1998**	**2002**
	%	%	%	%	%	%	%	%
Using method(s)*								
Non-surgical†								
Pill	23	22	23	25	25	24	24	26
IUD	7	5	5	5	4	5	5	4
Condom	13	15	16	17	18	18	19	19
Cap	2	1	1	1	1	1	1	1
Withdrawal	4	4	3	3	3	3	3	4
Safe period	1	1	1	1	1	1	1	1
Other	1	1	1	1	2	3	2	4
At least one	49	46	46	48	49	49	49	51
Surgical								
Female sterilisation	12 ⎤	11 ⎤	12 ⎤	12 ⎤	12 ⎤	11 ⎤	11 ⎤	9 ⎤
Male sterilisation	11 ⎦ 23	12 ⎦ 23	13 ⎦ 25	12 ⎦ 24	11 ⎦ 24	12 ⎦ 23	12 ⎦ 23	12 ⎦ 21
Total using at least one	71	69	70	72	73	72	72	72
Not using a method								
Sterile after another operation	3	5	3	2	3	2	2	2
Pregnant/wanting to get pregnant	7	7	9	8	8	6	6	6
Abstinence/no partner	15 ⎤	16 ⎤	16 ⎤	15 ⎤	14 ⎤	14 ⎤	14 ⎤	14 ⎤
Other	5 ⎦ 20	5 ⎦ 22	5 ⎦ 21	5 ⎦ 21	6 ⎦ 20	6 ⎦ 20	6 ⎦ 20	6 ⎦ 20
Total	29	31	30	29	28	28	28	28
*Weighted base (000's)=100%***							12,059	12,100
*Unweighted sample***	5866	5802	5571	5303	5067	4251		4218

*   Percentages add to more than 100 because some women used more than one non-surgical method or had more than one reason for not using a method.

†   Abstinence is not included here as a method of contraception.

**   Trend tables show unweighted and weighted figures for 1998 to allow direct comparison between 1998 and 2002 and to give an indication of the effect of the weighting. For the weighted data (1998 and 2002) the weighted base (000's) is the base for percentages. Unweighted data (up to 1998) are based on the unweighted sample.

Table 10.2　Current use of contraception by age

Women aged 16-49　　　　　　　　　　　　　　　　　　　　　　　　Great Britain: 2002

Current use of contraception	Age								
	16-17	18-19	20-24	25-29	30-34	35-39	40-44	45-49	Total
	%	%	%	%	%	%	%	%	%
Using method(s)*									
Non-surgical†:									
Pill**	24	46	53	45	28	20	10	9	26
Mini pill	6	10	8	7	6	6	4	5	6
Combined pill	14	29	41	35	21	13	5	3	19
(Mini + Combined subtotal)	20	39	49	42	27	18	9	8	25
IUD	0	1	2	6	6	5	5	4	4
Condom	23	21	30	24	22	17	13	12	19
Cap	0	1	0	0	1	1	1	1	1
Withdrawal	3	1	2	4	5	4	4	3	4
Safe period	0	1	1	1	2	1	2	1	1
Spermicides	0	0	0	0	0	0	0	0	0
Injection	3	8	6	5	4	3	2	1	3
At least one	39	61	74	75	60	47	34	27	51
Surgical:									
Sterilisation - female	0	0	0	2	6	12	17	21	9
- male	1	0	1	1	7	17	22	23	12
(Sterilisation subtotal)			1	3	13	29	40	44	21
Total - at least one	40	61	75	78	72	75	74	71	72
Not using a method									
Sterile after other operation	0	1	0	0	1	2	4	5	2
Pregnant now	3	4	6	6	5	2	1	1	4
Going without sex to avoid pregnancy	2	0	1	0	0	0	0	0	0
No sexual relationship	52	27	15	11	11	9	9	10	13
Wants to get pregnant	0	1	2	2	5	5	2	1	3
Unlikely to conceive because of menopause	0	0	0	0	0	0	1	7	1
Possibly infertile	0	0	0	0	1	3	3	3	2
Doesn't like contraception	0	1	0	0	1	0	1	1	1
Others	2	3	1	1	3	3	3	2	2
Total not using a method	60	39	25	22	28	25	26	29	28
Weighted base (000's) = 100%	606	482	1,506	1,582	2,007	2,121	2,001	1,794	12,100
Unweighted sample	215	161	485	539	704	754	689	671	4218

*　Percentages add to more than 100 because of rounding and because some women used more than one non-surgical method.

†　Abstinence is not included here as a method of contraception. Those who said that 'going without sex to avoid getting pregnant' was their only method of contraception are shown with others not using a method.

**　The total percentage using the pill includes those who did not know which type of pill.

Table 10.3   **Trends in use of the pill as a usual method of contraception by age: 1986 to 2002**

Women aged 16-49   **Great Britain**

Age	Unweighted						Weighted		Weighted base 2002 (000's) = 100%*	Unweighted sample* 2002
	1986	1989	1991	1993	1995	1998	1998	2002		
				Percentage of women who used the pill						
16-17	20	19	16	20	25	17	17	24	606	215
18-19	42	39	46	42	37	41	43	46	482	161
20-24	55	48	48	50	49	52	53	53	1,506	485
25-29	38	36	42	44	41	41	42	45	1,582	539
30-34	21	22	25	29	29	24	24	28	2,007	704
35-39	8	12	11	16	20	18	18	20	2,121	754
40-44	4	4	4	7	9	9	10	10	2,001	689
45-49	1	3	2	4	3	4	3	9	1,794	671
All aged 16-49	23	22	23	25	25	24	24	26	12,100	4218

*   Trend tables show unweighted and weighted figures for 1998 to allow direct comparison between 1998 and 2002 and to give an indication of the effect of the weighting. For the weighted data (1998 and 2002) the weighted base (000's) is the base for percentages. Unweighted data (up to 1998) are based on the unweighted sample. Bases for earlier years are of a similar size and can be found in GHS reports for each year.

Table 10.4   **Percentage of women and partners sterilised for contraceptive reasons: 1986 to 2002**

Women aged 16-49   **Great Britain**

Age	Unweighted						Weighted		Weighted base 2002 (000's) = 100%*	Unweighted sample* 2002
	1986	1989	1991	1993	1995	1998	1998	2002		
				Percentage of women and partners[†] sterilised						
16-24	1	1	1	0	0	1	1	1	2,594	861
25-29	6	7	8	5	7	5	4	3	1,582	539
30-34	25	23	21	21	18	17	17	13	2,007	704
35-39	42	40	38	34	32	33	33	29	2,121	754
40-44	48	47	50	47	45	42	43	40	2,001	689
45-49	35	37	47	47	46	50	50	44	1,794	671
All aged 16-49	23	23	25	24	24	23	23	21	12,100	4218

*   Trend tables show unweighted and weighted figures for 1998 to allow direct comparison between 1998 and 2002 and to give an indication of the effect of the weighting. For the weighted data (1998 and 2002) the weighted base (000's) is the base for percentages. Unweighted data (up to 1998) are based on the unweighted sample. Bases for earlier years are of a similar size and can be found in GHS reports for each year.

†   Refers to the woman's partner whether in the household or not.

Table 10.5   **Trends in use of the condom as a usual method of contraception by age: 1986 to 2002**

Women aged 16-49   **Great Britain**

Age	Unweighted						Weighted		*Weighted base 2002 (000's) = 100%**	*Unweighted sample* 2002*
	1986	1989	1991	1993	1995	1998	1998	2002		
				Percentage whose partners[†] used the condom						
16-17	6	6	10	17	13	18	18	23	606	215
18-19	6	12	15	22	26	21	22	21	482	161
20-24	9	14	14	18	21	23	24	30	1,506	485
25-29	13	17	19	21	20	23	24	24	1,582	539
30-34	15	19	17	18	20	20	20	22	2,007	704
35-39	15	16	20	17	16	18	18	17	2,121	754
40-44	14	16	13	14	16	15	14	13	2,001	689
45-49	16	15	12	12	14	10	10	12	1,794	671
All aged 16-49	13	15	16	17	18	18	19	19	12,100	4218

*   Trend tables show unweighted and weighted figures for 1998 to allow direct comparison between 1998 and 2002 and to give an indication of the effect of the weighting. For the weighted data (1998 and 2002) the weighted base (000's) is the base for percentages. Unweighted data (up to 1998) are based on the unweighted sample. Bases for earlier years are of a similar size and can be found in GHS reports for each year.

†   Refers to the woman's partner whether in the household or not.

Table 10.6   **Current use of contraception by marital status**

Women aged 16-49   **Great Britain: 2002**

Current use of contraception	Single	Married/ cohabiting	Widowed/divorced/ separated	Total
	%	%	%	%
Using method(s)				
Non-surgical*:				
Pill	37	23	17	27
IUD	3	5	7	5
Condom	23	19	10	19
Cap	0	1	1	1
Withdrawal	3	5	1	4
Safe period	1	2	0	1
Spermicides	0	0	0	0
Injection	5	3	4	3
At least one	57	51	35	51
Surgical:				
Sterilisation - female	2 ⎫ 4	11 ⎫ 29	19 ⎫ 24	9 ⎫ 21
- male	2 ⎭	18 ⎭	4 ⎭	12 ⎭
Total - at least one†	60	80	59	72
Not using a method				
Sterile after other operation	1	3	3	2
Pregnant now	2	5	1	4
Going without sex to avoid pregnancy	1	0	0	0
No sexual relationship	33	1	30	13
Wants to get pregnant	0	4	1	3
Unlikely to conceive because of menopause	0	2	1	1
Possibly infertile	0	2	2	2
Doesn't like contraception	0	1	0	1
Others	2	3	1	2
Total not using a method	40	20	41	28
*Weighted base (000's) = 100%***	3,603	7,318	1,145	12,066
*Unweighted sample***	1221	2579	405	4205

*   Abstinence is not included here as a method of contraception. Those who said that 'going without sex to avoid getting pregnant' was their only method of contraception are shown with others not using a method.

†   Includes a few cases where other methods of contraception were used.

**   Percentages add to more than 100 because of rounding and because some women used more than one non-surgical method.

Table 10.7   **Current use of contraception by age and marital status**

Women aged 16-49   Great Britain: 2002

Current use of contraception	Age						
	16-24	25-29	30-34	35-39	40-44	45-49	Total
	%	%	%	%	%	%	%
Single							
Pill	42	44	30	32	11	[18]	37
Male condom	27	24	21	18	11	[5]	23
Sterilised	0	2	6	4	8	[9]	2
Partner sterilised	1	1	2	6	4	[5]	2
No heterosexual relationship	34	27	29	33	46	[43]	33
Other†	16	17	27	12	26	[25]	18
Weighted base (000's) = 100%	*1,929*	*580*	*505*	*273*	*197*	*118*	*3,601*
Unweighted sample	*653*	*199*	*171*	*91*	*64*	*42*	*1220*
	%	%	%	%	%	%	%
Married or cohabiting							
Pill	55	46	27	17	10	8	23
Male condom	24	25	23	18	14	15	19
Sterilised	0	1	5	12	18	21	11
Partner sterilised	1	2	9	21	28	28	18
No heterosexual relationship	0	1	0	1	0	1	1
Other†	35	34	42	35	33	31	35
Weighted base (000's) = 100%	*631*	*947*	*1,331*	*1,547*	*1,495*	*1,357*	*7,308*
Unweighted sample	*197*	*321*	*469*	*554*	*520*	*515*	*2576*
	%	%	%	%	%	%	%
Widowed, divorced or separated							
Pill	**	**	31	21	9	8	17
Male condom	**	**	12	11	11	4	10
Sterilised	**	**	14	20	20	25	19
Partner sterilised	**	**	0	3	6	7	4
No heterosexual relationship	**	**	35	28	25	38	30
Other†	**	**	17	20	31	18	24
Weighted base (000's) = 100%	*25*	*52*	*167*	*280*	*307*	*314*	*1,145*
Unweighted sample	*8*	*18*	*62*	*101*	*104*	*112*	*405*

*   Percentages sum to more than 100 as respondents could give more than one answer.
†   Includes women using other methods of contraception and those not currently using a method.
**   Base too small for analysis.

Table 10.8   Current users of a contraceptive method: changes to and from condom use by woman's partner
during the two years prior to interview by woman's marital status and age

Women aged 16-49 with a partner* and currently using a method of contraception   Great Britain: 2002

Marital status and change to/from condom	Age								
	16-17	18-19	20-24	25-29	30-34	35-39	40-44	45-49	Total
	%	%	%	%	%	%	%	%	%
Single women									
Partner's use of condom									
Condom user throughout two years††	5	14	20	18	20	19	[22]	**	17
Changed to condom†	45	18	15	9	9	7	[2]	**	15
Changed from condom to current method	10	9	7	6	1	1	[7]	**	6
Condom has not been main method									
throughout two years	40	59	57	67	70	73	[68]	**	62
Weighted base (000's) = 100%	*235*	*233*	*667*	*402*	*324*	*164*	*85*	*44*	*2,154‡*
Unweighted sample	*83*	*79*	*220*	*138*	*110*	*55*	*28*	*15*	*728‡*
	%	%	%	%	%	%	%	%	%
Married/cohabiting women									
Partner's use of condom									
Condom user throughout two years†	**	**	6	12	19	16	14	16	15
Changed to condom†	**	**	17	15	9	4	3	1	7
Changed from condom to current method	**	**	6	6	5	6	3	1	4
Condom has not been main method									
throughout two years	**	**	70	67	68	74	81	82	74
Weighted base (000's) = 100%	*11*	*54*	*438*	*784*	*1,022*	*1,255*	*1,216*	*1,061*	*5,841‡*
Unweighted sample	*3*	*17*	*138*	*265*	*362*	*451*	*425*	*403*	*2064‡*
	%	%	%	%	%	%	%	%	%
All marital statuses of women††									
Partner's use of condom									
Condom user throughout two years†	5	13	14	14	19	16	14	15	15
Changed to condom†	43	17	16	13	9	5	3	1	9
Changed from condom to current method	10	10	7	6	4	5	4	1	5
Condom has not been main method									
throughout two years	42	60	62	67	69	75	79	83	72
Weighted base (000's) = 100%	*245*	*287*	*1,122*	*1,224*	*1,444*	*1,598*	*1,478*	*1,266*	*8,664‡*
Unweighted sample	*86*	*96*	*363*	*416*	*509*	*571*	*514*	*475*	*3030‡*

*   Refers to the woman's partner whether in the household or not.
†   Condom was main method at the time of interview.
**   Bases are too small to enable reliable analysis to be made.
††   'All marital statuses' includes widowed/divorced/separated.
‡   Total base includes groups not shown because of small bases.

Table 10.9   Percentage of women who
(a) had no sexual partner
(b) were sterilised or who had partners who were sterilised
by selected characteristics

Women aged 16-49   Great Britain: 2002

	(a) No sexual partner	(b) Self or partner* sterilised	Weighted base (000's) = 100%	Unweighted sample
	Percentages			
Age				
16-19	41	1	1,088	376
20-24	15	1	1,503	484
25-29	11	3	1,579	538
30-34	11	13	2,007	704
35-39	9	29	2,114	752
40-44	9	40	2,001	689
45-49	10	44	1,794	671
Marital status				
Single	33	4	3,601	1220
Married	1	34	5,584	1999
Cohabiting	0	13	1,723	577
Widowed/divorced/separated	30	24	1,145	405
Number of children born				
None	21	5	4,903	1650
One	11	12	2,236	764
Two	5	36	3,007	1094
Three or more	9	48	1,928	702
Opinion whether woman would have (more) children				
Yes, probably yes	17	1	4,404	1473
No, probably not	11	33	7,587	2709
Did not know/no answer	[20]	[6]	95	32
Highest qualification level attained				
GCE 'A' level or above	13	15	5,471	1900
GCSE grades A-C or equivalent	13	22	3,023	1064
Other	10	27	1,872	658
None	17	31	1,699	585
Socio-economic classification†				
Managerial and professional	10	19	3,520	1242
Intermediate	8	22	2,555	901
Routine and manual	11	28	4,153	1443
Total**	13	21	12,086	4214

*   Refers to the woman's partner, whether in the household or not.
†   From April 2001 the National Statistics Socio-economic Classification (NS-SEC) was introduced for all official statistics and surveys. It replaced Social Class based on Occupation and Socio-economic Groups (SEG). Full-time students, persons in inadequately described occupations, persons who have never worked and the long term unemployed are not shown as separate categories, but are included in the figure for all persons (see Appendix A).
** Total includes no answers to some of the selected characteristics.

Table 10.10   **Use of the pill and condom as a usual method of contraception by selected characteristics**

Women aged 16-49 (excluded if pregnant, self or partner* sterilised or no sexual relationship)   **Great Britain: 2002**

		1. Pill user†	2. Partner using condom†	3. User of both pill and condoms	4. Neither pill user nor condom user	*Weighted base (000's) = 100%*	*Unweighted sample*
Age							
16-19	%	39	18	23	21	*593*	*201*
20-24	%	46	16	22	17	*1,173*	*379*
25-29	%	47	21	9	23	*1,262*	*428*
30-34	%	34	25	5	35	*1,432*	*498*
35-39	%	30	25	3	42	*1,269*	*453*
40-44	%	19	26	0	54	*1,015*	*353*
45-49	%	18	25	2	55	*812*	*307*
Marital status							
Single	%	41	19	19	20	*2,212*	*745*
Married	%	25	27	3	45	*3,457*	*1235*
Cohabiting	%	46	20	6	29	*1,378*	*459*
Widowed/divorced/separated	%	31	15	7	47	*503*	*178*
Number of children born							
None	%	41	20	14	25	*3,448*	*1153*
One	%	32	23	6	39	*1,596*	*545*
Two	%	27	30	3	40	*1,687*	*618*
Three or more	%	25	17	2	56	*822*	*302*
Opinion whether woman would have (more) children							
Yes, probably yes	%	41	20	15	24	*3,408*	*1137*
No, probably not	%	28	25	3	44	*4,076*	*1458*
Did not know/no answer	%	[24]	[30]	[0]	[46]	*71*	*24*
Educational qualifications							
Formal qualifications	%	35	24	9	33	*6,721*	*2332*
No formal qualifications	%	31	12	4	53	*817*	*281*
Socio-economic classification**							
Managerial and professional	%	33	25	7	36	*2,378*	*836*
Intermediate	%	35	26	6	33	*1,682*	*589*
Routine and manual	%	34	19	8	38	*2,434*	*843*
Total††	%	34	23	8	35	*7,556*	*2619*

*   Refers to the woman's partner, whether in the household or not.
†   Includes women using the pill or condom in combination with other methods of contraception.
** From April 2001 the National Statistics Socio-economic Classification (NS-SEC) was introduced for all official statistics and surveys. It replaced Social Class based on Occupation and Socio-economic Groups (SEG). Full-time students, persons in inadequately described occupations, persons who have never worked and the long term unemployed are not shown as separate categories, but are included in the figure for all persons (see Appendix A).
†† Total includes no answers to some of the selected characteristics.

Table 10.11 Knowledge of emergency contraception by age

Women aged 16-49 (excluded if self or partner sterilised)   Great Britain: 2002

Emergency contraception	Age								Total
	16-17	18-19	20-24	25-29	30-34	35-39	40-44	45-49	
	Percentage who had heard of emergency contraception								
Hormonal emergency contraception	89	92	91	95	94	95	92	92	93
Emergency IUD	40	39	48	48	53	52	55	60	51
Weighted base (000)s = 100%	606	479	1,531	1,557	1,836	1,642	1,276	983	9,910
Unweighted base	215	160	492	530	640	582	442	369	3430

Table 10.12 Use of hormonal emergency contraception during the 12 months prior to interview by selected characteristics

Women aged 16-49 (excluded if partner* or self sterilised)   Great Britain: 2002

		Use of hormonal emergency contraception ('morning after pill')				Weighted base (000's) = 100%	Unweighted sample
		Used once	Used twice	Used more than twice	Not used†		
Age							
16-17	%	10	1	1	88	536	191
18-19	%	12	3	1	84	444	148
20-24	%	9	2	1	88	1,407	454
25-29	%	7	1	0	92	1,495	509
30-34	%	5	1	0	94	1,745	609
35-39	%	4	1	0	95	1,571	559
40-44	%	3	0	0	97	1,206	418
45-49	%	2	0	0	97	914	344
Marital status							
Single	%	9	2	1	89	3,279	1110
Married/cohabiting	%	4	1	0	95	5,129	1799
Widowed/divorced/separated	%	6	0	0	94	876	310
Number of children born							
None	%	7	2	0	91	4,441	1491
One	%	4	0	0	95	1,894	645
Two	%	4	1	0	95	1,930	709
Three or more	%	6	0	0	94	1,045	384
Opinion whether woman would have (more) children							
Yes, probably yes	%	8	2	0	90	4,155	1393
No, probably not	%	4	1	0	95	5,079	1811
Did not know/no answer	%	[0]	[4]	[0]	[96]	85	28
Highest qualification level attained							
GCE 'A' level or above	%	6	1	0	93	4,601	1584
GCSE grades A-C or equivalent	%	7	1	0	91	2,328	817
Other	%	5	1	0	93	1,312	459
None	%	3	1	0	96	1,070	369
Socio-economic classification**							
Managerial and professional	%	6	1	0	93	2,890	1012
Intermediate	%	5	0	0	95	1,966	687
Routine and manual	%	5	1	0	93	2,968	1027
Currently uses a method of contraception							
Yes	%	7	1	0	91	6,333	2195
No	%	3	0	0	97	2,874	997
Total††	%	6	1	0	93	9,318	3232

*   Refers to the woman's partner, whether in the household or not.
†   Includes women who had used the emergency IUD.
**   From April 2001 the National Statistics Socio-economic Classification (NS-SEC) was introduced for all official statistics and surveys. It replaced Social Class based on Occupation and Socio-economic Groups (SEG). Full-time students, persons in inadequately described occupations, persons who have never worked and the long term unemployed are not shown as separate categories, but are included in the figure for all persons (see Appendix A).
††   Total includes no answers to some of the selected characteristics.

Table 10.13 **Where hormonal emergency contraception (the 'morning after pill') obtained by selected characteristics**

Women aged 16-49, who had used the 'morning after pill' in the year prior to interview

Great Britain: 2002

		Where obtained*							Weighted	Unweighted
		Own GP/ practice nurse	Family planning clinic	Other GP/ practice nurse	Hospital Accident & Emergency	Chemist/ pharmacy	Walk-in centre/ minor injuries unit	Other	base (000's) = 100%	sample
Age										
16-29	%	43	21	3	0	26	5	2	423	143
30-49	%	56	11	3	5	18	0	6	238	84
Marital status										
Single	%	37	23	3	1	27	5	4	371	126
Married/cohabiting	%	60	11	3	4	18	1	3	236	82
Widowed/divorced/separated	%	†	†	†	†	†	†	†	52	19
Socio-economic classification**										
Managerial and professional	%	45	12	3	3	29	3	5	194	67
Intermediate	%	[41]	[24]	[2]	[0]	[24]	[3]	[6]	97	34
Routine and manual	%	61	12	2	3	16	3	3	207	72
Total††	%	47	17	3	2	23	3	4	661	227

*   Percentages sum to more than 100 as respondents could give more than one answer.
†   Base too small for analysis.
** From April 2001 the National Statistics Socio-economic Classification (NS-SEC) was introduced for all official statistics and surveys. It replaced Social Class based on Occupation and Socio-economic Groups (SEG). Full-time students, persons in inadequately described occupations, persons who have never worked and the long term unemployed are not shown as separate categories, but are included in the figure for all persons (see Appendix A).
†† Total includes no answers to some of the selected characteristics.

Chapter 11 Hearing and hearing aids

Questions about hearing difficulties and hearing aids were first asked on the GHS in 1977 and have been sponsored by the Department of Health. In 2002 all respondents aged 16 and over were asked if they had any difficulties with hearing. If so a further set of questions was asked about whether they wore aids and if they obtained them through the NHS or privately. The Department of Health also has an interest in whether people have NHS hearing aids that they no longer wear.

- Prevalence of hearing difficulties

- Hearing aids

- Hearing aids in use

- Working hearing aids that are not worn

Prevalence of hearing difficulties

Between 1979 and 2002 the percentage of adults with hearing difficulties increased from 13% to 16%. Most of this change occurred between 1979 and 1992, mainly among the older age groups, and has remained relatively stable since.

Men were significantly more likely than women to have hearing difficulties. In 2002 19% of men had a hearing difficulty compared with 13% of women. Having a hearing difficulty is a condition strongly associated with age. Figure 11A shows that men reported a higher prevalence of hearing difficulties than women at all ages. Although there was a big increase between 1979 and 1998 in the proportion of the oldest men reporting a hearing difficult (42% and 53% respectively) this has now levelled off at 52%. **Table 11.1, Figure 11A**

Hearing aids

The percentage of adults wearing a hearing aid has doubled from 2% in 1979 to 4% in 2002.

- In 2002 4% of men wore at least one hearing aid compared with 3% of women.
- Men aged 75 and over were significantly more likely to wear hearing aids than women in the same age group (23% compared with 17%).
- The likelihood of a man aged 75 and over wearing a hearing aid almost doubled between 1979 and 2002 (12% and 23% respectively).

Figure 11A **Percentage of men and women reporting some difficulty with hearing by age, 2002**

Great Britain
Percentage

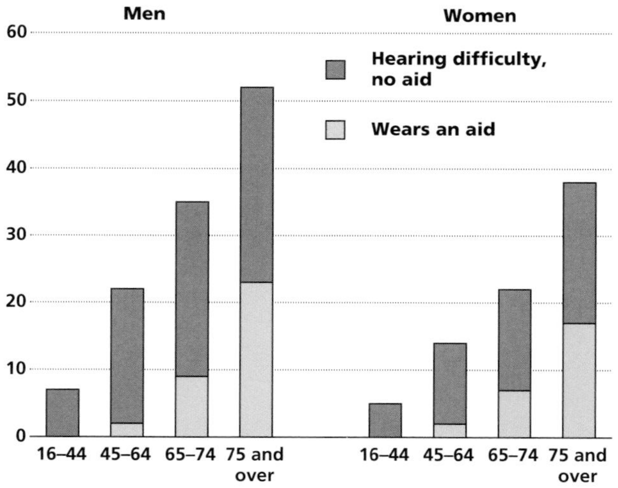

The proportion of people who had a hearing difficulty but did not wear an aid has shown a small but statistically significant increase, from 11% in 1979 to 12% in 2002.

- Overall, 15% of men had a hearing difficulty and no aid compared with 10% of women.
- There has been a decrease in the proportion of women aged 75 and over who have a hearing difficulty but do not wear an aid, from 27% in 1979 to 21% in 2002. **Table 11.2**

The use of a hearing aid does not necessarily overcome people's hearing problems. Table 11.3 shows that of the 4% of people who wear an aid, 62% reported continuing problems with their hearing. **Table 11.3**

Hearing aids in use

In 2002 28% of respondents who wore a hearing aid said they wore more than one. Out of all the hearing aids in use, 75% were obtained from the NHS and the rest (25%) were obtained privately (table not shown).

For each private hearing aid that a respondent was using, they were asked why they had bought it. Among the respondents who had bought an aid privately, 59% said it was for a better choice, 44% said it was not available on the NHS and 19% said they could obtain it more quickly. **Table 11.4**

Working hearing aids that are not worn

All respondents who said they had a hearing difficulty were asked if they had any hearing aids that worked, which they no longer wore. Just over one in ten (11%) of those currently wearing a hearing aid said they had at least one working aid that they did not wear. A further 8% of respondents who had a hearing difficulty but did not wear a hearing aid said they had at least one working aid in their possession. In 2002 1% of the GHS sample (206 respondents) said they had at least one working hearing aid that they did not wear (table not shown). Among all the working hearing aids that were not worn, 79% were obtained from the NHS and the rest (21%) were obtained privately (table not shown). **Table 11.5**

Respondents were asked why they were not using each working hearing aid they had and did not wear. Out of the respondents who said they had an aid they did not use, 55% said it did not help their hearing and 11% said

it was due to its appearance. A wide range of other
reasons were given by 56% of respondents, such as:

- it was badly fitting;
- it magnified background noise;
- it caused infections;
- it was a spare; or
- they had others they preferred to use. **Table 11.6**

Among all working hearing aids obtained from the NHS
that were not worn, half (50%) were not worn because
respondents reported that they did not help with their
hearing. This compared with 39% of those obtained
privately. **Table 11.7**

Table 11.1 Difficulty with hearing, by sex and age: 1979, 1992, 1995, 1998 and 2002

Persons aged 16 and over

Great Britain

Age	Unweighted				Weighted		Weighted base 2002 (000's) = 100%*	Unweighted sample 2002*
	1979	1992	1995	1998	1998	2002		
Percentage having a hearing difficulty								
Men								
16-44	6	6	7	7	7	7	9,703	3157
45-64	19	20	23	24	23	22	6,246	2297
65-74	31	36	39	34	34	35	2,193	829
75 and over	42	45	48	53	53	52	1,465	568
All aged 16 and over	15	17	18	20	18	19	19,607	6851
Women								
16-44	5	4	5	5	5	5	10,748	3699
45-64	12	11	13	13	13	14	6,705	2573
65-74	24	21	20	21	21	22	2,475	892
75 and over	39	41	41	40	40	38	2,352	803
All aged 16 and over	12	12	13	13	13	13	22,280	7967
All persons								
16-44	5	5	6	6	6	6	20,452	6856
45-64	15	15	18	18	18	18	12,949	4870
65-74	27	27	28	27	26	28	4,668	1721
75 and over	40	43	44	45	45	44	3,817	1371
All aged 16 and over	13	14	15	16	16	16	41,886	14818

* Trend tables show unweighted and weighted figures for 1998 to give an indication of the effect of the weighting. For the weighted data (1998 and 2002) the weighted base (000's) is the base for percentages. Unweighted data (up to 1998) are based on the unweighted sample. Bases for earlier years are of similar size and can be found in GHS reports for each year.

Table 11.2　**Percentage with a hearing difficulty: 1979, 1992, 1995, 1998 and 2002**
a) Wears an aid
b) Does not wear an aid

Persons aged 16 and over　　　　　　　　　　　　　　　　　　　　　　　　　Great Britain

Age		Unweighted								Weighted				Weighted base 2002 (000's) = 100%*	Unweighted sample 2002*
		1979		1992		1995		1998		1998		2002			
		Wears an aid	Hearing difficulty and no aid	Wears an aid	Hearing difficulty and no aid	Wears an aid	Hearing difficulty and no aid	Wears an aid	Hearing difficulty and no aid	Wears an aid	Hearing difficulty and no aid	Wears an aid	Hearing difficulty and no aid		
Men															
16-44	%	0	6	0	6	0	7	0	7	0	7	0	7	9,703	3157
45-64	%	2	17	3	17	2	21	2	22	2	21	2	20	6,246	2297
65-74	%	7	24	11	25	9	30	8	26	8	26	9	26	2,193	829
75 and over	%	12	30	20	25	21	27	23	30	23	30	23	29	1,465	568
All aged 16 and over	%	2	13	4	13	3	15	4	16	3	15	4	15	19,607	6851
Women															
16-44	%	0	5	0	4	0	5	0	5	0	5	0	5	10,748	3699
45-64	%	2	10	2	9	2	11	2	11	2	11	2	12	6,705	2573
65-74	%	6	18	6	15	6	14	5	16	5	16	7	15	2,475	892
75 and over	%	12	27	17	24	16	25	15	25	15	25	17	21	2,352	803
All aged 16 and over	%	2	10	3	9	3	10	3	10	3	10	3	10	22,280	7967
All persons															
16-44	%	0	5	0	5	0	6	0	6	0	6	0	6	20,452	6856
45-64	%	2	13	2	13	2	16	2	16	2	16	2	16	12,949	4870
65-74	%	6	21	8	19	7	21	6	21	6	20	8	20	4,668	1721
75 and over	%	12	28	18	25	18	26	18	27	18	27	20	24	3,817	1371
All aged 16 and over	%	2	11	3	11	3	12	3	13	3	13	4	12	41,886	14818

* Trend tables show unweighted and weighted figures for 1998 to give an indication of the effect of the weighting. For the weighted data (1998 and 2002) the weighted base (000's) is the base for percentages. Unweighted data (up to 1998) are based on the unweighted sample. Bases for earlier years are of similar size and can be found in GHS reports for each year.

Table 11.3　**Whether hearing problems continue when wearing hearing aid by age**

Persons aged 16 and over who wear a hearing aid　　　　　　　　　　　　　Great Britain: 2002

Age	Percentage who continue to have hearing problems when wearing an aid	Weighted base (000's) = 100%	Unweighted sample
16-44	[70]	60	20
45-64	65	281	106
65-74	60	381	142
75 and over	62	746	272
Total	62	1,468	540

Table 11.4   Reasons for buying a private hearing aid*

Persons aged 16 and over who have bought a private hearing aid   **Great Britain: 2002**

Reasons for buying private hearing aid†	
	%
Better choice	59
Not available on NHS	44
Obtain aid more quickly	19
Other	27
Weighted base (000's) = 100%	*355*
Unweighted sample	*135*

* Includes those only worn occasionally.

† The reasons given total more than 100%, because respondents could give more than one reason.

Table 11.5   Whether working hearing aids are no longer worn, by whether currently wears an aid

Persons aged 16 and over with hearing difficulties   **Great Britain: 2002**

Whether wears an aid	Percentage with working aids no longer worn	Weighted base (000's) = 100%	Unweighted sample
Wears an aid	11	*1,469*	*540*
Does not wear an aid	8	*5,151*	*1855*

Table 11.6   Reasons for not wearing a hearing aid

Persons aged 16 and over who no longer wear a working aid they possess   **Great Britain: 2002**

Reasons for not wearing aid*	
	%
Did not help hearing	55
Appearance	11
Other	56
Weighted base (000's) = 100%	*565*
Unweighted sample	*206*

*The reasons given total more than 100%, because respondents could give more than one reason.

Table 11.7   Working hearing aids not worn: where aid obtained by reasons for aid no longer being worn

All working hearing aids not worn   **Great Britain: 2002**

Reasons for aid no longer being worn*	Where obtained	
	NHS	**Private**
	%	%
Did not help hearing	50	39
Appearance	11	4
Other	45	62
Weighted base (000's) = 100%	*513*	*135*
Unweighted sample	*186*	*51*

* The reasons given total more than 100%, because more than one reason could be reported.

Appendix A Definitions and terms

Definitions and terms used in the report are listed in alphabetical order.

Acute sickness

See Sickness

Adults

Adults are defined as persons aged 16 or over in all tables except those showing dependent children where single persons aged 16 to 18 who are in full-time education are counted as dependent children.

Bedroom standard

This concept is used to estimate occupation density by allocating a standard number of bedrooms to each household in accordance with its age/sex/marital status composition and the relationship of the members to one another. A separate bedroom is allocated to each married couple, any other person aged 21 or over, each pair of adolescents aged 10 to 20 of the same sex, and each pair of children under 10. Any unpaired person aged 10 to 20 is paired if possible with a child under 10 of the same sex, or, if that is not possible, is given a separate bedroom, as is any unpaired child under 10. This standard is then compared with the actual number of bedrooms (including bedsitters) available for the sole use of the household, and deficiencies or excesses are tabulated. Bedrooms converted to other uses are not counted as available unless they have been denoted as bedrooms by the informants; bedrooms not actually in use are counted unless uninhabitable.

Central heating

Central heating is defined as any system whereby two or more rooms (including kitchens, halls, landings, bathrooms and WCs) are heated from a central source, such as a boiler, a back boiler to an open fire, or the electricity supply. This definition includes a system where the boiler or back boiler heats one room and also supplies the power to heat at least one other room.

Under-floor heating systems, electric air systems, and night storage heaters are included.

Where a household has only one room in the accommodation, it is treated as having central heating if that room is heated from a central source along with other rooms in the house or building.

Chronic sickness

See Sickness

Cohabitation

See Marital status

Co-ownership or equity sharing schemes

Co-ownership or equity sharing schemes are those where a share in the property is bought by the occupier under an agreement with the housing association. The monthly charges paid for the accommodation include an amount towards the repayment of the collective mortgage on the scheme. The co-owner never becomes the sole owner of the property, but on leaving the scheme usually receives a cash sum.

See also *Tenure*

Dependent children

Dependent children are persons aged under 16, or single persons aged 16 to 18 and in full-time education, in the family unit and living in the household.

Doctor consultations

Data on doctor consultations presented in this report relate to consultations with National Health Service general medical practitioners during the two weeks before interview. Visits to the surgery, home visits, and telephone conversations are included, but contacts only with a receptionist are excluded. Consultations with practice nurses were excluded prior to 2000, but since then are identified separately. The GHS also collects information about consultations paid for privately.

The average number of consultations per person per year is calculated by multiplying the total number of consultations within the reference period, for any particular group, by 26 (the number of two-week periods in a year) and dividing the product by the total number of persons in the sample in that group.

Economic activity

Economically active

People over the minimum school-leaving age of 16, who were working or unemployed (as defined below) in the week before the week of interview. These persons constitute the labour force.

Working persons

This category includes persons aged 16 and over who, in the week before the week of interview, worked for wages, salary or other form of cash payment such as commission or tips, for any number of hours. It covers persons absent from work in the reference week because of holiday, sickness, strike, or temporary lay-off, provided they had a job to return to with the same employer. It also includes persons attending an educational establishment during the specified week if they were paid by their employer while attending it, people on Government training schemes and unpaid family workers.

Persons are excluded if they worked in a voluntary capacity for expenses only, or only for payment in kind, unless they worked for a business, firm or professional practice owned by a relative.

Full-time students are classified as 'working', 'unemployed' or 'inactive' according to their own reports of what they were doing during the reference week.

Unemployed persons

The GHS uses the International Labour Organisation (ILO) definition of unemployment. This classifies anyone as unemployed if he or she was out of work and had looked for work in the four weeks before interview, or would have but for temporary sickness or injury, and was available to start work in the two weeks after interview.

The treatment of all categories on the GHS is in line with that used on the Labour Force Survey (LFS).

Economically inactive

People who are neither working nor unemployed by the ILO measure. For example, this would include those who were looking after a home or retired.

Ethnic group

The GHS introduced the current National Statistics ethnic classification in 2001. The classification has a separate category for people from mixed ethnic backgrounds. In the previous system, people with these backgrounds had to select a specific ethnic group or categorise themselves as 'other'.

Household members are classified by the person answering the Household Schedule as:

- British, other White background;
- White and Black Caribbean, White and Black African, White and Asian, Other Mixed background;
- Indian, Pakistani, Bangladeshi, Other Asian background;
- Black Caribbean, Black African, Other Black background; or
- Chinese, Other ethnic group.

Family

A GHS family unit is defined as:

(a) a married or opposite sex cohabiting couple on their own; or
(b) a married or opposite sex cohabiting couple, or a lone parent, and their never-married children (who may be adult), provided these children have no children of their own.

Persons who cannot be allocated to a family as defined above are said to be persons not in the family – i.e. as 'non-family units'.

In general, GHS family units cannot span more than two generations, i.e. grandparents and grandchildren cannot belong to the same family unit. The exception to this is where it is established that the grandparents are responsible for looking after the grandchildren (e.g. while the parents are abroad).

Adopted and stepchildren belong to the same family unit as their adoptive/step-parents. Foster-children, however, are not part of their foster-parents' family (since they are not related to their foster-parents) and are counted as separate non-family units.

See also Lone-parent family.

Full-time working

Full-time working is defined as more than 30 hours a week with the exception of occupations in education where more than 26 hours a week was included as full time.

Government Office Region (GOR)

Government Office Regions came into force in 1998. They replaced the Standard Statistical Regions as the primary classification for the presentation of English regional statistics. Standard Statistical Region was

retained for some long term trend tables up to 2000. See also NHS Regional Office.

GP consultations

See Doctor consultations

Hospital visits

Inpatient stays

Inpatient data relate to stays overnight or longer (in a twelve month reference period) in NHS or private hospitals. All types of cases are counted, including psychiatric and maternity, except babies born in hospital who are included only if they remained in hospital after their mother was discharged.

Outpatient attendances

Outpatient data relate to attendances (in a reference period of three calendar months) at NHS or private hospitals, other than as an inpatient. Consultative outpatient attendances, casualty attendances, and attendances at ancillary departments are all included and a separate count is made of attendances at a casualty department.

Day patient

Day patients are defined as patients admitted to a hospital bed during the course of a day or to a day ward where a bed, couch or trolley is available for the patient's use. They are admitted with the intention of receiving care or treatment which can be completed in a few hours so that they do not require to remain in hospital overnight. If a patient admitted as a day patient then stays overnight they are counted as an inpatient.

Household

A household is defined as:

a single person or a group of people who have the address as their only or main residence and who either share one meal a day or share the living accommodation. (See L McCrossan, *A Handbook for Interviewers*. HMSO, London 1991.)

A group of people is not counted as a household solely on the basis of a shared kitchen or bathroom.

A person is in general regarded as living at the address if he or she (or the informant) considers the address to be his or her main residence. There are, however, certain rules which take priority over this criterion.

(a) Children aged 16 or over who live away from home for purposes of either work or study and come home only for holidays are *not* included at the parental address under any circumstances.
(b) Children of any age away from home in a temporary job and children under 16 at

boarding school are *always* included in the parental household.
(c) Anyone who has been away from the address *continuously* for six months or longer is excluded.
(d) Anyone who has been living continuously at the address for six months or longer is included even if he or she has his or her main residence elsewhere.
(e) Addresses used only as second homes are never counted as a main residence.

Householder

The householder: the member of the household in whose name the accommodation is owned or rented, or is otherwise responsible for the accommodation.

Non-household members are never defined as householders even though they may own or rent the accommodation in question.

Household Reference Person (HRP)

For some topics it is necessary to select one person in the household to indicate the characteristics of the household more generally. In common with other government surveys, in 2000, the GHS replaced the Head of Household with the Household Reference Person for this purpose.

The household reference person is defined as follows:

- in households with a *sole* householder that person is the household reference person;
- in households with *joint* householders the person with the *highest income* is taken as the household reference person;
- if both householders have exactly the same income, the *older* is taken as the household reference person.

Note that this definition does not require a question about people's actual incomes; only a question about who has the highest income.

Main changes from the HOH definition

Female householders with the highest income are now taken as the HRP. In the case of joint householders, income then age, rather than sex then age is used to define the HRP. This means that in both cases more women are defined as HRP than were classified as the HOH.

Appendix A in 'Living in Britain 2000' suggested that in the GHS, the HRP would not be the same as the HOH in about 14% of households. Households consisting of a single adult or a sole male householder do not change and in many other households the new definition in practice results in the same person being selected. Part of the change is due to sole female householders who are living with a non-householder partner: they were not classified as the HOH, but they are the HRP, and they account for 4% of households. The remainder is due to the use of income (or age) to choose between joint householders resulting in a different person being selected.

Household type

There are many ways of grouping or classifying households into household types; most are based on the age, sex and number of household members.

The main classification of household type uses the following categories:

- 1 adult aged 16–59
- 2 adults aged 16–59
- small family
  - 1 or 2 persons aged 16 or over and 1 or 2 persons aged under 16
- large family
  - 1 or more persons aged 16 or over and 3 or more persons aged under 16, or 3 or more persons aged 16 or over and 2 persons aged under 16
- large adult household
  - 3 or more persons aged 16 or over, with or without 1 person aged under 16
- 2 adults, 1 or both aged 60 or over
- 1 adult aged 60 or over

The term 'family' in this context does not necessarily imply any relationship.

Chapter 3 also uses a modified version of household type which takes account of the age of the youngest household member. 'Small family', 'large family' and 'large adult household' are replaced by the following:

- youngest person aged 0–4
  - 1 or more persons aged 16 or over and 1 or more persons aged under 5
- youngest person aged 5–15
  - 1 or more persons aged 16 or over and 1 or more persons aged 5–15
- 3 or more adults
  - 3 or more persons aged 16 or over and no-one aged under 16

The first two categories above are combined in some tables.

In Chapter 3, households are also classified according to the family units they contain (see Family for definition), into the following categories:

- One family households* containing
  - married couple with dependent children
  - married couple with non-dependent children only
  - married couple with no children
  - cohabiting couple with dependent children
  - cohabiting couple with non-dependent children only
  - cohabiting couple with no children
  - lone parent with dependent children
  - lone parent with non-dependent children only
- Households containing two or more families
- Non-family households containing
  - 1 person only
  - 2 or more non-family† adults.

Some of the above categories are combined for certain tables and figures.

* Other individuals who were not family members may also have been present
† Individuals may, of course, be related without constituting a GHS family unit. A household consisting of a brother and sister, for example, is a non-family household of two or more non-family adults.

Income

Usual gross weekly income

Total income for an individual refers to income at the time of the interview, and is obtained by summing the components of earnings, benefits, pensions, dividends, interest and other regular payments. Gross weekly income of employees and those on benefits is calculated if interest and dividends are the only components missing.

If the last pay packet/cheque was unusual, for example in including holiday pay in advance or a tax refund, the respondent is asked for usual pay. No account is taken of whether a job is temporary or permanent. Payments made less than weekly are divided by the number of weeks covered to obtain a weekly figure.

Usual gross weekly household income is the sum of usual gross weekly income for all adults in the household. Those interviewed by proxy are also included.

Labour force

See Economic activity.

Lone-parent family

A lone-parent family consists of one parent, irrespective of sex, living with his or her never-married dependent children, provided these children have no children of their own.

Married or cohabiting women with dependent children, whose partners are not defined as resident in the household, are not classified as one-parent families because it is known that the majority of them are only temporarily separated from their husbands for a reason that does not imply the breakdown of the marriage (for example, because the husband usually works away from home). (See the GHS 1980 Report p.9 for further details.)

Longstanding conditions and complaints

See Sickness

Marriage and cohabitation

From 1971 to 1978 the Family Information section was addressed only to married women aged under 45 who were asked questions on their present marriage and birth expectations. In 1979 the section was expanded to include questions on cohabitation prior to marriage, previous marriages and all live births, and was addressed to all women aged 16 to 49 except non-married women aged 16 and 17. In 1986 the section was extended to cover all women and men aged 16 to 59. In 1998 all adults aged 16 to 59 were asked about any periods of cohabitation not leading to marriage. This section was extended in 2000.

Marital status

Since 1996 separate questions have been asked at the beginning of the questionnaire to identify the legal marital status and living arrangements of respondents in the household. The latter includes a category for cohabiting.

Cohabiting

Before 1996, unrelated adults of the opposite sex were classified as cohabiting if they considered themselves to be living together as a couple. From 1996, this has included a small number of same sex couples, unless otherwise stated in the table.

Married/non-married

In this dichotomy 'married ' generally includes cohabiting and 'non- married' covers those who are single, widowed, separated or divorced and not cohabiting.

Living arrangements (de facto marital status)

Before 1996, additional information from the Family Information section of the individuals' questionnaire has been used to determine living arrangements (previously known as 'defacto marital status') and the classification has only applied to those aged 16 to 59 who answer the marital history questions. For this population it only differed from the main marital status for those who revealed in the

Family Information section that they were cohabiting rather than having the marital status given at the beginning of the interview. 'Cohabiting' took priority over other categories. Since 1996, information on legal marital status and living arrangements, has been taken from the beginning of the interview where both are now asked.

Legal marital status

This classification applies to persons aged 16 to 59 who answer the marital history questions. Cohabiting people are categorised according to formal marital status. The classification differs from strict legal marital status in accepting the respondents' opinion of whether their marriage has terminated in separation rather than applying the criterion of legal separation.

Median

See Quantiles.

National Statistics Socio-economic classification (NS-SEC)

From April 2001 the National Statistics Socio-economic Classification (NS-SEC) was introduced for all official statistics and surveys. It replaced Social Class based on occupation and Socio-economic Groups (SEG). Full details can be found in *The National Statistics Socio-economic Classification User Manual 2002'* ONS 2002.

Descriptive definition	NS-SEC categories
Large employers and higher managerial occupations	L1, L2
Higher professional occupations	L3
Lower managerial and professional occupations	L4, L5, L6
Intermediate occupations	L7
Small employers and own account workers	L8, L9
Lower supervisory and technical occupations	L10, L11
Semi-routine occupations	L12
Routine occupations	L13
Never worked and long-term unemployed	L14

The three residual categories: L15 (full time students); L16 (occupation not stated or inadequately described) and L17 (not classifiable for other reasons) are excluded when the classification is collapsed into its analytical classes.

The categories can be further grouped into:

Managerial and professional occupations	L1–L6
Intermediate occupations	L7–L9
Routine and manual occupations	L10–L13

This results in the exclusion of those who have never worked and the long term unemployed, in addition to the groups mentioned above.

The main differences users need to be aware of are:

- the introduction of SOC2000 which includes various new technology occupations not previously defined in SOC90;
- definitional variations in employment status in particular with reference to the term 'supervisor';
- the inclusion of armed forces personnel in the appropriate occupation group;
- the separate classification of full-time students, whether or not they have been or are presently in paid employment; and
- the separate classification of long term unemployed who previously were classified by their most recent occupation.

This change has resulted in a discontinuity in time series data. The operational categories of NS-SEC can be aggregated to produce an approximated version of the previous Socio-economic Group. These approximations have been shown to achieve an overall continuity level of 87%. Some tables in Chapter 8 (Smoking) have used this approximation.

NHS Regional Office

Between 1996 and 2002, there were eight NHS Regional Offices in England, which together with Scotland and Wales were part of the administrative geography of the Department of Health. Each office was divided into approximately 100 District Health Authorities (DHAs) which, from April 1999, were in turn separated into Primary Care Organisations. In April 2002, Strategic Health Authorities (SHAs) replaced the DHAs, and between April 2002 and June 2003, four Directorates of Health and Social Care replaced the NHS Regional Offices in the organisational hierarchy. Since July 2003, the structure for health administration in England has consisted of 28 SHAs that report directly to the Department of Health, and manage the NHS locally.

Although NHS Regional Offices were no longer in use at the time of the survey, it has been used in this report for purposes of consistency and for comparison of results with previous years.

Pensions

The GHS asks questions about any pension scheme, either occupational or personal, that the respondent belonged to on the date of interview. It is quite possible that some respondents may have held entitlement in the occupational pension scheme of a previous employer or a personal pension scheme in the past. The GHS measures current membership and not the percentage of respondents who will get an occupational or personal pension when they retire.

In April 2002 the State Second Pension (SSP) was introduced. The new pension reformed the State Earnings Related Pension Scheme (SERPS) to provide a more generous additional pension for low and moderate earners, certain groups of carers and people with a longstanding illness or disability.

Since 1988, individual employees have had the option of contracting out of the SSP (formerly SERPS) by starting their own personal pension plan.

Some respondents may be contributing to both an occupational and personal pension scheme.

Qualification levels

Degree or degree equivalent, and above

- Higher degree and postgraduate qualifications
- First degree (including B.Ed.)
- Postgraduate Diplomas and Certificates (including PGCE)
- Professional qualifications at degree level e.g. graduate member of professional institute, chartered accountant or surveyor
- NVQ or SVQ level 4 or 5

Other higher education below degree level

- Diplomas in higher education & other higher education qualifications
- HNC, HND, Higher level BTEC
- Teaching qualifications for schools or further education (below degree level standard)
- Nursing, or other medical qualifications not covered above (below degree level standard)
- RSA higher diploma

A levels, vocational level 3 & equivalents

- A level or equivalent
- AS level
- SCE Higher, Scottish Certificate Sixth Year Studies or equivalent
- NVQ or SVQ level 3
- GNVQ Advanced or GSVQ level 3
- OND, ONC, BTEC National, SCOTVEC National Certificate
- City & Guilds advanced craft, Part III (& other names)
- RSA advanced diploma

Trade apprenticeships
GCSE/O Level grade A–C, vocational level 2 & equivalents

- NVQ or SVQ level 2
- GNVQ intermediate or GSVQ level 2
- RSA Diploma
- City & Guilds Craft or Part II (& other names)
- BTEC, SCOTVEC first or general diploma etc.
- O level or GCSE grade A–C, SCE Standard or Ordinary grades 1–3

Qualifications at level 1 and below
- NVQ or SVQ level 1
- GNVQ Foundation level, GSVQ level 1
- GCSE or O level below grade C, SCE Standard or Ordinary below grade 3
- CSE below grade 1
- BTEC, SCOTVEC first or general certificate
- SCOTVEC modules
- RSA Stage I, II, or III
- City and Guilds part 1
- Junior certificate
- YT Certificate/ YTP

Other qualifications: level unknown
- Other vocational or professional or foreign qualifications

No qualifications
- Excludes those who never went to school (omitted from the classification altogether).

This is not a complete listing of all qualifications. In particular, it does not give all the names which have been used by BTEC or City and Guilds. Neither does it give names for vocational qualifications from other awarding bodies besides BTEC, City and Guilds, RSA and SCOTVEC, although it should cover the majority of vocational qualifications awarded.

The qualification levels do not in all cases correspond to those used in statistics published by the Department for Education and Skills.

Quantiles

The quantiles of a distribution, eg of household income, divide it into equal parts.

Median: the median of a distribution divides it into two equal parts. Thus half the households in a distribution of household income have an income higher than the median, and the other half have an income lower than the median.

Quartiles: the quartiles of a distribution divide it into quarters. Thus the upper quartile of a distribution of household income is the level of income that is expected by 25% of the households in the distribution; and 25% of the households have an income less than the lower quartile. It follows that 50% of the households have an income between the upper and lower quartiles.

Quintiles: the quintiles of a distribution divide it into fifths. Thus the upper quintile of a distribution of household income is the level of income that is expected by 20% of the households in the distribution; and 20% of the households have an income less than the lower quintile. It follows that 60% of the households have an income between the upper and lower quintiles.

Relatives in the household

The term 'relative' includes any household member related to the head of household by blood, marriage, or adoption. Foster-children are therefore not regarded as relatives.

Rooms

These are defined as habitable rooms, including (unless otherwise specified) kitchens, whether eaten in or not, but excluding rooms used solely for business purposes, those not usable throughout the year (eg conservatories), and those not normally used for living purposes such as toilets, cloakrooms, store rooms, pantries, cellars and garages.

Sickness

Acute sickness

Acute sickness is defined as restriction of the level of normal activity, because of illness or injury, at any time during the two weeks before interview. Since the two-week reference period covers weekends, normal activities include leisure activities as well as school attendance, going to work, or doing housework. Anyone with a chronic condition that caused additional restriction during the reference period is counted among those with acute sickness.

The average number of restricted activity days per person per year is calculated in the same way as the average number of doctor consultations.

Chronic sickness

Information on chronic sickness was obtained from the following two-part question:

'Do you have any longstanding illness, disability or infirmity? By longstanding I mean anything that has troubled you over a period of time or that is likely to affect you over a period of time.

IF YES
Does this illness or disability limit your activities in any way?'

'Longstanding illness' is defined as a positive answer to the first part of the question, and 'limiting longstanding illness' as a positive answer to both parts of the question.

The data collected are based on people's

subjective assessment of their health, and therefore changes over time may reflect changes in people's expectations of their health as well as changes in incidence or duration of chronic sickness. In addition, different sub-groups of the population may have varying expectations, activities and capacities of adaptation.

Longstanding conditions and complaints

The GHS collects information about the nature of longstanding illness. Respondents who report a longstanding illness are asked 'What is the matter with you?' and details of the illness or disability are recorded by the interviewers and coded into a number of broad categories. Interviewers are instructed to focus on the symptoms of the illness, rather than the cause, and code what the respondent said was currently the matter without probing for cause. This approach has been used in 1988, 1989, 1994 to 1996, 1998 and 2000 to 2002.

The categories used when coding the conditions correspond broadly to the chapter headings of the International Classification of Diseases (ICD). However, the ICD is used mostly for coding conditions and diseases according to cause whereas the GHS coding is based only on the symptoms reported. This gives rise to discrepancies in some areas between the two classifications.

Socio-economic classification

See National Statistics Socio-economic classification.

Step family

See Family.

Tenure

From 1981, households who were buying a share in the property from a housing association or co-operative through a shared ownership (equity sharing) or co-ownership scheme are included in the category of owner-occupiers. In earlier years such households were included with those renting from a housing association or co-operative.

Renting from a council includes renting from a local authority or New Town corporation or commissions or Scottish Homes (formerly the Scottish Special Housing Association).

Renting from a housing association also includes co-operatives and charitable trusts. It also covers fair rent schemes. Since 1996, housing associations are more correctly described as Registered Social Landlords (RSLs). RSLs are not-for-profit organisations which include: charitable housing associations, industrial and provident societies and companies registered under the Companies Act 1985.

Social sector renters include households renting from a local authority or New Town corporation or commission or Scottish Homes and those renting from housing associations, cooperatives and charitable trusts.

Private renters include those who rent from a private individual or organisation and those whose accommodation is tied to their job even if the landlord is a local authority, housing association or Housing Action Trust, or if the accommodation is rent free. Squatters are also included in this category.

Unemployed

See Economic activity

Working

See Economic activity

Appendix B Sample design and response

The GHS samples around 13,000 addresses each year and aims to interview all adults aged 16 or over at every household at the sampled address[1]. It uses a probability, stratified two-stage sample design. The Primary Sampling Units (PSUs) are postcode sectors, which are similar in size to wards and the secondary sampling units are addresses within those sectors.

Sample design

The revised 2000 survey design introduced new stratifiers[2]. Stratification involves the division of the population into sub-groups, or strata, from which independent samples are taken. This ensures that a representative sample will be drawn with respect to the stratifiers (i.e. the proportion of units sampled from any particular stratum will equal the proportion in the population with that characteristic). Stratification of a sample can lead to substantial improvements in the precision of survey estimates. Optimal precision is achieved where the factors used as strata are those that correlate most highly with the survey variables. From 2000, the stratification factors were based on an area classifier and selected indicators from the 1991 census. Details of how these were selected were reported in the January 2000 edition of the ONS Survey Methodology Bulletin[3].

Initially, postcode sectors were allocated to 30 major strata. These were based on the 10

Government Office Regions in England, 5 subdivisions in Scotland and 2 in Wales. The English regions were divided between the former Metropolitan and non-Metropolitan counties. In addition London was subdivided into quadrants (Northwest, Northeast, Southwest and Southeast) with each quadrant being divided into inner and outer areas[4]. Using a finer division of London in the regional stratifier had a large effect on the increase in precision.

Within each major stratum, postcode sectors were then stratified according to the selected indicators taken from the 1991 Census. Sectors were initially ranked according to the proportion of households with no car, then divided into three bands containing approximately the same number of households. Within each band, sectors were re-ranked according to the proportion of households with household reference person in socio-economic groups 1 to 5 and 13, and these bands were then sub-divided into three further bands of approximately equal size. Finally, within each of these bands, sectors were re-ranked according to the proportion of people who were pensioners. In order to minimise the difference between one band and the next, the ranking by the pensioners and socio-economic group criteria were in the reverse order in consecutive bands, as shown in Figure B.A.   **Figure B.A**

Major strata were then divided into minor strata with equal numbers of addresses, the number of minor strata per major strata being proportionate to the size of the major stratum. Since 1984 the frame has been divided into 576 minor strata and one PSU has been selected from each per year. Of the 576 PSUs selected, 48 are randomly allocated to each month of the year. Each PSU forms a quota of work for an interviewer. Within each PSU 23 addresses are randomly selected.

Conversion of multi-occupancy addresses to households

Most addresses contain just one private household, a few - such as institutions and purely business addresses[5] - contain no private households, while others contain more than one private household. For addresses containing more than one household, set procedures are laid down in order to give each household one and only one chance of selection.

As the Postcode Address File (PAF) does not give names of occupants of addresses, it is not possible to use the number of different surnames at an address as an indicator of the number of households living there. A rough guide to the number of households at an address is provided on the PAF by the multi-occupancy (MO) count. The MO count is a fairly accurate indicator in Scotland but is less accurate in England and Wales, so it is used only when sampling at addresses in Scotland.

Figure BA

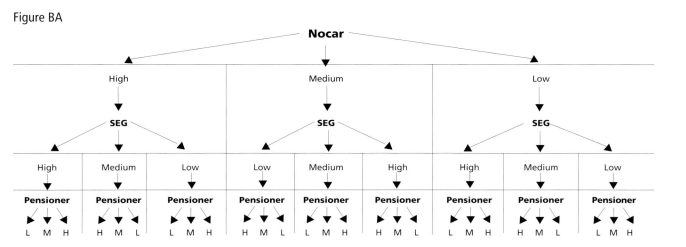

All addresses in England and Wales, and those in Scotland with an MO count of two or less, are given only one chance of selection for the sample. At such addresses, interviewers interview all the households they find up to a maximum of three. If there are more than three households at the address, the interviewer selects the households for interview by listing all households at the address systematically then making a random choice by referring to a household selection table.

Addresses in Scotland with an MO count of three or more, where the probability that there is more than one household is fairly high, are given as many chances of selection as the value of the MO count. When the interviewer arrives at such an address, he or she checks the actual number of households and interviews a proportion of them according to instructions. The proportion is set originally by the MO count and adjusted according to the number of households actually found, with a maximum of three households being interviewed at any address. The interviewer selects the households for interview by listing all households at the address systematically and making a random choice, as above, by means of a table.

No addresses are deleted from the sample to compensate for the extra interviews that may result from these multi-household addresses but a maximum of four extra interviews per quota of addresses is allowed. Once four extra interviews have been carried out in an interviewer's quota, only the first household selected at each multi-occupancy address is included. As a result of the limits on additional interviews, households in concealed multi-occupied addresses may be slightly under-represented in the GHS sample.

Data collection

Information for the GHS is collected week by week throughout the year[6] by personal interview. In 2002, interviews took place from April 2002 to March 2003. Since 1994 the survey has been carried out using Computer Assisted Personal Interviewing (CAPI) on laptop computers and Blaise software by face-to-face interviewers. Since 2000, telephone interviewers have converted GHS proxy interviews to full interviews using Computer Assisted Telephone Interviewing (CATI) from a central unit. Interviews are sought with all adult members (aged 16 or over) of the sample of private households and some information about children in the household is also collected.

A letter is sent in advance of an interviewer calling at an address. The letter briefly describes the purpose and nature of the

survey and prepares the recipient for a visit by an interviewer. Since 2001, postage stamps have been included in the advance letter (see 'Improving response').

Data quality

The face-to-face and telephone interviewers who work on the GHS are recruited only after careful selection procedures after which they take part in an initial training course. Before working on the GHS they attend a briefing and new recruits are always supervised either by being accompanied in the field by a Field Manager or monitored by a Telephone Interviewing Unit (TIU) supervisor. All interviewers who continue to work on the GHS are observed regularly in their work.

Proxy interviews and the proxy conversion exercise

On occasion it may prove impossible, despite repeated calls, to contact a particular member of a household in person and, in strictly controlled circumstances, interviewers are permitted to conduct a proxy interview with a close household member. In these cases opinion-type questions and questions on smoking and drinking behaviour, qualifications, health, family information and income are omitted.

During the review of the GHS[7] the conversion of proxy interviews to full interviews was examined in order to improve the quality of data (a full interview is one in which the respondent has answered all sections of the questionnaire in person, either face to face or by telephone). This was achieved by re-contacting the household member, who was unavailable during the initial face-to-face interview, to answer the questions that were not asked of the proxy respondent on his/her behalf. The most efficient way of re-contacting these respondents was by employing Telephone Interviewing Unit (TIU) interviewers who could contact a widely

dispersed population more efficiently than would be possible by conducting face-to-face interviews. Table B.1 shows the percentages of the types of interview taken for all persons in the co-operating households since 2000 before and after the proxy conversion exercise. In 2002 the TIU increased the percentage of full interviews conducted on the GHS from 71% before proxy conversion to 74%. The process of proxy conversion allows the GHS to provide more information on the topics not asked in detail of proxy respondents **Table B.1**

Response

The GHS is conducted with people who volunteer their time to answer questions about themselves. The voluntary nature of the survey means that people who do not wish to take part in the survey can refuse to do so. Reasons for not participating in the survey vary from a dislike of surveys to poor health that prevents them from taking part. The sample is designed to ensure that the results of the survey represent the population of Great Britain. The risk of the survey not being representative is likely to increase with every refusal or noncontact with a sampled household (survey nonresponse). One measure of the quality of survey results is therefore the response rate.

Harmonised outcome codes and survey response rates

Harmonised outcome codes and definitions of response rates[8] were introduced for the first time on the 2002/3 GHS and other large household surveys, following recommendations from National Statistics and the National Centre for Social Research joint working group on standard survey outcomes. The harmonised outcome codes are categorised as complete and partial interviews, non-contact, refusals and other non-responders, and unknown and known ineligibility.

Table B1   Type of interview taken by proxy conversion status

All persons   **Great Britain**

	Proxy conversion status					
	2000		2001		2002	
	Before	After	Before	After	Before	After
	%	%	%	%	%	%
Full interview	72	73	71	73	71	74
Proxy	6	5	7	5	7	5
Child/Other	22	22	22	22	22	22
Unweighted sample	*19266*	*19266*	*21180*	*21180*	*20149*	*20149*

The joint working group also recommended that surveys include an estimate of the proportion of cases of unknown eligibility that are eligible from 2002 onwards. It is assumed that the proportion of eligible cases amongst those cases where eligibility is unknown is the same as that amongst cases where eligibility has been established.

Four new response rates are now calculated for the GHS on the basis of the new outcome codes:

1. Overall response rate
This indicates how many full and partial interviews were achieved as a proportion of those eligible for the survey. In order to obtain the most conservative response rate measures, the denominator includes an estimate of the proportion of cases of unknown eligibility that would in fact be eligible for interview. In 2002 the overall response rate for the GHS was 71%.

2. Full response rate
This is similar to the overall response rate calculated above, but only full interviews are included in the numerator. The full response rate for the GHS in 2002 was 63%.

3. Co-operation rate
This indicates the number of achieved interviews as a proportion of those contacted during the fieldwork period. The co-operation rate for the GHS in 2002 was 72%.

4. Contact rate
The contact rate measures the proportion of cases in which some household members were reached by the interviewer even though they might then have refused or been unable to give further information about the household or to participate in the survey. In 2002 the contact rate for the GHS was 96%.

Table B.2 shows the outcome of visits to the addresses selected for the 2002 sample and the resultant number of households interviewed. Out of the 13,248 addresses that were selected, 12,069 were eligible and this yielded a sample of 12,159 eligible households. In 8,420 households, interviews (including proxy interviews) were achieved with every member of the household. In a further 200 households interviews were achieved with some but not all members of the household. This produced a total of 8,620 full or partial interviews.

In total, 25% of households selected for interview in 2002 were lost to the sample altogether, because they did not wish to take part (21%) or because they could not be contacted (4%) (table not shown).

Table B.3 shows annual response by interview outcome category. **Tables B.2–B.3**

Trends in response

In order to continue to measure the response to the survey over time, a 'middle' response rate has been calculated since 1971. The middle response rate includes full interviews and accepts some of the partial household interviews as response – that is, it includes households where information has been collected by proxy and is therefore missing certain sections (category 2 in Table B.3), but does not include those where information is missing altogether for one or more household members (categories 3 and 4 in Table B.3). In other words, this middle rate can be thought of as the proportion of the sample of households known to be eligible from whom all or nearly all the information was obtained.

For the purposes of comparison, the middle response rate has been calculated as in previous years, although it is not in itself a classification of the harmonised response rates.

Table B2   The sample of addresses and households

	Great Britain: 2002
Selected addresses	13248
Ineligible addresses:	
Demolished or derelict	
Used wholly for business purposes	
Empty	
Institutions	1179
Other ineligible	
No sample selected at address	
Address not traced	
Eligible addresses	12069
Number of households at eligible addresses	12159
Number of households where all individual interviews achieved (including proxies) 8420	8620
Number of households where some but not all individual interviews achieved 200	

Table B3   Annual response

Households   Great Britain: 2002

Outcome category	No.	%
1 Complete household co-operation	7680	63.2
2 Non-interview of one or more household members, proxy taken	740	6.1
3 Non-contact with one or more household members	93	0.8
4 Refusal by at least one household member	107	0.9
5 Non-contact with household/resident	621	5.1
6 HQ refusal	367	3.0
7 Other refusal	2543	20.9
8 Other non-response	8	0.1
Unweighted sample base = 100%	*12159*	

In 2002 the middle response rate was 69%. Table B.4 shows middle response rates by Government Office Region. Trends in the middle response rate are shown in Table B.5. Since 1971, the middle response rate has shown some fluctuation. The decline in response rate since the early 1990s is due to an increase in the proportion of households refusing to participate (12% in 1991 rising to 23% in 2002) rather than failure to contact people. This decline reflects a general trend in decreasing response experienced by all survey organisations. **Tables B.4–B.5**

Improving response

The GHS uses a number of methods to try to improve response. One ongoing method of improving response has been to reissue addresses to interviewers where there is a possibility of obtaining a better outcome, for example if there was initially a non-contact or a circumstantial refusal.

During the 2001/2 survey the advance letter was changed to mention that the survey had been running for 30 years. This change shows respondents how important the survey is and

Table B5 **Trends in the middle response rate: 1971 to 2002**

Households   Great Britain

Year	Response rate
	%
1971	83
1972	81
1973	81
1974	83
1975	84
1976	84
1977	83
1978	82
1979	83
1980	82
1981	84
1982	84
1983	82
1984	81
1985	82
1986	84
1987	85
1988	85
1989	84
1990	81
1991	84
1992	83
1993	82
1994	80
1995	80
1996	76
1998	72
2000	67
2001	72
2002	69

Table B4 Middle response rates by Government Office Region

Households   Great Britain: 2002

Government Office Region	%	Rank
North East	72.3	2
North West	68.5	9
Yorkshire and the Humber	71.6	4
East Midlands	70.5	5
West Midlands	66.6	10
East of England	68.9	6
London	63.3	11
South East (excluding Greater London)	72.8	1
South West	71.9	3
Wales	68.7	8
Scotland	68.7	7
Great Britain	68.9	

Table B6 Unweighted bases: number of household reference persons in GHS 2002 by age, sex, region and country

Household reference persons   Great Britain: 2002

	Age							All
	16-24	25-34	35-44	45-54	55-64	65-74	75+	
Sex								
Male	139	759	1160	1069	977	753	536	5393
Female	192	551	578	521	434	399	552	3227
Government Office Region								
North East	30	50	59	87	71	61	56	414
North West	39	161	177	192	175	161	117	1022
Yorkshire and the Humber	36	112	152	115	123	106	116	760
East Midlands	18	87	123	121	119	78	86	632
West Midlands	28	108	170	119	117	116	101	759
East of England	19	99	168	168	154	102	106	816
London	44	221	230	165	102	112	97	971
South East	35	173	258	248	235	159	156	1264
South West	30	123	165	145	117	103	108	791
Country								
England	279	1134	1502	1360	1213	998	943	7429
Wales	15	67	90	87	71	61	58	449
Scotland	37	109	146	143	127	93	87	742
Total	331	1310	1738	1590	1411	1152	1088	8620

Shaded figures also show the number of households in each region and country.

Table B7　　**Unweighted bases: number of people in GHS 2002 by age, sex, region and country**

All persons　　　　　　　　　　　　　　　　　　　　　　　　　　　　　　　　　　　　　Great Britain: 2002

	Age									All
	0-4	5-15	16-24	25-34	35-44	45-54	55-64	65-74	75+	
Sex										
Male	606	1521	982	1204	1455	1318	1176	849	595	9706
Female	562	1488	1041	1353	1567	1444	1231	917	840	10443
Government Office Region										
North East	48	117	118	104	111	148	115	89	71	921
North West	150	318	244	306	339	320	298	241	153	2369
Yorkshire and the Humber	112	249	170	222	266	201	217	157	144	1738
East Midlands	83	241	143	170	221	207	208	115	117	1505
West Midlands	97	292	188	208	307	209	208	181	138	1828
East of England	109	265	154	199	286	298	259	171	142	1883
London	158	363	252	441	365	279	176	163	124	2321
South East	153	451	289	364	453	452	387	241	214	3004
South West	115	306	182	230	283	262	213	157	146	1894
Country										
England	1025	2602	1740	2244	2631	2376	2081	1515	1249	17463
Wales	65	169	97	119	155	139	119	105	75	1043
Scotland	78	238	186	194	236	247	207	146	111	1643
Total	1168	3009	2023	2557	3022	2762	2407	1766	1435	20149

how many people take part. For subsequent years a similar advance letter has been produced informing respondents that the survey has been running for over 30 years.

Another method was introduced in 2002 following the findings of the Response Working Group of the Office for National Statistics. The Group found that response was higher among households who received postage stamps with their advance letters. Since August 2002, stamps have been included in all advance letters.

Sample sizes

Tables B.6 and B.7 show the numbers of households and individuals interviewed on the 2002 GHS by age, sex, region and country. **Tables B.6–B.7**

Notes and references

1　A limit is put on the number of households that are contacted per address. This is explained in detail at the 'Conversion of multi-occupancy addresses to households' section of Appendix B.

2　From 1984 to 1998, the stratifiers used were a regional variable (based on the standard statistical region until 1996 and on the Government Office Region in 1998) and variables that measured the prevalence of privately rented accommodation, local authority accommodation and people in professional and managerial socio-economic groups.

3　Insalaco F Choosing stratifiers for the General Household Survey *ONS Social Survey Division, Survey Methodology Bulletin*, No. 46, January 2000.

4　The GOR regional stratifier:
1.　North East Met
2.　North East Non Met
3.　North West Met
4.　North West Non Met
5.　Merseyside
6.　Yorks and Humberside Met
7.　Yorks and Humberside Non Met
8.　East Midlands
9.　West Midlands Met
10.　West Midlands Non Met
11.　Eastern Outer Met
12.　Eastern Other
13.　Inner London North-East
14.　Inner London North-West
15.　Inner London South-East
16.　Inner London South-West
17.　Outer London North-East
18.　Outer London North-West
19.　Outer London South-East
20.　Outer London South-West
21.　South East Outer Met
22.　South East Other
23.　South West
24.　Wales 1 – Glamorgan, Gwent
25.　Wales 2 – Clwydd, Gwenneyd, Dyfed, Powys
26.　Highlands, Grampian, Tayside
27.　Fife, Central, Lothian
28.　Glasgow Met
29.　Strathclyde (excl. Glasgow)
30.　Borders, Dumfries, Galloway

5　Most institutions and business addresses are not listed on the small-user PAF. If an address was found in the field to be non-private (e.g. boarding house containing four or more boarders at the time the interviewer calls), the interviewer was instructed not to take an interview. However, a household member in hospital at the time of interview was included in the sample provided that he or she had not been away from home for more than six months and was expected to return. In this case a proxy interview was taken.

6　From 1988, the GHS interviewing year was changed from a calendar year to a financial year basis.

7　Walker A et al *Living in Britain Results from the 2000 General Household Survey: Appendix E*. TSO London 2002. Also available on the web: www.statistics.gov.uk/lib

8　Lynn P, Beerten R, Laiho J and Martin J. Recommended Standard Final Outcome Categories and Standard Definition of Response Rate for Social Surveys. ISER Working Papers. Number 2001 – 23. http://www.iser.essex.ac.uk/pubs/ workpaps/pdf/2001-23.pdf

Appendix C Sampling errors

Tables in this appendix present estimates of sampling errors for some of the main variables used in this report, taking into account the complex sample design of the survey.

Sources of error in surveys

Survey results are subject to various sources of error. The total error in a survey estimate is the difference between the estimate derived from the sample data collected and the true value for the population. The total error is made up of two main types: systematic and random error.

Systematic error

Systematic error occurs when data are consistently biased in a certain way, such that the variation from the true values for the population will not average to zero over repeats of the survey. For example, if a certain section of the population is excluded from the sampling frame, estimates may be biased because non-respondents to the survey have different characteristics to respondents. Another cause of bias may be that interviewers systematically influence responses in one way or another. Substantial efforts have been made to avoid systematic errors, for example, through extensive interviewer training and by weighting the data collected for non-response.

Random error

Random error, or bias, is the variation in sample data from the true values for the population, which occurs by chance. This type of error is expected to average to zero over a number of repeats of the survey. Random error may result from sources such as variation in respondents' interpretation of the survey questions, or interviewer variation. Efforts are made to minimise these effects through pilot work and consistent interviewer training.

Sampling errors

An important component of random error is sampling error, which arises because the variable estimates are based on a sample rather than a full census of the population. The results obtained for any single sample would be likely to vary slightly from the true values for the population. The difference between the estimates derived from the

sample and the true population values is referred to as the sampling error. The amount of variation can generally be reduced by increasing the size of the sample, and by improving the sample design. Sampling errors have been measured for estimates derived from the General Household Survey (GHS), and these may be used to assess the accuracy of the estimates presented in this report.

Calculating standard errors

Unlike non-sampling errors, it is possible to estimate the size of sampling error, by calculating the standard error of the survey estimates. The standard error (se) of a percentage p, based on a simple random sample of size n is calculated by the formula,

$$se(p)_{srs} = \sqrt{(p(100-p)/n)}$$

The GHS uses a multi-stage sample design, which involves both clustering and stratification (see Appendix B). The complexity of the design means that sampling errors calculated on the basis of a simple random sample design will not reflect the true variance in the survey estimates. Clustering can lead to a substantial increase in sampling error if the households or individuals within the primary sampling units (PSUs) are relatively homogenous but the PSUs differ from one another. By contrast, stratification tends to reduce sampling error and is particularly effective when the stratification factor is related to the characteristics of interest on the survey.

Because of the complexity of the GHS sample design, the size of the standard error depends on how the characteristic of interest is spread within and between the PSUs and strata. The method used to calculate the standard errors for the survey takes this into account. It explicitly allows for the fact that the estimated values (percentages and means) are ratios of two survey estimates: the number with the characteristic of interest is the numerator (y) and the sample size is the denominator (x), both of which are subject to random error.

The standard error of a survey estimate is found by calculating the positive square root of the estimated variance of the ratio. The formula used to estimate the variance of a ratio estimator r (where $r = y/x$) is shown below.

$$var(r) = \frac{1}{x^2} [var(y) + r^2 var(x) - 2r\ cov(y,x)]$$

$Var(r)$ is the estimate of the variance of the ratio, r, expressed in terms of $var(y)$ and $var(x)$ which are the estimated variances of y and x, and $cov(y,x)$ which is their estimated covariance. The resulting estimate is only valid if the denominator (x) does not vary too greatly. The method compares the differences between totals for adjacent PSUs (postal sectors) in the characteristic of interest. The ordering of PSUs reflects the ranking of postal sectors on the stratifiers used in the sample design.

Design factors

The design factor, or deft, of an estimate p is the ratio of the complex standard error of p to the standard error of p that would have resulted had the survey design been a simple random sample of the same size.

$$deft(p) = \frac{se(p)}{se_{srs}(p)}$$

This is often used to give a broad indication of the effect of the clustering on the reliability of estimates. The size of the design factor varies between survey variables reflecting the degree to which a characteristic of interest is clustered within PSUs, or is distributed between strata. For a single variable the size of the design factor also varies according to the size of the subgroup on which the estimate is based, and on the distribution of that subgroup between PSUs and strata. Design factors below 1.0 show that the complex sample design improved on the estimate that we would have expected from a simple random sample, probably due to the benefits of stratification. Design factors greater than 1.0 show less reliable estimates than might be gained from a simple random sample, due to the effects of clustering. Design factors equal to 1.0 indicate no difference in the survey design on the reliability of the estimate.

The formula to calculate the standard error of the difference between two percentages for a complex sample design is:

$$se(p_1-p_2) = \sqrt{[deft^2_1(p_1(100-p_1)/n_1) + deft2^2 (p_2(100-p_2)/n_2)]}$$

where p_1 and p_2 are observed percentages for the two sub-samples and n_1 and n_2 are the sub-sample sizes.

Confidence intervals

The estimate produced from a sample survey will rarely be identical to the population value, but statistical theory allows us to measure its accuracy. A confidence interval can be calculated around the estimated value, which gives a range in which the true value for the population is likely to fall. The standard error measures the precision with which the estimates from the sample approximate to the true population values and is used to construct the confidence interval for each survey estimate.

The 95% confidence intervals have been calculated for each estimated value presented. These are known as such, because if it were possible to repeat the survey under the same conditions a number of times, we would expect 95% of the confidence intervals calculated in this way to contain the true population value for that estimate. When assessing the results of a single survey, it is usual to assume that there is only a 5% chance that the true population value falls outside the 95% confidence interval calculated for each survey estimate. To construct the bounds of the confidence interval, 1.96 times the standard error is subtracted from, and added to, the estimated value, since under a normal distribution, 95% of values lie within 1.96 standard errors of the mean value. The confidence interval is then given by:

$$p +/- 1.96 \times se(p)$$

The 95% confidence interval for the difference between two percentages is given by:

$$(p_1 - p_2) +/- 1.96 \times se\ (p_1 - p_2)$$

If this confidence interval includes zero then the observed difference is considered to be a result of chance variation in the sample. If the interval does not include zero then it is unlikely (less than 5% probability) that the observed difference could have occurred by chance.

Standard errors for the 2002 GHS

The standard errors were calculated on weighted data using STATA [1]. Weighting for different sampling probabilities results in larger sampling errors than for an equal-probability sample without weights. However, weighting which uses population totals to control for differential non-response tends to lead to a reduction in the errors. The method used to calculate the sampling errors correctly allows for the inflation in the sampling errors caused by the first type of weighting but, in treating the second type of weighting in the same way as the first, incorrectly inflates the estimates further. Therefore the standard

errors and defts presented are likely to be slight over-estimates. Weighted data were used so that the values of the percentages and means were the same as those in the substantive chapters of the report.

Tables C.1 to C.12 show the standard error, the 95% confidence interval and the deft for selected survey estimates. The tables do not cover all the topics discussed in the report but show a selection of estimates.

For the design factors of household based estimates, one in ten (10%) were below 1, close to a half (45%) were below 1.1 and 72% of the defts were less than 1.2. There were four cases (10% of all the household-based estimates) where the deft was 1.5 or greater. The higher defts were for tenure and accommodation type (Table C.1) where the effects of clustering lead to a loss of precision compared with that of a simple random sample. The defts that were below 1 were in part for the number of persons in the household and household type, indicating that stratification has increased the precision of the sample over a simple random sample for these estimates of household variables.

Only 6% of the design factors of person based estimates were below 1, but over two fifths (42%) were below 1.1 and almost three-quarters (74%) were less than 1.2. Four per cent of the defts were 1.5 or greater, including many of those for estimates of ethnicity, shown in Table C.6. As well as clustering in the same sectors, people from the same ethnic backgrounds will generally cluster within the same households, and so estimates have high sampling errors and high defts. In contrast, estimates broken down by gender will generally have lower sampling errors because there is often one man and one woman in a household; for example, the estimates of males and females in the population have defts of 0.82 (Table C.4).

Estimating standard errors for other survey measures

The standard errors of survey measures, which are not presented in the tables and for sample subgroups may be estimated by applying an appropriate value of deft to the sampling error. The choice of an appropriate value of deft will vary according to whether the basic survey measure is included in the tables. Since most deft values are relatively small (1.2 or less) the absolute effect of adjusting sampling errors to take account of the survey's complex design will be small. In most cases it will result in an increase of less than 20% over the standard error assuming a simple random sample. Whether it is considered necessary to use deft or to use the basic estimates of standard errors assuming a

simple random sample is a matter of judgement and depends chiefly on the use to which the survey results are to be put.

Notes and references

[1] STATA is a statistical analysis software package. For further details of the method of calculation see: Elliot D. A comparison of software for producing sampling errors on social surveys. *SSD Survey Methodology Bulletin* 1999; 44: 27–36.

Table C.1   **Standard errors and 95% confidence intervals for household tenure, household type and accommodation type**

Base	Characteristic	% (p)	Unweighted sample size	Standard error of p	95% confidence intervals		Deft
All households	**Household type**						
	1 adult aged 16-59	15.9	8620	0.46	15.0	- 16.8	1.17
	2 adults aged 16-59	16.4	8620	0.44	15.5	- 17.3	1.10
	Youngest person aged 0-4	11.0	8620	0.36	10.2	- 11.7	1.07
	Youngest person aged 5-15	16.2	8620	0.44	15.4	- 17.1	1.11
	3 or more adults	10.3	8620	0.35	9.6	- 11.0	1.07
	2 adults, 1 or both aged 60 or over	15.4	8620	0.38	14.6	- 16.1	0.98
	1 adult aged 60 or over	14.9	8620	0.38	14.1	- 15.6	0.99
	Tenure						
	Owner occupied, owned outright	28.7	8613	0.55	27.6	- 29.7	1.13
	Owner occupied, with mortgage	40.3	8613	0.58	39.1	- 41.4	1.10
	Rented from council	13.9	8613	0.60	12.7	- 15.1	1.61
	Rented from housing association	6.5	8613	0.39	5.8	- 7.3	1.47
	Rented privately, unfurnished	3.0	8613	0.28	2.4	- 3.5	1.53
	Rented privately, furnished	7.7	8613	0.35	7.0	- 8.3	1.22
	Accommodation type						
	Detached house	21.9	8619	0.53	20.8	- 22.9	1.19
	Semi-detached house	31.6	8619	0.66	30.3	- 32.9	1.32
	Terraced house	26.6	8619	0.69	25.3	- 28.0	1.45
	Purpose-built flat or maisonette	16.1	8619	0.61	14.9	- 17.3	1.54
	Converted flat or maisonette/rooms	3.8	8619	0.30	3.2	- 4.4	1.46
	With business premises/other	0.1	8619	0.03	0.0	- 0.1	1.14

Table C.2   **Standard errors and 95% confidence intervals for number of persons and cars at each household**

Base	Characteristic	% (p)	Unweighted sample size	Standard error of p	95% confidence intervals	Deft
All households	**Number of persons**					
	1	30.7	8620	0.54	29.7 - 31.8	1.09
	2	34.9	8620	0.51	33.9 - 35.9	0.99
	3	15.8	8620	0.39	15.0 - 16.5	0.99
	4	12.7	8620	0.36	12.0 - 13.4	1.00
	5	4.2	8620	0.22	3.8 - 4.6	1.02
	6 or more	1.7	8620	0.15	1.4 - 2.0	1.09
	Number of cars/light vans					
	1	43.9	8614	0.54	42.8 - 45.0	1.01
	2 or more	29.6	8614	0.51	28.6 - 30.6	1.04
	none	26.5	8614	0.51	25.5 - 27.5	1.07

Table C.3   **Standard errors and 95% confidence intervals for households' ownership of selected consumer durables**

Base	Characteristic	% (p)	Unweighted sample size	Standard error of p	95% confidence intervals	Deft
All households	**Selected consumer durables**					
	Satellite	25.6	8620	0.59	24.5 - 26.8	1.25
	Cable	14.4	8620	0.62	13.2 - 15.6	1.64
	Digital	27.8	8611	0.61	26.6 - 29.0	1.26
	Video recorder	89.4	8619	0.37	88.6 - 90.1	1.11
	DVD player	32.0	8619	0.55	31.0 - 33.1	1.09
	Compact disc (CD) player	82.6	8620	0.43	81.7 - 83.4	1.05
	Home computer	53.8	8620	0.59	52.7 - 55.0	1.10
	Microwave oven	87.4	8620	0.38	86.6 - 88.1	1.06
	Deep freezer/fridge freezer	95.4	8620	0.24	94.9 - 95.9	1.06
	Washing machine	93.4	8620	0.28	92.9 - 94.0	1.05
	Tumble drier	54.5	8620	0.61	53.3 - 55.7	1.14
	Dishwasher	28.3	8620	0.55	27.2 - 29.4	1.13

Table C.4   **Standard errors and 95% confidence intervals for age and sex**

Base	Characteristic	% (p)	Unweighted sample size	Standard error of p	95% confidence intervals	Deft
All persons	**Sex**					
	Male	48.7	20149	0.29	48.1 - 49.2	0.82
	Female	51.4	20149	0.29	50.8 - 51.9	0.82
All persons	**Age**					
	0-4	5.8	20149	0.19	5.4 - 6.2	1.15
	5-15	14.2	20149	0.30	13.6 - 14.8	1.22
	16-44	40.1	20149	0.42	39.3 - 40.9	1.22
	45-64	24.3	20149	0.39	23.6 - 25.1	1.29
	65-74	8.5	20149	0.25	8.0 - 9.0	1.27
	75 and over	7.1	20149	0.22	6.6 - 7.5	1.22
All males	0-4	6.2	9706	0.27	5.7 - 6.7	1.10
	5-15	14.8	9706	0.42	13.9 - 15.6	1.17
	16-44	40.6	9706	0.55	39.5 - 41.7	1.10
	45-64	24.6	9706	0.47	23.7 - 25.6	1.07
	65-74	8.2	9706	0.28	7.6 - 8.7	1.01
	75 and over	5.6	9706	0.25	5.1 - 6.1	1.07
All females	0-4	5.4	10443	0.24	4.9 - 5.9	1.08
	5-15	13.7	10443	0.34	13.1 - 14.4	1.01
	16-44	39.6	10443	0.48	38.7 - 40.6	1.00
	45-64	24.0	10443	0.43	22.2 - 24.9	1.03
	65-74	8.8	10443	0.29	8.2 - 9.3	1.05
	75 and over	8.5	10443	0.29	7.9 - 9.0	1.06

Table C.5   **Standard errors and 95% confidence intervals for marital status**

Base	Characteristic	% (p)	Unweighted sample size	Standard error of p	95% confidence intervals	Deft
All persons aged 16 and over	**Marital status**					
	Married	52.4	15972	0.53	51.4 - 53.4	1.34
	Cohabiting	9.2	15972	0.34	8.5 - 9.9	1.49
	Single	22.8	15972	0.47	21.9 - 23.7	1.42
	Widowed	7.6	15972	0.22	7.1 - 8.0	1.05
	Divorced	5.7	15972	0.20	5.3 - 6.1	1.09
	Separated	2.1	15972	0.13	1.8 - 2.3	1.16
Men aged 16 and over	Married	54.5	7579	0.64	53.2 - 55.7	1.12
	Cohabiting	9.6	7579	0.36	8.9 - 10.3	1.07
	Single	25.9	7579	0.60	24.7 - 27.0	1.19
	Widowed	3.6	7579	0.21	3.2 - 4.0	0.98
	Divorced	4.6	7579	0.25	4.2 - 5.1	1.03
	Separated	1.5	7579	0.16	1.2 - 1.8	1.14
Women aged 16 and over	Married	50.5	8393	0.59	49.3 - 51.6	1.08
	Cohabiting	8.9	8393	0.34	8.2 - 9.6	1.09
	Single	20.0	8393	0.54	18.9 - 21.0	1.24
	Widowed	11.2	8393	0.35	10.6 - 11.9	1.02
	Divorced	6.7	8393	0.29	6.1 - 7.3	1.06
	Separated	2.6	8393	0.18	2.2 - 2.9	1.05
All persons aged 16 to 24	Married	4.3	2023	0.57	3.2 - 5.4	1.26
	Cohabiting	13.0	2023	0.99	11.1 - 15.0	1.32
	Single	81.7	2023	1.13	79.4 - 83.9	1.31
	Widowed	0.0	2023	-	- -	-
	Divorced	0.1	2023	0.06	0.0 - 0.2	0.95
	Separated	0.6	2023	0.17	0.2 - 0.9	1.02
All persons aged 25 to 34	Married	40.4	2557	1.23	38.0 - 42.8	1.27
	Cohabiting	20.8	2557	1.02	18.8 - 22.8	1.27
	Single	34.2	2557	1.08	32.1 - 36.3	1.15
	Widowed	0.1	2557	0.08	0.0 - 0.3	1.08
	Divorced	1.6	2557	0.23	1.1 - 2.1	0.93
	Separated	2.7	2557	0.33	2.0 - 3.3	1.03
All persons aged 35 to 44	Married	63.8	3022	1.08	61.7 - 65.9	1.24
	Cohabiting	11.1	3022	0.74	9.6 - 12.5	1.30
	Single	14.1	3022	0.74	12.7 - 15.6	1.17
	Widowed	0.4	3022	0.11	0.1 - 0.6	1.01
	Divorced	7.1	3022	0.48	6.1 - 8.0	1.03
	Separated	3.1	3022	0.35	2.4 - 3.7	1.12
All persons aged 45 to 54	Married	70.6	2762	0.99	68.7 72.6	1.14
	Cohabiting	6.4	2762	0.55	5.3 - 7.4	1.18
	Single	8.0	2762	0.52	7.0 - 9.0	1.01
	Widowed	1.5	2762	0.23	1.0 - 1.9	1.00
	Divorced	10.2	2762	0.63	9.0 - 11.5	1.09
	Separated	2.9	2762	0.34	2.2 - 3.6	1.07
All persons aged 55 to 64	Married	73.9	2407	1.06	71.8 - 75.9	1.18
	Cohabiting	3.4	2407	0.43	2.6 - 4.3	1.16
	Single	5.3	2407	0.49	4.3 - 6.3	1.07
	Widowed	6.1	2407	0.52	5.1 - 7.1	1.06
	Divorced	9.4	2407	0.68	8.0 - 10.7	1.15
	Separated	1.9	2407	0.31	1.3 - 2.5	1.12
All persons aged 65 to 74	Married	66.3	1766	1.22	63.9 - 68.7	1.08
	Cohabiting	1.5	1766	0.35	0.8 - 2.1	1.23
	Single	4.6	1766	0.57	3.5 - 5.7	1.15
	Widowed	19.4	1766	0.98	17.4 - 21.3	1.04
	Divorced	7.1	1766	0.59	5.9 - 8.2	0.97
	Separated	1.2	1766	0.30	0.6 - 1.8	1.16
All persons aged 75 and over	Married	41.3	1435	1.50	38.3 - 44.2	1.15
	Cohabiting	0.9	1435	0.31	0.3 - 1.5	1.26
	Single	4.5	1435	0.56	3.4 - 5.6	1.03
	Widowed	49.0	1435	1.52	46.0 - 52.0	1.15
	Divorced	3.7	1435	0.52	2.7 - 4.7	1.05
	Separated	0.7	1435	0.22	0.3 - 1.2	0.99

Table C.6 **Standard errors and 95% confidence intervals for ethnic origin***

Base†	Characteristic	% (p)	Unweighted sample size	Standard error of p	95% confidence intervals	Deft
All persons aged 16 and over	**Ethnic origin**					
	White	92.4	15906	0.61	91.2 - 93.6	2.91
	Mixed race	0.7	15906	0.09	0.5 - 0.9	1.38
	Asian-Indian	1.5	15906	0.31	0.9 - 2.1	3.19
	Asian-Pakistani, Bangladeshi, other	1.9	15906	0.30	1.3 - 2.5	2.77
	Black Caribbean	1.1	15906	0.16	0.8 - 1.4	1.91
	Black African	0.9	15906	0.14	0.6 - 1.1	1.91
	Other	1.5	15906	0.17	1.1 - 1.8	1.78

* Other includes other Black groups. Information on those giving no answer has not been presented

† These estimates are based on 2002 data only, whereas in the report estimates are based on 2002 and 2001 data combined. We would expect the defts to be very similar for the combined years estimates.

Table C.7 **Standard errors and 95% confidence intervals for education level**

Base	Characteristic	% (p)	Unweighted sample size	Standard error of p	95% confidence intervals	Deft
All persons aged 16 to 69	**Education level**					
	Higher education	26.7	12636	0.61	25.5 - 27.9	1.55
	Other qualifications	51.1	12636	0.59	49.9 - 52.2	1.33
	None	22.2	12636	0.52	21.2 - 23.2	1.41
All men aged 16 to 69	Higher education	27.7	5899	0.77	26.2 - 29.2	1.32
	Other qualifications	51.5	5899	0.77	50.0 - 53.0	1.18
	None	20.8	5899	0.65	19.5 - 22.1	1.23
All women aged 16 to 69	Higher education	25.8	6737	0.65	24.5 - 27.1	1.22
	Other qualifications	50.7	6737	0.67	49.4 - 52.0	1.10
	None	23.5	6737	0.59	22.3 - 24.6	1.14

Table C.8 **Standard errors and 95% confidence intervals for socio-economic classification and employment status of adults**

Base	Characteristic	% (p)	Unweighted sample size	Standard error of p	95% confidence intervals	Deft
All persons aged 16 and over	**Socio-economic classification**					
	Higher managerial and professional	9.9	14910	0.31	9.3 - 10.5	1.27
	Lower managerial and professional	21.8	14910	0.39	21.0 - 22.5	1.15
	Intermediate	12.6	14910	0.29	12.1 - 13.2	1.07
	Small employers and own account	8.0	14910	0.27	7.5 - 8.5	1.22
	Lower supervisory and technical	10.5	14910	0.27	10.0 - 11.0	1.08
	Semi-routine	17.8	14910	0.34	17.1 - 18.4	1.09
	Routine	14.7	14910	0.37	14.0 - 15.5	1.27
	Never worked and long-term unemployed	4.7	14910	0.26	4.2 - 5.2	1.50
All men aged 16 and over	Higher managerial and professional	15.3	7063	0.48	14.4 - 16.8	1.12
	Lower managerial and professional		7063	0.52	20.1 - 20.9	1.07
	Intermediate	5.9	7063	0.30	5.3 - 7.0	1.07
	Small employers and own account	11.5	7063	0.41	10.7 - 11.8	1.08
	Lower supervisory and technical	15.5	7063	0.48	14.5 - 16.3	1.12
	Semi-routine	12.0	7063	0.41	11.2 - 13.1	1.06
	Routine	15.9	7063	0.51	14.9 - 16.4	1.17
	Never worked and long-term unemployed	2.9	7063	0.24	2.4 - 4.3	1.20
All women aged 16 and over	Higher managerial and professional	5.0	7847	0.29	4.4 - 5.6	1.18
	Lower managerial and professional	22.4	7847	0.50	21.4 - 22.4	1.06
	Intermediate	18.9	7847	0.49	17.9 - 19.8	1.11
	Small employers and own account	4.8	7847	0.25	4.3 - 5.3	1.04
	Lower supervisory and technical	5.9	7847	0.28	5.4 - 6.5	1.05
	Semi-routine	23.1	7847	0.49	22.1 - 24.0	1.03
	Routine	13.7	7847	0.44	12.8 - 14.5	1.13
	Never worked and long-term unemployed	6.3	7847	0.38	5.6 - 7.1	1.38
All persons aged 16 to 44	Higher managerial and professional	11.1	6626	0.49	10.2 - 12.1	1.27
	Lower managerial and professional	22.8	6626	0.54	21.7 - 23.8	1.05
	Intermediate	13.3	6626	0.43	12.4 - 14.1	1.03
	Small employers and own account	6.6	6626	0.34	5.9 - 7.2	1.12
	Lower supervisory and technical	9.8	6626	0.38	9.0 - 10.5	1.04
	Semi-routine	17.3	6626	0.50	16.3 - 18.3	1.08
	Routine	13.1	6626	0.49	12.1 - 14.1	1.18
	Never worked and long-term unemployed	6.1	6626	0.36	5.4 - 6.8	1.22
All persons aged 45 to 64	Higher managerial and professional	10.6	5095	0.49	9.7 - 11.6	1.13
	Lower managerial and professional	23.3	5095	0.66	22.0 - 22.6	1.11
	Intermediate	11.3	5095	0.45	10.4 - 12.2	1.02
	Small employers and own account	10.7	5095	0.48	9.8 - 11.7	1.11
	Lower supervisory and technical	10.4	5095	0.46	9.5 - 11.3	1.08
	Semi-routine	17.2	5095	0.55	16.1 - 18.3	1.04
	Routine	13.8	5095	0.54	12.7 - 14.8	1.12
	Never worked and long-term unemployed	2.7	5095	0.31	2.1 - 3.3	1.37
All persons aged 65 to 74	Higher managerial and professional	6.3	1761	0.58	5.2 - 7.5	1.00
	Lower managerial and professional	17.5	1761	0.99	15.6 - 19.5	1.09
	Intermediate	12.6	1761	0.79	11.1 - 14.2	1.00
	Small employers and own account	7.9	1761	0.69	6.5 - 9.2	1.08
	Lower supervisory and technical	11.6	1761	0.76	10.1 - 13.1	1.00
	Semi-routine	20.4	1761	0.98	18.5 - 22.3	1.02
	Routine	20.5	1761	0.98	18.5 - 22.4	1.02
	Never worked and long-term unemployed	3.2	1761	0.46	2.3 - 4.1	1.10
All persons aged 75 and over	Higher managerial and professional	6.0	1428	0.68	4.6 - 7.3	1.08
	Lower managerial and professional	16.7	1428	1.19	14.4 - 19.1	1.20
	Intermediate	14.3	1428	1.04	12.2 - 16.3	1.12
	Small employers and own account	5.7	1428	0.64	4.5 - 7.0	1.04
	Lower supervisory and technical	12.9	1428	1.01	11.0 - 14.9	1.14
	Semi-routine	19.0	1428	1.08	16.8 - 21.1	1.04
	Routine	19.2	1428	1.16	17.0 - 21.5	1.11
	Never worked and long-term unemployed	6.2	1428	0.69	4.8 - 7.5	1.08
All persons aged 16 and over	**Employment status**					
	In employment	60.3	15729	0.50	59.4 - 61.3	1.28
	Unemployed	2.5	15729	0.13	2.3 - 2.8	1.04
	Economically inactive	37.1	15729	0.49	36.2 - 38.1	1.27
All men aged 16 and over	In employment	67.9	7433	0.62	66.7 - 69.1	1.15
	Unemployed	3.1	7433	0.21	2.7 - 3.5	1.04
	Economically inactive	29.0	7433	0.61	27.8 - 30.2	1.16
All women aged 16 and over	In employment	53.4	8296	0.59	52.2 - 54.5	1.08
	Unemployed	2.0	8296	0.16	1.7 - 2.3	1.04
	Economically inactive	44.6	8296	0.59	43.5 - 45.8	1.08

Table C.9   **Standard errors and 95% confidence intervals for health measures**

Base	Characteristic	% (p)	Unweighted sample size	Standard error of p	95% confidence intervals	Deft
All persons	**Self-reported sickness**					
	Longstanding illness	34.6	19875	0.47	33.7 - 35.5	1.39
	Limiting longstanding illness	20.7	19908	0.35	20.1 - 21.4	1.22
	Restricted activity in the last 14 days	14.8	19879	0.33	14.1 - 15.4	1.31
All males	Longstanding illness	34.2	9542	0.58	33.0 - 35.3	1.19
	Limiting longstanding illness	19.8	9560	0.47	18.9 - 20.7	1.15
	Restricted activity in the last 14 days	13.7	9543	0.41	12.9 - 14.5	1.16
All females	Longstanding illness	35.0	10333	0.58	33.9 - 36.1	1.24
	Limiting longstanding illness	21.6	10348	0.46	20.7 - 22.5	1.14
	Restricted activity in the last 14 days	15.8	10336	0.42	14.9 - 16.6	1.17
All persons aged 0 to 4	Longstanding illness	14.8	1164	1.06	12.7 - 16.8	1.02
	Limiting longstanding illness	4.0	1167	0.57	2.9 - 5.1	0.99
	Restricted activity in the last 14 days	9.7	1165	0.97	7.8 - 11.6	1.12
All persons aged 5 to 15	Longstanding illness	20.0	2999	0.86	18.4 - 21.7	1.18
	Limiting longstanding illness	8.3	3009	0.57	7.2 - 9.4	1.13
	Restricted activity in the last 14 days	9.0	3000	0.61	7.8 - 10.2	1.17
All persons aged 16 to 44	Longstanding illness	23.7	7408	0.61	22.5 - 24.9	1.23
	Limiting longstanding illness	13.2	7419	0.46	12.3 - 14.1	1.17
	Restricted activity in the last 14 days	12.4	7405	0.44	11.5 - 13.3	1.15
All persons aged 45 to 64	Longstanding illness	44.9	5117	0.78	43.4 - 46.4	1.12
	Limiting longstanding illness	28.2	5122	0.64	26.9 - 29.4	1.02
	Restricted activity in the last 14 days	18.3	5120	0.60	17.1 - 19.4	1.11
All persons aged 65 to 74	Longstanding illness	63.0	1758	1.26	60.5 - 65.4	1.09
	Limiting longstanding illness	40.6	1761	1.28	38.1 - 43.1	1.09
	Restricted activity in the last 14 days	19.7	1760	1.05	17.6 - 21.7	1.11
All persons aged 75+	Longstanding illness	71.5	1429	1.30	69.0 - 74.1	1.09
	Limiting longstanding illness	52.6	1430	1.48	49.8 - 55.6	1.12
	Restricted activity in the last 14 days	26.1	1429	1.35	23.4 - 28.7	1.16

Table C.10 **Standard errors and 95% confidence intervals for cigarette smoking**

Base	Characteristic	% (p)	Unweighted sample size	Standard error of p	95% confidence intervals	Deft
All persons aged 16 and over	**Cigarette smoking**					
	Current cigarette smoker	25.9	14788	0.45	25.0 - 26.7	1.25
	Ex-regular cigarette smoker	24.0	14788	0.44	23.1 - 24.8	1.25
	Never regularly smoked cigarettes	50.2	14788	0.55	49.1 - 51.2	1.34
All men aged 16 and over	Current cigarette smoker	26.8	6837	0.61	25.6 - 28.0	1.14
	Ex-regular cigarette smoker	27.5	6837	0.58	26.4 - 28.7	1.07
	Never regularly smoked cigarettes	45.6	6837	0.71	44.2 - 47.0	1.18
All women aged 16 and over	Current cigarette smoker	25.0	7951	0.56	23.9 - 26.1	1.15
	Ex-regular cigarette smoker	20.9	7951	0.51	19.9 - 21.9	1.12
	Never regularly smoked cigarettes	54.1	7951	0.67	52.8 - 55.5	1.20

Table C.11 **Standard errors and 95% confidence intervals for alcohol consumption (maximum daily amount)**

Base	Characteristic	% (p)	Unweighted sample size	Standard error of p	95% confidence intervals	Deft
All men aged 16 and over	**Alcohol consumption* (maximum daily amount)**					
	Drank nothing last week	26.6	6827	0.76	25.1 - 28.0	1.42
	Drank up to 4 units	36.4	6827	0.68	35.0 - 37.7	1.17
	Drank more than 4 and up to 8 units	16.5	6827	0.49	15.5 - 17.4	1.09
	Drank more than 8 units	20.6	6827	0.60	19.4 - 21.8	1.23
All women aged 16 and over	Drank nothing last week	40.7	7938	0.72	39.3 - 42.1	1.31
	Drank up to 3 units	36.6	7938	0.66	35.3 - 37.9	1.22
	Drank more than 3 and up to 6 units	13.1	7938	0.42	12.3 - 13.9	1.11
	Drank more than 6 units	9.6	7938	0.42	8.8 - 10.4	1.27
All aged 16 to 24	Drank nothing last week	35.9	1668	1.75	32.5 - 39.4	1.49
	Drank up to 4/3 units	19.1	1668	1.01	17.2 - 21.1	1.05
	Drank more than 4/3 and up to 8/6 units	14.4	1668	0.93	12.6 - 16.2	1.08
	Drank more than 8/6 units	30.6	1668	1.48	27.6 - 33.5	1.31
All aged 25 to 44	Drank nothing last week	29.8	5141	0.82	28.2 - 31.4	1.29
	Drank up to 4/3 units	33.7	5141	0.72	32.3 - 35.1	1.09
	Drank more than 4/3 and up to 8/6 units	16.9	5141	0.59	15.7 - 18.0	1.13
	Drank more than 8/6 units	19.6	5141	0.70	18.3 - 21.0	1.26
All aged 45 to 64	Drank nothing last week	31.0	4865	0.91	29.2 - 32.8	1.37
	Drank up to 4/3 units	41.2	4865	0.88	39.5 - 42.9	1.25
	Drank more than 4/3 and up to 8/6 units	17.0	4865	0.59	15.8 - 18.1	1.10
	Drank more than 8/6 units	10.8	4865	0.50	9.9 - 11.8	1.12
All aged 65 and over	Drank nothing last week	45.2	3091	1.01	43.3 - 47.2	1.13
	Drank up to 4/3 units	45.0	3091	1.04	43.0 - 47.1	1.16
	Drank more than 4/3 and up to 8/6 units	7.3	3091	0.49	6.4 - 8.3	1.04
	Drank more than 8/6 units	2.4	3091	0.29	1.8 - 3.0	1.06
All aged 16 and over	Drank nothing last week	34.1	14765	0.64	32.8 - 35.3	1.64
	Drank up to 4/3 units	36.5	14765	0.55	35.4 - 37.6	1.39
	Drank more than 4/3 and up to 8/6 units	14.7	14765	0.34	14.0 - 15.3	1.17
	Drank more than 8/6 units	14.8	14765	0.41	14.0 - 15.6	1.40

* The first of each pair of figures shown relates to men, and the second, to women.

Table C.12 **Standard errors and 95% confidence intervals for number of cohabitations**

Base	Characteristic	% (p)	Unweighted sample size	Standard error of p	95% confidence intervals	Deft
All women aged 16 to 59	**Numbers of cohabitations**					
	None	84.4	5627	0.54	83.3 - 85.5	1.12
	One	12.2	5627	0.49	11.2 - 13.2	1.12
	Two or more	3.4	5627	0.26	2.9 - 3.9	1.07
All men aged 16 to 59	None	85.2	4893	0.56	84.1 - 86.3	1.10
	One	10.9	4893	0.50	9.9 - 11.8	1.12
	Two or more	4.0	4893	0.28	3.4 - 4.5	1.00
All people aged 16 to 24	None	91.1	1645	0.79	89.6 - 92.7	1.13
	One	7.6	1645	0.75	6.2 - 9.1	1.15
	Two or more	1.3	1645	0.26	0.7 - 1.8	0.95
All people aged 25 to 34	None	74.3	2312	0.99	72.4 - 76.2	1.09
	One	19.9	2312	0.93	18.0 - 21.7	1.12
	Two or more	5.8	2312	0.47	4.9 - 6.8	0.96
All people aged 35 to 44	None	80.7	2761	0.79	79.1 - 82.2	1.05
	One	13.9	2761	0.71	12.5 - 15.3	1.08
	Two or more	5.5	2761	0.44	4.6 - 6.3	1.02
All people aged 45 to 54	None	90.2	2536	0.63	88.9 - 91.4	1.07
	One	7.3	2536	0.55	6.2 - 8.4	1.06
	Two or more	2.5	2536	0.32	1.9 - 3.2	1.02
All people aged 55 to 59	None	95.4	1266	0.64	94.1 - 96.6	1.08
	One	3.6	1266	0.57	2.5 - 4.7	1.09
	Two or more	1.1	1266	0.31	0.4 - 1.7	1.08
All people aged 16 to 59	None	84.8	10520	0.42	83.9 - 85.6	1.20
	One	11.6	10520	0.37	10.8 - 12.3	1.19
	Two or more	3.7	10520	0.19	3.3 - 4.1	1.04

Appendix D **Weighting and grossing**

All surveys accept that there will be some degree of nonresponse, although great efforts are made to keep nonresponse to a minimum[1]. During the review of the GHS in 1999, two methods of compensating for nonresponse were examined with the aim of improving the quality of data. The method adopted to compensate for **total nonresponse** (where all survey information for a sampled household is missing) will be described here. The method adopted to reduce **item nonresponse** (where information for particular questions is missing as the result of conducting proxy interviews) is discussed in Appendix B.

The 2002 GHS is weighted using a two-step approach. In the first step, the data are weighted to compensate for nonresponse (sample-based weighting). The second step weights the sample distribution so that it matches the population distribution in terms of region, age group and sex (population-based weighting).[2]

Weighting for nonresponse

Weighting for total nonresponse involves giving each respondent a weight so that they represent the non-respondents who are similar to them in terms of survey characteristics. To be able to use this method, information about non-respondents is needed. By their very nature, non-responding households yield little information. Although some surveys collect information about the characteristics of these households, information about non-respondents is not routinely collected on the GHS. An alternative approach to gaining information about the GHS's non-responding households was needed to carry out such a weighting procedure.

Sample-based weighting using the Census

The decennial Census was found to be the most appropriate source of information about non-responding addresses on the GHS. Unlike the GHS, which relies upon voluntary co-operation from respondents, the Census is mandatory therefore nonresponse is kept to an absolute minimum.

After the 1991 Census, methodological work was conducted to match Census addresses with the sampled addresses of some of the large continuous surveys, of which the GHS

was one. In this way it was possible to match the address details of the GHS respondents as well as the non-respondents with corresponding information gathered from the Census for the same address. It was then possible to identify any types of household that were being under-represented in the survey. Similar work is currently being conducted using the 2001 Census.

The information collected during the 1991 Census/GHS matching work was used to weight the 2002/2003 GHS data by identifying types of households that differed in terms of response rates. A combination of household variables such as household type, social class, region and car ownership were analysed using the software package *Answer Tree* (using the chi-squared statistic CHAID)[3] to identify which characteristics were most significant in distinguishing between responding and non-responding households. These characteristics were sorted by the program to produce the weighting classes shown in Figure D.A. The variables used to identify the weighting classes were restricted to those that appear annually on the GHS.

Figure D.A

Population-based weighting (grossing)

Population-based weighting schemes address deficiencies in the data due to sample-non coverage.

Source of the population totals

The GHS sample is based on private households, which means that the population totals used in the weighting need to relate to people in private households. These totals are taken from the Labour Force Survey (LFS). The LFS derives household population estimates by excluding residents of institutions initially from population projections based on mid-year estimates[4]. Since the LFS estimates are based on population projections as well as estimates, they are generally available earlier than the mid-year estimates, so in 2000, when weighting was introduced, the GHS was able to use the 2000 LFS household population estimates.

Weighting of the 2001 and 2002 GHS was less straightforward. It was decided that the LFS 2000 population estimates would be used for weighting the 2001 GHS because

they were the most accurate figures available at the time. However, during the writing of the 2001 report the results from the 2001 Census were published. They indicated that previous mid-year estimates of the total UK population were around 900,000 too high, with disparities being most apparent among men aged 25 to 39. These disparities were larger than expected, so it was decided that the 2001 weighting would need to be revisited post-publication as the reporting process was too far advanced for them to be incorporated[5].

In February 2003, the Office for National Statistics (ONS) published revised mid-year population estimates for 1991 to 2000. This brought them into line with the post 2001 Census-based mid-2001 population estimates (hereafter referred to as the 2001-based intermediate population estimates). In September 2003, there was a relatively small upward revision to the mid-2001 population estimate of men (mainly in those aged 25-34 of around 190,000). There have been, and will be, further revisions to some local authority population estimates[6].

ONS examined the effects on GHS estimates of re-weighting the data using the 2001-based intermediate population estimates. The effects were sufficiently small that it was decided that it was unnecessary to issue revised weighting post-publication. However, GHS 2001 data with weights derived from the 2001-based intermediate population estimates are available from the UK Data Archive (www.data-archive.co.uk).

The population-based (grossing) method

The population information and GHS data were grouped into 12 age and sex categories within 6 region categories to form weighting classes as shown in Figure D.B[7]. The population data and the GHS data were then matched using the age and sex categories within region, giving corresponding population totals for each of these sub-groups within the GHS sample.

The population-based and sample-based weighting methods were used in conjunction to produce the final weight. The final weight was created using a calibration procedure called CALMAR (a SAS-based Macro)[7] which used the pre-weighted (sample-based weighted) data.

Figure D.B

Figure D.A   **Weighting classes formed in the CHAID analysis**

Level 1 split	Level 2 split	Level 3 split	Level 4 split	Weight class
Region North East Merseyside Yorks & Humbs W Midlands South East Scotland	**No. of Cars** 0 or 1	**No. of dependent children** 0 or 1	**Household type 1** 1 adult 16-59 Youngest 5-15 3+ adults no child	1
			Household type 1 2 adults 16-59 Youngest 0-4 2 adults, 1 or 2 60+ 1 adult only 60+	2
		No. of dependent children 2 or more		3
	No. of Cars 2 or more			4
Region North West E Midlands Eastern South West	**Pensioner HH** Pensioner in HH	**SEG (grouped)** Skilled Manual Partly-skilled manual Unskilled manual & others Not employed in last 10 years		5
		SEG (grouped) Professional Manager/employer Intermediate/jnr		6
	Pensioner HH No pensioner in HH	**No. of adults** 1		7
		No. of adults 2 or more	**Social Class** I, II, IV Not employed in last 10 years	8
			Social Class IIInm, IIIm, V Other	9
Region London	**Type of building** Detached Semi-detached Terraced Converted flat/other			10
	Type of building Purpose built flat			11
Region Wales				12

Effects of re-weighting the 2001/2 GHS data

Tables D1 and D2 compare the effects of weighting a selection of GHS variables with the original 2001 weight and the revised 2001 weight. Overall, none of the effects of re-weighting the data were large[8].

- The proportion of households containing one adult increased by 1.2 percentage points when weighted with the original 2001 weight, whereas the revised weight increased it by 1.8 percentage points. Conversely, the proportion of households containing two adults decreased by 2.0 percentage points when weighted with the original 2001 weight, whereas the revised weight decreased it by 2.6 percentage points.
- The original 2001 weight increased the proportion of those buying on a mortgage by 0.5 percentage points whereas the revised weight decreased the same group by 0.3 percentage points.
- The original 2001 weight decreased the proportion of ex-regular male cigarette smokers by 1.4 percentage points whereas the revised 2001 weight decreased the same group by 0.9 percentage points.

Tables D.1 and D.2

It should be noted that the weighted bases used in this report are not recommended as a source for population estimates. They should primarily be used as bases for the percentages shown in tables rather than estimates of population size[9].

Presentation and interpretation of weighted data

Weighted data cannot be meaningfully compared to unweighted data from previous years without knowledge of how the weighting changes the estimates. In the trend tables in the 2002 report, weighted and unweighted data are presented for 1998 and weighted data are only shown from 2000 to 2002. Care should be taken when interpreting trend data or individual tables compared with other years as part of a time series.

Tables D.3 and D.4 identify the effects of weighting by comparing unweighted and weighted data for 2002. They also show the differences between the weighted and unweighted estimates for 1998 and 2000 to 2002, on a selection of household and individual level variables. **Tables D.3 and D.4**

Effects of weighting on data

A comparison of the characteristics recorded on the 1991 Census forms of respondents and non-respondents in the 1991 GHS sample showed that households comprising one adult aged 16 to 59 or a couple with non-dependent children were under-represented[10]. Households containing dependent children were over-represented in the responding sample. As would be expected, weighting has changed the value of the estimate for some variables, but the overall changes have been relatively small.

For the 2002 estimates, the most marked effect of weighting was seen in the following variables. None of the effects are large.

Increase in value of estimate.
- 1 person households from 29% to 31%.
- 1 adult households from 35% to 37%.
- male current cigarette smokers from 26% to 27%.

Decrease in value of estimate.
- 2 person households from 37% to 35%.
- 2 adult households from 51% to 48%.
- households containing a married couple with no children from 25% to 23%.
- households who own two cars or vans from 24% to 22%.
- male ex-regular cigarette smokers from 29% to 28%.

The differences between the weighted and unweighted data for 1998, 2000 and 2001 are also shown in Tables D.3 and D.4. It can be seen that the differences produced by weighting in 2002 were similar to those in 1998, 2000 and 2001 for the same variables.

Tables D.3 and D.4

Notes and references

1. Appendix B describes the variation in response for the GHS since it began in 1971.
2. Barton, J. Developing a Weighting and Grossing System for the General Household Survey: *Social Survey Methodology Bulletin* (Issue 49 July 2001).
3. CHAID is an acronym that stands for Chi-squared Automatic Interaction Detection. As is suggested by its name, CHAID uses chi-squared statistics to identify optimal splits or groupings of independent variables in terms of predicting the outcome of a dependent variable, in this case response.
4. These estimates are revised when mid-year estimates for the appropriate year are available.
5. For more details about the revised census-based estimates see http://www.statistics.gov.uk/census2001/implications.asp
6. For more details about revisions to local authority population estimates see www.statistics.gov.uk/about/methodology_by_theme/sape/default.asp
7. CALMAR uses a calibration procedure, also known as raking ratio or rim weighting, which divides the sample into weighting classes which in this case will have known population totals. The weighting classes used were those that were recommended for the GHS by Elliot, D. (1999) "Report of the Task Force on Weighting and Estimation", *GSS Methodology Series*.
8. See Chapters 8 and 9 for more detailed analysis on the effects of re-weighting the 2001/2 GHS data in relation to smoking and drinking variables.
9. Missing answers are excluded from the tables and in some cases this is reflected in the weighted bases, i.e. these numbers vary between tables. For this reason, the bases themselves are not recommended as a source for population estimates. Recommended data sources for population estimates for most socio-demographic groups are: ONS mid-year estimates, the Labour Force Survey, or Housing Statistics from the Office of the Deputy Prime Minister.
10. Foster K et al. *General Household Survey 1993*. HMSO 1995. Appendix C.

Figure D.B   **Weighting classes used for CALMAR analysis**

Age/sex	Region
0-4	London
5-15	Scotland
16-24 Male	Wales
16-24 Female	Other metropolitan
25-44 Male	Other non-metropolitan
25-44 Female	South East
45-64 Male	
45-64 Female	
65-74 Male	
65-74 Female	
75+ Male	
75+ Female	

221

Table D1   **Comparison of the original weight used for the 2001 report and the revised 2001 weight**

Household level variables

% of households	Effect of weighting	
	Original weighted 2001 - Unweighted 2001	Revised weighted 2001 - Unweighted 2001
Household size		
1 person	1.7	1.8
2 persons	-1.4	-1.3
3 persons	0.2	0.1
4 persons	-0.3	-0.5
5 persons	-0.1	-0.2
6 or more persons	-0.1	-0.1
Number of adults		
1 adult	1.2	1.8
2 adults	-2.0	-2.6
3 adults	0.4	0.4
4 or more adults	0.3	0.4
Number of children		
No children	1.1	1.0
1 child	0.1	0.3
2 children	-0.7	-0.8
3 or more children	-0.4	-0.5
Household type		
1 person only	1.8	1.9
2 or more unrelated adults	0.2	0.2
Married couple, dependent children	-0.6	-0.9
Married couple, independent children	0.3	0.2
Married couple, no children	-1.7	-1.5
Lone parent, dependent children	-0.5	0.0
Lone parent, independent children	0.2	0.1
2 or more families (inc. same sex cohab)	0.0	0.0
Cohabiting couple, with children	0.0	-0.1
Cohabiting couple, no children	0.3	0.0
Tenure - harmonised		
Owns outright	-1.7	-1.2
Buying on mortgage	0.5	-0.3
Rents from LA	0.2	0.6
Rents from HA	0.1	0.2
Rents privately - unfurnished/nk	0.4	0.2
Rents privately - furnished	0.6	0.5
Ownership of consumer durables		
Video	-0.2	-0.3
Freezer	-0.3	-0.3
Washing machine	-0.6	-0.6
Drier	-0.9	-1.0
Dishwasher	-1.0	-1.2
Microwave oven	-0.4	-0.3
Telephone	-0.1	-0.2
CD player	0.4	0.0
Home computer	0.4	-0.2
Central heating	-0.2	-0.3
Car or van ownership		
No car or van	0.6	1.2
One car or van	0.4	0.6
Two cars or vans	-1.0	-1.6
Three or more cars or vans	0.0	-0.2

Table D2   **Comparison of the original weight used for the 2001 report and the revised 2001 weight**

Individual level variables

% of individuals	Effect of weighting	
	Original weighted 2001 - Unweighted 2001	Revised weighted 2001 - Unweighted 2001
Limiting longstanding illness		
Male	-0.3	0.0
Female	-0.1	0.1
Total	-0.2	0.0
Non-limiting longstanding illness		
Male	-0.2	-0.2
Female	-0.1	0.0
Total	-0.2	-0.1
No longstanding illness		
Male	0.6	0.2
Female	0.2	-0.1
Total	0.4	0.1
General health		
Good		
Male	0.0	-0.3
Female	-0.1	-0.3
Total	0.0	-0.3
Fairly good		
Male	0.0	0.2
Female	0.0	0.1
Total	0.1	0.2
Not good		
Male	-0.1	0.1
Female	0.1	0.2
Total	0.0	0.1
Restricted activity in the last 14 days		
Male	0.1	0.1
Female	0.0	0.1
Total	0.1	0.1
Cigarette smoking by sex		
Men		
Current cigarette smokers	1.2	0.9
Ex-regular cigarette smokers	-1.4	-0.9
Never or (only occasionally) smoked	0.1	0.0
Women		
Current cigarette smokers	0.3	0.5
Ex-regular cigarette smokers	0.0	-0.2
Never or (only occasionally) smoked	-0.3	-0.4
Total		
Current cigarette smokers	0.8	0.8
Ex-regular cigarette smokers	-0.6	-0.5
Never or (only occasionally) smoked	-0.2	-0.3
Weekly alcohol consumption by sex		
Men		
Non-drinker	0.0	0.2
Under 1 unit	-0.2	-0.1
1-10 units	-0.4	-0.5
11-21 units	0.0	-0.1
22-35 units	0.0	0.0
36-50 units	0.2	0.1
51 + units	0.4	0.3
Women		
Non-drinker	0.1	0.3
Under 1 unit	-0.1	0.0
1-7 units	0.0	-0.3
8-14 units	-0.1	-0.1
15-25 units	0.0	0.0
26 -35 units	0.0	0.0
36+ units	0.0	0.0

Table D3   **Weighted versus unweighted data for years 1998 to 2002**

Household level variables

% of households		2002		Effect of weighting			
		Unweighted (a)	Weighted (b)	Weighted 1998 - Unweighted 1998	Weighted 2000 - Unweighted 2000	Weighted* 2001 - Unweighted 2001	Weighted 2002 - Unweighted 2002 (b-a)
Household size							
1 person		28.7	30.7	1.9	2.4	1.7	2.0
2 persons		36.6	34.9	-1.3	-1.6	-1.4	-1.7
3 persons		15.7	15.8	0.2	0.2	0.2	0.1
4 persons		13.0	12.7	-0.4	-0.5	-0.3	-0.3
5 persons		4.3	4.2	-0.1	-0.3	-0.1	-0.1
6 or more persons		1.7	1.7	-0.2	-0.1	-0.1	0.0
	Base	8620	24,529,000				
Number of adults							
1 adult		34.5	36.6	1.4	1.9	1.2	2.1
2 adults		51.0	48.5	-2.7	-2.6	-2.0	-2.5
3 adults		10.2	10.3	0.5	0.3	0.4	0.1
4 or more adults		4.2	4.7	0.7	0.5	0.3	0.5
	Base	8620	24,529,000				
Number of children							
No children		72.2	72.8	1.5	1.5	1.1	0.6
1 child		12.9	13.2	0.3	0.0	0.1	0.3
2 children		10.8	10.2	-1.0	-0.8	-0.7	-0.6
3 or more children		4.2	3.8	-0.8	-0.6	-0.4	-0.4
	Base	8620	24,529,000				
Household type							
1 person only		28.8	30.8	1.9	2.4	1.8	2.0
2 or more unrelated adults		2.2	2.5	0.4	0.3	0.2	0.3
Married couple, dependent children		18.4	17.8	-0.8	-0.8	-0.6	-0.6
Married couple, independent children		5.5	5.6	0.3	0.4	0.3	0.1
Married couple, no children		24.8	22.8	-2.0	-2.3	-1.7	-2.0
Lone parent, dependent children		7.2	7.2	-0.4	-0.5	-0.5	0.0
Lone parent, independent children		2.3	2.4	0.1	0.2	0.2	0.1
2 or more families (inc. same sex cohab)		3.1	3.1	-0.1	0.0	0.0	0.0
Cohabiting couple, with children		3.0	3.1	0.0	0.0	0.0	0.1
Cohabiting couple, no children		4.7	4.8	0.4	0.4	0.3	0.1
	Base	8598	24,464,000				
Tenure - harmonised							
Owns outright		30.3	28.7	-1.7	-1.9	-1.7	-1.6
Buying on mortgage		40.3	40.3	0.3	0.1	0.5	0.0
Rents from LA		13.2	13.9	0.1	0.5	0.2	0.7
Rents from HA		6.3	6.5	0.1	0.1	0.1	0.2
Rents privately - unfurnished/nk		7.3	7.7	0.4	0.5	0.4	0.4
Rents privately - furnished		2.5	3.0	0.6	0.6	0.6	0.5
	Base	8613	24,508,000				
Ownership of consumer durables							
Video		89.8	89.4	-0.4	-0.6	-0.2	-0.4
Freezer		95.7	95.4	-0.6	-0.6	-0.3	-0.3
Washing machine		93.9	93.4	-0.8	-0.7	-0.6	-0.5
Drier		55.6	54.5	-1.1	-1.4	-0.9	-1.1
Dishwasher		29.7	28.3	-1.0	-1.4	-1.0	-1.4
Microwave oven		87.7	87.4	-0.6	-0.6	-0.4	-0.3
Telephone		98.7	98.7	-0.3	-0.2	-0.1	0.0
CD player		82.6	82.5	0.8	0.0	0.4	-0.1
Home computer		54.2	53.8	0.4	-0.1	0.4	-0.4
	Base	8619	24,526,000				
Central heating		93.3	93.0	-0.3	-0.3	-0.2	-0.3
	Base	8619	24,526,000				
Car or van ownership							
No car or van		25.7	27.1	0.6	1.1	0.6	1.4
One car or van		44.9	45.4	0.6	0.4	0.4	0.5
Two cars or vans		23.9	22.2	-1.1	-1.3	-1.0	-1.7
Three or more cars or vans		5.5	5.3	-0.1	-0.2	0.0	-0.2
	Base	8620	24,529,000				

* Original 2001 weighting (based on LFS 2000 population estimates).

Table D4   **Weighted versus unweighted data for years 1998 to 2002**

Individual level variables

% of individuals	2002 Unweighted (a)	2002 Weighted (b)	Weighted 1998 - Unweighted 1998	Weighted 2000 - Unweighted 2000	Weighted* 2001 - Unweighted 2001	Weighted 2002 - Unweighted 2002 (b-a)
Limiting longstanding illness						
Male	20.2	19.8	-0.2	-0.3	-0.3	-0.4
Female	21.6	21.7	-0.1	0.0	-0.1	0.1
Total	21.0	20.8	-0.1	-0.1	-0.2	-0.2
Non-limiting longstanding illness						
Male	14.7	14.3	-0.4	-0.2	-0.2	-0.4
Female	13.5	13.3	0.0	-0.1	-0.1	-0.2
Total	14.1	13.8	-0.2	-0.1	-0.2	-0.3
No longstanding illness						
Male	65.0	65.8	0.6	0.5	0.6	0.8
Female	64.9	65.0	0.1	-0.1	0.2	0.1
Total	65.0	65.4	0.4	0.2	0.4	0.4
General health						
Good						
Male	63.1	63.2	0.6	0.1	0.0	0.1
Female	60.1	59.7	0.2	-0.3	-0.1	-0.4
Total	61.5	61.4	0.5	-0.1	0.0	-0.1
Fairly good						
Male	25.7	25.7	-0.2	0.0	0.0	0.0
Female	27.4	27.6	0.0	0.2	0.0	0.2
Total	26.6	26.7	-0.2	0.0	0.1	0.1
Not good						
Male	11.1	11.1	-0.4	0.0	-0.1	0.0
Female	12.6	12.7	-0.2	0.1	0.1	0.1
Total	11.9	11.9	-0.3	0.0	0.0	0.0
Restricted activity in the last 14 days						
Male	13.8	13.7	0.0	-0.1	0.1	-0.1
Female	15.7	15.8	0.0	0.1	0.0	0.1
Total	14.8	14.8	-0.1	-0.1	0.1	0.0
Cigarette smoking by sex						
Men						
Current cigarette smokers	25.8	26.8	1.4	1.2	1.2	1.0
Ex-regular cigarette smokers	29.0	27.5	-2.0	-2.0	-1.4	-1.5
Never or (only occasionally) smoked	45.1	45.6	0.6	0.8	0.1	0.5
Women						
Current cigarette smokers	24.6	25.0	0.4	0.2	0.3	0.4
Ex-regular cigarette smokers	21.1	20.9	-0.3	-0.3	0.0	-0.2
Never or (only occasionally) smoked	54.3	54.1	-0.1	0.1	-0.3	-0.2
Total						
Current cigarette smokers	25.2	25.9	0.8	0.8	0.8	0.7
Ex-regular cigarette smokers	24.7	24.0	-0.9	-1.0	-0.6	-0.7
Never or (only occasionally) smoked	50.1	50.2	0.1	0.3	-0.2	0.1
Weekly alcohol consumption by sex						
Men						
Non-drinker	9.0	9.2	-0.1	0.3	0.0	0.2
Under 1 unit	8.2	8.0	-0.2	-0.2	-0.2	-0.2
1-10 units	34.4	34.0	-0.6	-0.6	-0.4	-0.4
11-21 units	22.7	22.6	0.1	-0.1	0.0	-0.1
22-35 units	13.5	13.6	0.1	0.2	0.0	0.1
36-50 units	6.0	6.2	0.2	0.1	0.2	0.2
51 + units	6.2	6.6	0.5	0.3	0.4	0.4
Women						
Non-drinker	14.5	14.8	0.0	0.4	0.1	0.3
Under 1 unit	16.5	16.4	0.0	-0.1	-0.1	-0.1
1-7 units	37.8	37.5	-0.1	-0.3	0.0	-0.3
8-14 units	15.2	15.0	0.0	-0.1	-0.1	-0.2
15-25 units	9.8	9.8	0.1	0.1	0.0	0.0
26 -35 units	3.2	3.3	0.0	0.0	0.0	0.1
36+ units	3.1	3.2	0.1	0.0	0.0	0.1

* Original 2001 weighting (based on the LFS 2000 population estimates).

Appendix E
Household and Individual Questionnaires
General Household Survey 2002/03
Household Questionnaire

Areacode	Information already entered
Address	Information already entered
	1..30
HHold	Information already entered
	1..4
StartDat	ENTER DATE INTERVIEW WITH THIS HOUSEHOLD WAS STARTED

DateChk IS THIS...

The first time you've opened this
questionnaire 1
or the second or later time? 2
EMERGENCY CODE IF
COMPUTER'S DATE IS WRONG
AT LATER CHECK 5

IntEdit CODE WHETHER THIS IS THE
INTERVIEW STAGE, A PROXY
CONVERSION OR THE EDIT STAGE

Interview 1
Proxy Conversion by telephone
(TELEPHONE INTERVIEW UNIT
ONLY) 2
OFFICE ONLY - EDIT 7

HOUSEHOLD INFORMATION

Information to be collected for all persons in all households

Name Who normally lives at this address?

RECORD THE NAME (OR A UNIQUE IDENTIFIER)
FOR HOH, THEN A NAME / IDENTIFIER FOR EACH
MEMBER OF THE HOUSEHOLD

ENTER TEXT OF AT MOST 12 CHARACTERS → Q2

Sex Male .... 1 ⎤
Female .... 2 ⎦ → Q3

Birth What is your date of birth?

FOR DAY NOT GIVEN...........ENTER 15 FOR DAY.
FOR MONTH NOT GIVEN.....ENTER 6
FOR MONTH   → See Q4

AgeIf **Ask those who did not know, or refused to give their date of birth (Birth = DK OR REFUSAL)**

What was your age last birthday?

98 or more = CODE 97

0..97   → See Q5

MarStat **Ask if respondent is aged 16 or over (DVAge > 15)**

ASK OR RECORD, CODE FIRST THAT APPLIES

Are you

single, that is, never married? .... 1 → See Q6
married and living with your
husband/wife? .... 2 → See Q7
married and separated from your
husband/wife? .... 3 ⎤
divorced? .... 4 ⎥ → See Q6
or widowed? .... 5 ⎦

6. LiveWith **Ask if there is more than one person in the household AND respondent is aged 16 or over AND is single, separated, divorced or widowed (Household size > 1 & DVAge > 15 & Marstat = 1, 3, 4 or 5)**

ASK OR RECORD

May I just check, are you living with someone in the household as a couple?

Yes .... 1 ⎤
No .... 2 ⎥ → See Q7
SPONTANEOUS ONLY -   ⎥
same sex couple .... 3 ⎦

7. Hhldr **Ask if there is more than one person in the household, AND the respondent is aged 16 or over (Household size > 1 & DVAge > 15)**

In whose name is the accommodation owned or rented?
ASK OR RECORD

This person alone .... 1 ⎤
This person jointly .... 3 ⎥ → See Q8
NOT owner/renter .... 5 ⎦

8. HiHNum **Ask if there is more than one person in the household, AND the accommodation is jointly owned (Household size > 1 & Hhldr = 3)**

You have told me that...jointly own or rent the accommodation. Which of you/ who has the highest income (from earnings, benefits, pensions and any other sources)?

INTERVIEWER: THESE ARE THE JOINT HOUSEHOLDERS

ENTER PERSON NUMBER - IF TWO OR MORE HAVE SAME INCOME, ENTER 15
1..14   → See Q9

9. JntEldA **Ask if there is more than one person in the household, AND the joint householders have the same income (Household size > 1 & HiHNum = 15)**

ENTER PERSON NUMBER OF THE ELDEST JOINT HOUSEHOLDER FROM THOSE WITH THE SAME HIGHEST INCOME

ASK OR RECORD

1..14   →See Q10

10. JntEldB   Ask if household size is greater than one, AND the joint householders do not know, or refuse to say who has the greatest income *(Household size > 1 & HiHNum = Don't know or Refusal)*

ENTER PERSON NUMBER OF THE ELDEST JOINT HOUSEHOLDER

ASK OR RECORD

1..14   → Q11

11. HRPnum   Ask all households

PERSON NUMBER OF HRP.
(Computed in Blaise)   →See Q12

12. HRPprtnr   Ask if the HRP is married or cohabiting *(HRPnum = 1..14 & Marstat = 2 or LiveWith = 1)*

THE HRP IS (HRP's NAME)

ENTER THE PERSON NUMBER OF THE HRP's SPOUSE/PARTNER
NO SPOUSE/PARTNER = 15

1..15   → Q13

13. R   Ask all households

I would now like to ask how the people in your household are related to each other

CODE RELATIONSHIP - ... IS ...'S ....

Spouse...1
Cohabitee 2
Son/daughter (inc. adopted) 3
Step-son/daughter 4
Foster child 5
Son- in -law/daughter - in -law 6
Parent/Guardian.....................................7
Step-parent 8
Foster parent..9   → Q14
Parent- In - law......................................10
Brother/sister (inc. adopted)..................11
Step-brother/sister 12
Foster brother/sister.............................13
Brother/sister-in-law 14
Grand-child 15
Grand-parent...16
Other relative 17
Other non-relative 18

ACCOMMODATION TYPE

14. IntroAcc   The next section looks at the standard of people's housing.

15. Accom   All households

IS THE HOUSEHOLD'S ACCOMMODATION:

N.B. MUST BE SPACE USED BY HOUSEHOLD

a house or bungalow.............................1 → Q16
a flat or maisonette ....2 → Q17
a room/rooms..3 → Q19
or something else? ....4 → Q18

16. HseType   Ask if respondents live in a house or bungalow *(Accom = 1)*

IS THE HOUSE/BUNGALOW:

detached 1
semi-detached ....2 → Q2
or terraced/end of terrace? 3

17. FltTyp   Ask if respondents live in a flat or maisonette *(Accom = 2)*

IS THE FLAT/MAISONETTE:

a purpose-built block..............................1
a converted house/some other
kind of building? ....2 → Q1

18. AccOth   Ask if respondents said their accommodation wa 'something else' *(Accom = 4)*

IS THE ACCOMMODATION A:

caravan, mobile home or houseboat......1 → Q2
or some other kind of accommodation?. 2 → Q2

19. Storey   Ask if respondents live in a flat, maisonette, OR a room or rooms *(Accom = 2 or 3)*

What is the floor level of the main living part of the accommodation?

ASK OR RECORD.

Basement/semi-basement 1
Ground floor/street level........................2
1st floor 3
2nd floor..4 → Q2
3rd floor...5
4th to 9th floor.......................................6
10th floor or higher................................7

20. HasLift   Ask if respondents live in a flat, maisonette, OR a room or rooms *(Accom = 2 or 3)*

INTERVIEWER CODE: IS THERE A LIFT?

Yes ....1 → Q2
No 2

21. DateBlt   Ask all households, EXCEPT those living in a caravan, mobile house or houseboat *(AccOth ≠1)*

When was this building first built?

PROMPT IF NECESSARY - IF DK CODE YOUR ESTIMATE

before 1919...1
between 1919 and 1944 2
between 1945 and 1964 ....3 →See Q2
between 1965 and 1984 4
1985 or later..5
DK but after 1944..................................6

22. ShareH   Ask if living in a house, bungalow OR a converted flat/maisonette OR 'something else' *(Accom = 1, or FltTyp = 2)*

INTERVIEWER ASK OR RECORD

May I just check, does anyone else live in this building apart from the people in your household?

(I.E. IS THERE ANYONE ELSE IN THE BUILDING WITH WHOM THE HOUSEHOLD COULD SHARE ROOMS OR FACILITIES?)

Yes .... 1 ⎤→
No .... 2 ⎦   Q23

3. ShareE   INTERVIEWER ASK OR RECORD

Is there any empty living accommodation in this building outside your household's accommodation?

Yes .... 1 ⎤→ SeeQ24
No .... 2 ⎦

4. Share2   **Ask if other people live in the building, apart from the household, OR respondents live in a flat, maisonette or room(s) (ShareH = 1 or Accom = 2 or 3)**

Does your household (do you) have the whole accommodation to yourselves (yourself) or do you share any of it with someone outside your household?

Have the whole accommodation............ 1 ⎤
Share with someone else outside   ⎬→See Q25
the household .... 2 ⎦

5. Share3   **Ask if there is empty living accommodation in the building outside the household's accommodation, AND the accommodation is not shared with someone outside the household (ShareE = 1 & Share2 ≠ 2)**

If all the empty accommodation in this building were occupied, would your household (you) have to share any part of your accommodation with anyone who had moved in?

Yes .... 1 ⎤→See Q26
No .... 2 ⎦

6. Rooms1   **Ask if household shares part of its accommodation with someone else outside the household OR would have to share part of the accommodation if someone moved in to an empty part of the accommodation (Share2 = 2 or Share3 = 1)**

I want to ask you about all the rooms you have in your household's accommodation. Please include any rooms you sublet to other people and any rooms you share with people who are not in your household (or would share if someone moved into the empty accommodation).   →See Q27

7. Rooms2   **Ask if household does not share part of the accommodation (Share2 ≠ 2 OR Share3 ≠ 1)**

I want to ask you about all the rooms you have in your household's accommodation (including any rooms you sublet to other people). (How many of the following rooms do you have in this house/flat ...)   → Q28

28. Bedrooms   **Ask all households**

How many bedrooms do you have?

INCLUDE BEDSITTERS, BOXROOMS, ATTIC BEDROOMS

0..20   →See Q29

29. BedCook   **Ask those who have at least one bedroom (Bedrooms > 0)**

Are any of them used by your household for cooking in - like a bedsitter for example?

Yes .... 1 ⎤→
No .... 2 ⎦   Q30

30. KitOver   **Ask all households**

How many kitchens over 6.5 feet wide do you have?

NARROWEST SIDE MUST BE AT LEAST 6.5 FEET FROM WALL TO WALL

0..20   →   Q31

31. KitUnder   **How many kitchens under 6.5 feet do you have?**

0..20   →See Q32

32. ShareKit   **Ask those who have a kitchen AND share accommodation (KitOver > 0 or KitUnder > 0) AND (Share2 =2 or Share3=1)**

Do you share the kitchen with any other household?

Yes .... 1 ⎤→
No .... 2 ⎦   Q 33

33. Living   **Ask all households**

How many LIVING ROOMS do you have?

INCLUDE DINING ROOMS, SUNLOUNGE OR CONSERVATORY USED ALL YEAR ROUND.

0..20   →   Q34

34. Bathrooms   How many BATHROOMS do you have with PLUMBED IN BATH/SHOWER?

0..20   →   Q35

35. Utility   How many UTILITY and other rooms do you have?

0..20   →   Q36

36. GHSCentH   ASK OR RECORD

Do you have any form of central heating, including electric storage heaters, in your (part of the) accommodation?

Yes .... 1 →   Q37
No .... 2 →   Q38

37. GHSCHFuel **Ask if the household has some form of central heating (GHSCentH = 1)**

Which type of fuel does it use?

CODE MAIN METHOD ONLY PROBE 'Hot Air' FOR FUEL

Solid fuel: incl. coal, coke, wood, peat ... 1 ⎤
Electricity: storage heaters..................... 2 ⎥
Electricity: other (incl. oil filled radiators) 3 ⎥ → Q38
Gas/Calor gas .... 4 ⎥
Oil .... 5 ⎥
Other... 6 ⎦

CONSUMER DURABLES

38. IntroDur **Ask all households**

Now I'd like to ask you about various household items you may have - this gives us an indication of how living standards are changing.

39. HasDur Does your household have any of the following items in your (part of the)accommodation?

40. TVcol ...Colour TV set?

Yes .... 1 → Q41
No .... 2 → Q44

41. Tvnum **Ask those who have a colour TV set (Tvcol = 1)**

How many colour TVs do you have?

PROMPT AS NECESSARY TO PROBE FOR NUMBER OF TVS

1..7   →   Q42

42. UseColTV **Ask those who have a colour TV set (Tvcol = 1)**

ASK OR RECORD

Is this/are any of these colour TV set(s) currently in use?

Yes .... 1 → Q44
No .... 2 → Q43

43. BrkColTV **Ask if no colour TV sets currently in use (UseColTV = 2)**

Is this/are any of these colour TV set(s) broken but due to be repaired within 7 days?

Yes .... 1 ⎤ → Q44
No .... 2 ⎦

44. TVbw **Ask all households**

Black and white TV set?

Yes .... 1 → Q45
No .... 2 → Q48

45. Tvbwnum **Ask those who have a black and white TV set (Tvbw = 1)**

How many black and white TVs do you have?

PROMPT AS NECESSARY TO PROBE FOR NUMBER OF TVS

1..7   →See Q4

46. UseBwTV **Ask if NO colour TV set in use and none intended for repair AND has black and white TV (Tvcol = 2 or BrkColTV = 2) & (TvBw = 1)**

ASK OR RECORD

Is this/are any of these black and white TV set(s) currently in use?

Yes .... 1 → Q4
No .... 2 → Q4

47. BrkBwTV **If no black and white TV sets currently in use (UseBwTV = 2)**

Is this/are any of these black and white TV set(s) broken but due to be repaired within 7 days?

Yes .... 1 ⎤ Q4
No .... 2 ⎦

48. Satell **Ask all households**

Satellite TV?

Yes .... 1 ⎤ Q4
No .... 2 ⎦

49. Cable Cable TV?

Yes .... 1 ⎤ Q5
No .... 2 ⎦

50. Digital A digital TV receiver?

Yes .... 1 ⎤ Q5
No .... 2 ⎦

51. Video Video recorder?

Yes .... 1 ⎤ Q5
No .... 2 ⎦

52. Freezer Deep freezer or fridge freezer?

EXCLUDE FRIDGE ONLY

Yes .... 1 ⎤ Q5
No .... 2 ⎦

53. WashMach Washing machine?

Yes .... 1 ⎤ Q5
No .... 2 ⎦

54. Drier Tumble drier?

IF COMBINED WASHING MACHINE AND TUMBLE DRIER, CODE 1 FOR BOTH

Yes .... 1 ⎤
No .... 2 ⎦ → Q55

5. DishWash   Dish washer?

Yes .... 1 ⎤
No .... 2 ⎦ → Q56

6. MicroWve   Microwave oven?

Yes .... 1 ⎤
No .... 2 ⎦ → Q57

7. Telephon   Telephone?

SHARED TELEPHONES LOCATED IN PUBLIC
HALLWAYS TO BE INCLUDED ONLY IF THIS
HOUSEHOLD IS RESPONSIBLE FOR PAYING THE
ACCOUNT

Yes, fixed telephone .... 1 ⎤
Yes, mobile telephone .... 2 ⎥
Yes, fixed and mobile telephone............. 3 ⎥ → Q58
No .... 4 ⎦

8. CDplay   Compact disc (CD) player?

Yes .... 1 ⎤
No .... 2 ⎦ → Q59

9. DVD   DVD player?

Yes .... 1 → Q60
No .... 2 → Q61

10. DVDHow   Ask if DVD = 1

Can I just check, is that a stand-alone DVD player, or
do you have a DVD player as part of a computer or
games console such as Play Station 2?

CODE ALL THAT APPLY

Stand-alone DVD player........................ 1 ⎤
DVD as part of a computer .... 2 ⎥ → Q61
DVD as part of a games console .... 3 ⎦

11. Computer   Home computer?

EXCLUDE: VIDEO GAMES

Yes .... 1 ⎤
No .... 2 ⎦ → Q62

12. Internet   Does your household have access to the internet at
home?

Yes .... 1 → Q63
No .... 2 → Q64

13. Access   **Ask if has home access to the internet (Internet =
1)**

How does your household access the internet from
home?

CODE ALL THAT APPLY

Home computer .... 1 ⎤
Digital television .... 2 ⎥
Mobile phone .... 3 ⎥ → Q64
Games console .... 4 ⎥
Other.. 5 ⎦

64. UseVcl   **Ask all households**

Do you, or any members of your household, at
present own or have continuous use of any motor
vehicles?

INCLUDE COMPANY CARS (IF AVAILABLE FOR
PRIVATE USE)

Yes .... 1 → Q65
No .... 2 → Q68

65. TypeVcl   **Ask if the household has use of any motor
vehicles (If UseVcl =1)**

FOR EACH VEHICLE IN TURN:
I would now like to ask about the (Nth) vehicle. Is it...

CAR INCLUDES MINIBUSES, MOTOR CARAVANS,
'PEOPLE CARRIERS' AND 4- WHEEL DRIVE
PASSENGER VEHICLES

LIGHT VAN INCLUDES PICKUPS AND THOSE
4-WHEEL DRIVE VEHICLES, LAND ROVERS AND
JEEPS THAT DO NOT HAVE SIDE WINDOWS
BEHIND THE DRIVER

a car.. 1 ⎤
a light van.. 2 ⎥ → Q66
a motor cycle .... 3 ⎥
or some other motor vehicle? .... 4 ⎦

66. PrivVcl   FOR EACH VEHICLE IN TURN:
Is the [vehicle]...

privately owned .... 1 ⎤ → Q67
or is it a company vehicle?..................... 2 ⎦

67. AnyMore   Do (any of) you at present own or have continuous
use of any more motor vehicles?

INCLUDE COMPANY CARS - UNLESS NO PRIVATE
USE ALLOWED

Yes .... 1 ⎤ → Q68
No .... 2 ⎦

TENURE

68. Ten1   **Ask all households**

In which of these ways do you occupy this
accommodation?

SHOW CARD A
MAKE SURE ANSWER APPLIES TO HRP

Own outright .... 1 ⎤
Buying it with the help of a mortgage ⎥ → Q77
or loan... 2 ⎦

Pay part rent and part mortgage
(shared ownership) .... 3 → Q73
Rent it .... 4 ⎤
Live here rent-free (including ⎥→ Q69
rent-free in relative's/friend's property; ⎥
excluding squatting) .... 5 → ⎦
Squatting.. 6 → Q77

69. Tied   **Ask if household rents the accommodation, or lives there rent-free (Ten1 = 4 or 5)**

Does the accommodation go with the job of anyone in the household?

Yes .... 1 ⎤→
No .... 2 ⎦ Q70

70. LLord   Who is your landlord?...

CODE FIRST THAT APPLIES

the local authority/council/New Town
Development/Scottish Homes................ 1 ⎤
a housing association or co-operative ⎥
or charitable trust .... 2 ⎥
employer (organisation) of a ⎥
household member .... 3 ⎥
another organisation .... 4 → Q71
relative/friend (before you lived here) ⎥
of a household member .... 5 ⎥
employer (individual) of a household ⎥
member.. 6 ⎥
another individual private landlord? .... 7 ⎦

71. Furn   Is the accommodation provided: ...

furnished .... 1 ⎤
partly furnished (e.g. carpets and ⎥→See Q72
curtains only) .... 2 ⎥
or unfurnished?...................................... 3 ⎦

72. LandLive   **Ask if rented from an individual (Llord = 5, 6 or 7)**

Does the landlord live in this building?

Yes .... 1 ⎤→See Q73
No .... 2 ⎦

73. HB   **Ask if 'shared ownership' or 'rents' or 'rents free' (Ten1 = 3, 4 or 5)**

Some people qualify for Housing Benefit, that is a rent rebate or allowance.

Are you (or HRP) receiving Housing Benefit from your local authority or local Social Security office?

Yes .... 1 →See Q76
No .... 2 → Q74

74. HbWait   **Ask if not receiving Housing Benefit (HB = 2)**

Are you (or HRP) waiting to receive Housing Benefit or to hear the outcome of a claim?

Yes .... 1 →See Q76
No .... 2 → Q75

75. HbChk   **Ask if not waiting to receive Housing Benefit or to hear the outcome of a claim (HBWait = 2)**

May I just check, does the local authority or local Social Security office pay any part of your rent?

Yes .... 1 ⎤→See Q7
No .... 2 ⎦

76. HbOthr   **Ask if there is someone aged 16 and over, apart from HRP and partner, in the household**

Is anyone (else) in the household receiving a rent rebate, a rent allowance or Housing Benefit?

Yes .... 1 ⎤→
No .... 2 ⎦ Q7

MIGRATION

77. Reslen   **Ask All**

How many years have you /has(...) lived at this address?

IF UNDER 1, CODE AS 0

0..97   →See Q7

78. Hmnths   **Ask if respondent has lived at the address for less than a year (Reslen = 0)**

How many months have you/has (...) lived here?

1..12   →See Q7

79. Nmoves   **Ask if respondent has lived at the address for less than five years (Reslen < 5 years)**

How many moves have you /has (...) made in the last 5 years, not counting moves between places outside Great Britain?

0..97   → Q8

80. Cry1   **All persons**

In what country were you/was (...) born? ...

UK, British... 1 → Q8
Irish Republic .... 6 ⎤
Jamaica...26 ⎥
Bangladesh...33 → Q8
India ....34 ⎥
Pakistan ....56 ⎦
Other...59 → Q8

81. CrySpec   **Ask if country of birth was 'other' (Cry1 = 59)**

TYPE IN COUNTRY

ENTER TEXT OF AT MOST 40 CHARACTERS→ Q8

82. CryCode   **Ask if country of birth was 'other' (Cry1 = 59)**

CHOOSE COUNTRY FROM CODING FRAME

	1..116	→	Q83

8. Arruk **Ask if not born in the UK** *(Cry1 ≠ 1)*

In what year did you (...) first arrive in the United Kingdom? ...

ENTER IN 4 DIGIT FORMAT E.G.: 2000

1900..2005   →   Q84

4. FathCob **All persons**

ASK OR RECORD

In what country was your / (...'s) father born?

UK, British.. 1 ⌉	
Irish Republic 6	
Jamaica...26	
Bangladesh ....33 → Q87	
India 34	
Pakistan ....56 ⌋	
Other...59 → Q85	

5. CrySpec1 **Ask if father's country of birth was 'other'** *(FathCob = 59)*

TYPE IN COUNTRY

ENTER TEXT OF AT MOST 40 CHARACTERS → Q86

6. CryCode1 **Ask if father's country of birth was 'other'** *(FathCob = 59)*

CHOOSE COUNTRY FROM CODING FRAME

1..116   →   Q87

7. MothCob **Ask all persons**

ASK OR RECORD

In what country was your/ (...'s) mother born?

UK, British.. 1 ⌉	
Irish Republic 6	
Jamaica...26	
Bangladesh ....33 → Q90	
India 34	
Pakistan ....56 ⌋	
Other...59 → Q88	

8. CrySpec2 **Ask if mother's country of birth was 'other'** *(MothCob = 59)*

TYPE IN COUNTRY

ENTER TEXT OF AT MOST 40 CHARACTERS → Q89

9. CryCode2 **Ask if mother's country of birth was 'other'** *(MothCob = 59)*

CHOOSE COUNTRY FROM CODING FRAME

1..116   →   Q90

90. Nation [*] **All persons**

SHOWCARD NAT(E) in England, NAT(S) in Scotland, NAT(W) in Wales

What do you consider your national identity to be? Please choose your answer from this card, choose as many or as few as apply.

English .... 1 ⌉	
Scottish 2	
Welsh .... 3 → Q92	
Irish 4	
British .... 5 ⌋	
Other answer .... 6 → Q91	

91. NatSpec [*] **If answered other** *(Nation = 6)*

How would you describe your national identity?

ENTER DESCRIPTION OF NATIONAL IDENTITY→Q92

92. Ethnic [*] **All persons**

SHOW CARD B

To which of these ethnic groups do you consider you belong?

White - British... 1 ⌉	
White - Any other White background...... 2	
Mixed - White and Black Caribbean....... 3	
Mixed - White and Black African 4	
Mixed - White and Asian 5	
Mixed - Any other Mixed background...... 6	
Asian or Asian British - Indian................ 7	
Asian or Asian British - Pakistani 8	
Asian or Asian British - Bangladeshi...... 9 →See Q93	
Asian or Asian British - Any other Asian background 10	
Black or Black British - Caribbean 11	
Black or Black British - African..............12	
Black or Black British - Any other Black background 13	
Chinese...14	
Any other ethnic group..........................15 ⌋	

93. Ethdes **Ask those who describe themselves as:**
Any other White background
Any other Mixed background
Any other Asian background
Any other Black background
Any other ethnic group *(Ethnic = 2, 6, 10, 13 or 15)*

Please can you describe your ethnic group
ENTER DESCRIPTION OF ETHNIC GROUP

END OF HOUSEHOLD QUESTIONNAIRE

General Houshold Survey 2002/03
Individual Questionnaire

1. Iswitch   **Ask this section of all adults**

THIS IS WHERE YOU START RECORDING
ANSWERS FOR INDIVIDUALS
DO YOU WANT TO RECORD ANSWERS FOR
(name) NOW OR LATER?

Yes, now 1
Later... 2
or is there no interview with this person? 3

2. PersProx   **Ask if answers are to be recorded now (Iswitch = 1)**

INTERVIEWER: IS THE INTERVIEW ABOUT (name)
BEING GIVEN:

In person 1
or by someone else? 2

3. ProxyNum   **Ask if answers are to be recorded now, but are being answered by someone else (Iswitch = 1 & PersProx = 2)**

ENTER PERSON NUMBER OF PERSON GIVING
THE INFORMATION

1..14

EMPLOYMENT

1. Wrking   **Ask this section of all adults**

Did you do any paid work in the 7 days ending Sunday
the (n), either as an employee or as self-employed?

Yes .... 1 →   Q15
No .... 2 → See Q2

2. SchemeET   **Ask if respondent is not in paid work and is a man aged 16-64, or a woman aged 16-62 (Wrking = 2 & man aged 16-64 or woman aged 16-62)**

Were you on a government scheme for employment
training?

Yes .... 1 →   Q3
No .... 2 → See Q4

3. Trn   **Ask those on a government scheme for employment training (SchemeET = Yes)**

Last week were you ...

CODE FIRST THAT APPLIES

with an employer, or on a project
providing work experience
or practical training? .... 1 ⎤
or at a college or training centre? .... 2 ⎦ → Q15

4. JbAway   **Ask if not in paid work AND not on a government scheme for employment training (Wrking = 2 & (SchemeET = 2 or not asked SchemeET because not in the age bracket asked))**

Did you have a job or business that you were away
from?

Yes .... 1 →   Q1
No .... 2 ⎤
Waiting to take up a new job/   ⎦ →   C
business already obtained 3

5. OwnBus   **Ask if not in paid work AND not on a government scheme for employment training AND not away from a job (JbAway = 2 or 3)**

Did you do any unpaid work in that week for any
business that you own?

Yes .... 1 → See C
No .... 2 →   C

6. RelBus   **Ask if the respondent did not do any unpaid work for a business that they own (OwnBus = 2)**

...or that a relative owns?

Yes .... 1 → See C
No .... 2 → See C

7. Looked   **Ask if not in paid work AND not on a government scheme for employment training AND not doing unpaid work (Wrking = 2 & (SchemeET = 2 or not asked SchemeET because not in the age bracket asked) & (RelBus = 2 OR JbAway = 2))**

Thinking of the 4 weeks ending Sunday the (date last
Sunday), were you looking for any kind of paid work
or government training scheme at any time in those
weeks?

Yes .... 1 →   C
No .... 2 →   Q1
Waiting to take up a new job or
business already obtained .... 3 →   C

8. LKTime   **Ask if looking for paid work OR waiting to take up a new job or business already obtained (Looked 1 or 3 OR JbAway = 3)**

How long have you been/were you looking for paid
work/a place on a government training scheme?

Not yet started .... 1 ⎤
Less than 1 month .... 2 ⎪
1 month but less than 3 months............. 3 ⎬ → See C
3 months but less than 6 months.......... 4 ⎪
6 months but less than 12 months .... 5 ⎪
12 months or more............................... 6 ⎦

9. StartJ   **Ask if looking for paid work OR waiting to take up a new job or business already obtained (Looked 1 or 3 OR JbAway = 3)**

If a job or a place on a government scheme had been
available in the week ending Sunday the (n), would
you have been able to start within 2 weeks?

Yes .... 1 →See Q1
No .... 2 →   Q1

10. Yinact   **Ask if not looking for paid work, and would not be able to start work or training within 2 weeks (Looked = 2 or StartJ = 2)**

What was the main reason you did not seek any work in the last 4 weeks/would not be able to start in the next 2 weeks?

Student 1	
Looking after the family/home 2	
Temporarily sick or injured 3	→See Q11
Long-term sick or disabled 4	
Retired from paid work 5	
None of these 6	

1. Everwk   Ask if not in paid work

Have you ever had a paid job, apart from casual or holiday work?

Yes .... 1 → Q12
No .... 2 →See Q13

2. DtJbL   Ask if not in paid work, but has worked before (Everwk = 1)

When did you leave your last PAID job?

FOR DAY NOT GIVEN .... ENTER 15 FOR DAY
FOR MONTH NOT GIVEN .... ENTER 6 FOR MONTH

DATE   →See Q13

3. WantaJob   Ask if respondent is aged 16-68 and male, or 16-64 and female, and is not working because is a student, is looking after the family/home, is retired, or is at a college or training centre ((DVAge = 16-68 & Sex =1) or (DVAge = 16-64 & Sex = 2) & YInAct = 1, 2, 5 or 6 or Trn = 2)

Even though you were not looking for work (last week) would you like to have a regular paid job at the moment - either a full or part-time job?

Yes .... 1 → Q14
No .... 2 →See Q15

4. NablStrt   Ask if respondent would like a job (WantaJob = 1)

If a job or a place on a government scheme had been available last week, would you have been able to start within 2 weeks?

Yes 1
No .... 2 →See Q15

5. IndD   Ask those who are in current employment or have had a job in the past

CURRENT OR LAST JOB

What did the firm/organisation you worked for mainly make or do (at the place where you worked)?

DESCRIBE FULLY - PROBE MANUFACTURING or PROCESSING or DISTRIBUTING ETC. AND MAIN GOODS PRODUCED, MATERIALS USED WHOLESALE or RETAIL ETC.

ENTER TEXT AT MOST 80 CHARACTERS →   Q16

16. OccT   JOBTITLE CURRENT OR LAST JOB

What was your (main) job (in the week ending Sunday the (n))?

ENTER TEXT AT MOST 30 CHARACTERS →   Q17

17. OccD   CURRENT OR LAST JOB

What did you mainly do in your job?
CHECK SPECIAL QUALIFICATIONS/TRAINING NEEDED TO DO THE JOB

ENTER TEXT AT MOST 80 CHARACTERS →   Q18

**18. Stat   ** Were you working as an employee or were you self-employed?

Employee .... 1 → Q19
Self-employed .... 2 → Q22

19. Svise   Ask if employee (Stat = 1)

In your job, did you have formal responsibility for supervising the work of other employees?
DO NOT INCLUDE PEOPLE WHO ONLY SUPERVISE:
- children, e.g. teachers, nannies, childminders
- animals
- security of buildings, e.g. caretakers, security guards

Yes 1
No .... 2 → Q21

20. *Question number not used*

**21. NEmplee   ** How many people worked for your employer at the place where you worked?

1-2 1	
3-24 2	
25-99 3	
100-499 4	
500-999 5	→ Q24
1000 or more 6	
DK, but less than 25 7	
DK, but between 25 and 499 ... 8	
DK, but 500 or more 9	

22. Solo   Ask if self-employed (Stat = 2)

Were you working on your own or did you have employees?

on own/with partner(s) but no employees . 1 → Q24
with employees .... 2 → Q23

23. SNEmplee   Ask if self-employed with employees (Solo = 2)

How many people did you employ at the place where you worked?

1-5 1	
6-24 2	
25-499 3	→ Q24
500 or more 4	
DK but has/had employees 5	

24. FtPtWk   **Ask those who are in current employment or have had a job in the past**

In your (main) job were you working:

full time .... 1 ⎤ See Qs
or part time?... 2 ⎦ 24 &25

25. EmpStY   **Ask if employee (Stat = 1)**

In which year did you start working continuously for your current employer?

1900..2005   →See Q27

26. SempStY   **Ask if self-employed (Stat = 2)**

In which year did you start working continuously as a self-employed person?

1900..2005   →See Q27

27. JobstM   **If less than or equal to 8 years since started working continuously for current employer/ as a self-employed person (EmpStY ≤ 8 less than the present date or SEmpStY ≤ 8 less than the present date)**

and which month in (YEAR) was that?

0..12   →See Q28

28. Tothrs   **Ask all working (Working = 1 or JbAway = 1 or SchemeET = 1)**

How many hours a week do you usually work in your (main) job/business? Please exclude mealbreaks but include any paid or unpaid overtime that you usually work.

HOURS IN MAIN JOB ONLY

97 OR MORE = 97

0.00..99.00   →See Q29

29. UnPaidHr   **Ask if did unpaid work for a business (OwnBus = 1 or RelBus = 1)**

Thinking of the business that you did unpaid work for how many hours unpaid work did you do for that business in the 7 days ending last Sunday?

1..97   →   Q30

30. UnPaidHm   Did you do this work mainly...

somewhere quite separate from
home, ... 1 ⎤
in different places using home as
a base, ... 2 ⎥
or in your own home or in the same   ⎬→Pensions
grounds or buildings as your home?, ... 3 ⎥
SPONTANEOUSLY ONLY: some
days at home, other days somewhere
quite separate from home .... 4 ⎦

PENSIONS

The whole section on pensions (apart from the last question) is only asked of those in paid work (including those temporarily away from job or on a government scheme), but excluding unpaid family workers. ((Wrking = 1 OR JbAway = 1 O SchemeET = 1) & (OwnBus = 2 & RelBus = 2)) The routing instructions above each question apply only to those who meet the above criteria.

1. PenSchm   **If employee or on a government scheme (Stat = 1 or SchemeET = 1), others see Q14**

(Thinking now of your present job,) some people (will receive a pension from their employer when they retire, as well as the state pension.
Does your present employer run an occupational pension scheme or superannuation scheme for any employees?
INCLUDE CONTRIBUTORY AND NON-CONTRIBUTORY SCHEMES
EXCLUDE EMPLOYER SPONSORED GROUP PERSONAL PENSION AND STAKEHOLDER PENSIONS

Yes .... 1 →   C
No .... 2 →   C

2. Eligible   **Ask if employer runs an occupational pension scheme (PenSchm = 1)**

Are you eligible to belong to your employer's occupational pension scheme?

Yes .... 1 →   C
No .... 2 →   C

3. EmPenShm   **Ask if eligible for employer's pension scheme (Eligible = 1)**

Do you belong to your employer's occupational pension scheme?

Yes .... 1 ⎤
No .... 2 ⎦→ C

4. PschPoss   **Ask if did not know or refused to say whether the employer offered an occupational pension scheme, or whether they were eligible, or whether they belonged to one (PenSchm or Eligible or EmPenShm = DK / refusal)**

So do you think it's possible that you belong to an occupational pension scheme run by your employer, do you definitely not belong to one?

Possibly belongs .... 1 ⎤
Definitely not .... 2 ⎦→ C

5. PersPnt1   **Ask if employee OR (under pensionable age and not self-employed) - this is to select those who may have answered don't know, or refused to answer Stat (Stat = 1 OR (under pensionable age & Stat ≠ 2))**

INTERVIEWER - INTRODUCE IF NECESSARY.
Now I would like to ask you about personal pensions and stakeholder pensions (rather than employers' occupational pension schemes).

PersPens People can now save for retirement by contracting out of the State Second Pension (formally known as SERPS) and arranging their own personal pension or stakeholder pension. Part of your National Insurance contributions are then repaid into your chosen pension plan by the Inland Revenue (or formerly by the DSS).

Do you at present have any such arrangements?

Yes .... 1 →   Q7
No .... 2 →   Q10

OutSERPS If contracted out of SSP *(PersPens = 1)*

Is the arrangement you use to contract out of SSP...

CODE ONE ONLY (most recent arrangement)

a personal pension .... 1 →   Q8
or a stakeholder pension?...................... 2 →   Q9

PersCont Ask if contracted out of SSP with a personal pension *(OutSERPS = 1)*

Do you make any extra contributions over and above any rebated National Insurance contributions made by the Inland Revenue (or formerly by the DSS) on your behalf?

Yes .... 1 ⎤→   Q9
No .... 2 ⎦

EmpCont Ask if employee and has arranged own contracted out pension scheme *(Stat = 1 & PersPens = 1)*

Does your employer contribute to the scheme?

Yes .... 1 ⎤→   Q11
No .... 2 ⎦

10. EverPers Ask if employee and has not, or does not know if they have arranged own pension scheme *(Stat =1 & PersPens = 2 or DK)*

Have you ever had any such arrangements?

Yes .... 1 ⎤→   Q11
No .... 2 ⎦

11. OthPers Ask if employee OR (under pensionable age and not self-employed) - this is to select those who may have answered don't know, or refused to answer Stat *(Stat = 1 OR (under pensionable age & Stat ≠ 2))*

SHOW CARD PEN1

Please look at card PEN1. (Apart from the contributions you've already told me about,) do you have any other pension arrangements, such as those listed on the card, on which you receive income tax relief?

Yes .... 1 →   Q12
No .... 2 →See Q13

12. OtDetail Ask if respondent has other arrangements *(OthPers = 1)*

What arrangements do you have?

SHOW CARD PEN1

CODE ALL THAT APPLY

Personal pension .... 1 ⎤
Stakeholder pension .... 2 ⎥
Additional Voluntary Contribution .... 3 ⎥→See Q13
Free-Standing Additional Voluntary ⎥
Contribution .... 4 ⎥
Retirement annuities .... 5 ⎦

13. EmpConOt Ask if employee, and does not belong to employer's occupational scheme, and has other pension arrangements *(Stat = 1 & EmPenShm = 2 & OthPers = 1)*

Does your employer contribute to (any of) the arrangement(s)?

Yes .... 1 ⎤→See Q14
No .... 2 ⎦

14. PersPnt2 Ask if self-employed *(Stat = 2)*, others see Q18

INTERVIEWER - INTRODUCE IF NECESSARY.
Now I would like to ask you about personal pension schemes.

15. SePrsPen Self-employed people may arrange pensions for themselves and get tax relief on their contributions. These schemes include personal pensions, stakeholder pensions and 'self-employed pensions' (sometimes called 'Section 226 Retirement Annuities').

Do you at present contribute to one of these schemes?

Yes .... 1 →   Q16
No .... 2 →   Q17

16. SePrshp Ask if contributes to one of the schemes *(SePrsPen = 1)*

Which types of scheme are you contributing to - personal pension, stakeholder pension, or some other scheme?

CODE ALL THAT APPLY

Personal pension .... 1 ⎤
Stakeholder pension .... 2 ⎥→See Q18
Other... 3 ⎦

17. SeEvPers Ask if does not, or does not know if they contribute to one of the above schemes *(SePrsPen = 2 or DK)*

Have you ever contributed to one of these schemes?

Yes .... 1 ⎤→See Q18
No .... 2 ⎦

18. NewShp **This question is asked of anyone under pensionable age who is not currently in paid work** *(Under pensionable age AND (Wrking ≠ 1 OR JbAway ≠ 1 OR SchemeET ≠ 1))*

Since April 2001, anyone can arrange a stakeholder pension for themselves and get tax relief on the contribution.

Do you at present have a stakeholder pension?

Yes 1
No .... 2 →Education

EDUCATION

Ask this section of those aged 16-69 (it is not asked of proxies) *(DVAge = 16-69)*

1. QualCh I would now like to ask you about education and work-related training. Do you have any qualifications from school, college or university, connected with work or from government schemes?

Yes .... 1 → Q2
No .... 2 → Q20
Don't know .... 3 → Q2

2. Quals **Ask if respondent has a qualification, or answers don't know** *(QualCh = 1 or 3)*

Which qualifications do (you think) you have, starting with the highest qualifications?

SHOW CARD C

CODE ALL THAT APPLY - PROMPT AS NECESSARY

Degree level qualifications including graduate membership of a professional institute or PGCE or higher..................... 1 → Q3
Diploma in higher education 2
HNC/HND 3
ONC/OND... 4
BTEC, BEC OR TEC............................... 5
SCOTVEC, SCOTEC OR SCOTBEC 6
Teaching qualification (excluding PGCE) 7
Nursing or other medical qualification not yet mentioned 8
Other higher education qualification below degree level 9
A level or equivalent..............................10
SCE highers...11
NVQ/SVQ...12
GNVQ/GSVQ 13
AS level...14   → See Qs 5-19
Certificate of sixth year studies (CSYS) or equivalent 15
O level or equivalent..............................16
SCE STANDARD/ORDINARY (O) GRADE..17
GCSE...18
CSE 19
RSA 20
City and Guilds 21
YT Certificate/YTP 22
Any other professional/vocational qualifications/foreign qualifications 23
Don't know 24

3. Degree **Ask if highest qualification is a degree level qualification** *(Quals = 1 AND does NOT have a higher qualification)*

Is your degree...

a higher degree (including PGCE)? ...... 1 →   C
a first degree? 2
other (eg graduate member of a professional institute or chartered accountant)? 3   →See Q1
Don't know 4

4. HighO **Ask if has a higher degree** *(Degree = 1)*

ASK OR RECORD

Was your higher degree...

CODE FIRST THAT APPLIES

a Doctorate? 1
a Masters? 2
a Postgraduate Certificate in Education? 3   →See Q
or some other postgraduate degree or professional qualification? 4
Don't know 5

5. BTEC **Ask if highest qualification is BTEC, BEC or TEC** *(Quals = 5 AND does NOT have a higher qualification)*

Is your highest BTEC qualification...

CODE FIRST THAT APPLIES

at higher level?..................................... 1
at National Certificate or National Diploma level? 2
a first diploma or general diploma?........ 3   →See Q
a first certificate or general certificate? .. 4
Don't know 5

6. SCTVEC **Ask if highest qualification is SCOTVEC** *(Quals = AND does NOT have a higher qualification)*

Is your highest SCOTVEC qualification...

CODE FIRST THAT APPLIES

higher level?... 1
full National Certificate?........................ 2
a first diploma or general diploma?........ 3
a first certificate or general certificate? .. 4   →See Q
modules towards a National Certificate? 5
Don't know 6

7. Teach **Ask if highest qualification is a teaching qualification excluding PGCE** *(Quals = 7 AND does NOT have a higher qualification)*

Was your teaching qualification for...

Further education.................................. 1
Secondary education 2
or primary education? 3   →See Q
Don't know 4

NumAL   Ask if highest qualification is A levels *(Quals = 10 AND does NOT have a higher qualification)*

Do you have...

one A level or equivalent........................ 1
or more than one? .... 2 →See Q10
Don't know 3

NumSCE   Ask if highest qualification is Scottish highers *(Quals = 11 AND does NOT have a higher qualification)*

Do you have...

1 or 2 SCE highers 1
3 or more highers.................................. 2 →See Q10
Don't know 3

NVQlev   Ask if has NVQ/SVQ *(Quals = 12)*

What is your highest level of full NVQ/SVQ?

Level 1 1
Level 2 2
Level 3 3
Level 4 .... 4 →See Q11
Level 5 5
Don't know 6

GNVQ   Ask if highest qualification is GNVQ\GSVQ *(Quals = 13 AND does NOT have a higher qualification)*

Is your highest GNVQ/GSVQ at...

CODE FIRST THAT APPLIES

advanced level? 1
intermediate level?................................. 2
foundation level? .... 3 →See Q17
Don't know 4

NumAS   Ask if highest qualification is AS levels *(Quals = 14 AND does NOT have a higher qualification)*

Do you have...

one AS level... 1
2 or 3 AS levels..................................... 2
or 4 or more passes at this level? .... 3 →See Q17
Don't know 4

RSA   Ask if highest qualification is RSA *(Quals = 20 AND does NOT have a higher qualification)*

Is your highest RSA...

CODE FIRST THAT APPLIES

a higher diploma? 1
an advanced diploma or advanced
certificate? 2
a diploma? .... 3 →See Q17
or some other RSA (including Stage
I,II & III)? 4
Don't know 5

14. CandG   Ask if highest qualification is City and Guilds *(Quals = 21 AND does NOT have a higher qualification)*

Is your highest City and Guilds qualification....

CODE FIRST THAT APPLIES

advanced craft/part 3? 1
craft/part 2?... 2
foundation/part 1? .... 3 →See Q17
Don't know 4

15. GCSE   Ask if highest qualification is SCE Standard/ Ordinary Grade or GCSE *(Quals = 17 OR Quals = 18 AND does NOT have a higher qualification)*

Do you have any (GCSEs at grade C or above) (SCE Standard grades 1-3/ O grades at grade C or above)?

Yes 1
No .... 2 →See Q17
Don't know 3

16. CSE   Ask if highest qualification is CSE *(Quals = 19 AND does NOT have a higher qualification)*

Do you have any CSEs at grade 1?

Yes 1
No .... 2 →See Q17
Don't know 3

17. NumOL   Ask if passes at GCSE at Grade C or above OR CSE Grade 1 or O level or equivalent OR SCE level or equivalent *(CSE = 1 or GCSE = 1 or Quals = 16 or Quals = 17)*

ASK OR RECORD

You mentioned that you have passes at (GCSE at Grade C or above) (CSE Grade 1) (O level or equivalent) (SCE level or equivalent). Do you have...

fewer than 5 passes, ... 1
or 5 or more passes at this level? .... 2 →See Q18
Don't know 3

18. EngMath   Ask if has O levels, SCE Standard/Ordinary (O) Grade or GCSEs or CSEs *(Quals = 16 or GCSE = 1 or CSE = 1 or Quals = 19)*

Do you have (GCSEs at Grade C or above) (CSE Grade 1) (O levels or equivalent) in English or Mathematics?

EXCLUDE ENGLISH LITERATURE

English 1
Maths 2
Both .... 3 →See Q19
Neither 4

19. Appren   Ask if highest qualification is 'any other professional/vocational qualifications/foreign qualifications', or the respondent answered 'don't know' *(Quals = 23 or 24 or Qualch = 3 AND does NOT have a higher qualification)*

Are you doing or have you completed, a recognised trade apprenticeship?

Yes, (completed).................................... 1
Yes, (still doing) .... 2 ⌉→ Q20
No (including apprenticeships begun
but discontinued).................................... 3 ⌋

20. Enroll Are you at present (at school or sixth form college or) enrolled on any full-time or part-time education course excluding leisure classes? (Include correspondence courses and open learning as well as other forms of full-time or part-time education course.)

Yes .... 1 ⌉→ Q21
No .... 2 ⌋→ Q23
Don't know 3

21. Attend Ask if enrolled on a education course (Enroll = 1)

And are you ...

Still attending .... 1 ⌉→ Q22
Waiting for term to (re)start .... 2 ⌋
Or have you stopped going? .... 3 → Q23

22. Course Ask if respondent is still attending school or college, or waiting for term to [re]start (Attend = 1 or 2)

Are you (at school or 6th form college), on a full or part-time course, a medical or nursing course, a sandwich course, or some other kind of course?

CODE FIRST THAT APPLIES

School/full-time (age < 20 years only) ... 1
School/part-time (age < 20 years only).. 2
sandwich course 3
studying at a university or college
including sixth form college FULL-TIME 4
training for a qualification in nursing,
physiotherapy, or a similar medical
subject .... 5 → Q23
on a part-time course at university or
college INCLUDING day release and
block release.. 6
on an Open College Course 7
on an Open University Course.............. 8
any other correspondence course 9
any other self/open learning course.......10

23. EdAge Asked of all aged 16-69 (DVAge = 16-69)

How old were you when you finished your continuous full-time education?

CODE AS 97 IF NO EDUCATION
CODE AS 96 IF STILL IN EDUCATION

1..97   →   Q24

24. EducPres Are you at present attending any sort of leisure or recreation classes during the day, in the evenings or at weekends?

Yes .... 1 → Q25
No .... 2 → Adult
                                                              Health

25. EdTyp Ask if respondent is attending a leisure or recreation class (EducPres = 1)

What type of college or organisation runs these classes?

CODE ALL THAT APPLY
(Enter at most 4 codes)

Evening institute/Local Education
Authority/College or Centre of
Adult Education .... 1 ⌉
College of Further Education/
Technical College................................... 2 ⌐→ Ad
University Extra-Mural Department........ 3 │ Hea
Other.. 4 ⌋

ADULT HEALTH

Ask this section of all adults (except GenHlth which excludes proxy respondents)

1. Genhlth [*] **Ask all (except proxy informants)**

Over the last twelve months would you say your hea has on the whole been good, fairly good, or not goo

Good .... 1 ⌉
Fairly Good .... 2 ⌐→   (
Not Good .... 3 ⌋

2. Illness [*] **Ask all**

Do you have any long-standing illness, disability or infirmity? By long-standing, I mean anything that ha troubled you over a period of time or that is likely to affect you over a period of time?

Yes .... 1 →   (
No .... 2 →   (

3. Lmatter [*] **Ask if has a long-standing illness (Illness = 1)**

What is the matter with you?

RECORD ONLY WHAT RESPONDENT SAYS.

ENTER TEXT OF AT MOST 100 CHARACTERS→ (

4. LMatNum HOW MANY LONGSTANDING ILLNESSES OR INFIRMITIES DOES RESPONDENT HAVE?

ENTER NUMBER OF LONGSTANDING COMPLAINTS MENTIONED, IF MORE THAN 6 - TAKE THE SIX THAT THE RESPONDENT CONSIDERS THE MOST IMPORTANT

1..6   →   (

5. Lmat **For each illness mentioned above**

WHAT IS THE MATTER WITH RESPONDENT?

ENTER THE (FIRST/SECOND/etc.) CONDITION/ SYMPTOM RESPONDENT MENTIONED

ENTER TEXT OF AT MOST 40 CHARACTERS→ (

CD   CODE FOR COMPLAINT AT LMAT

ENTER TEXT OF AT MOST 12 CHARACTERS → Q7

LimitAct   Does this illness or disability (Do any of these illnesses or disabilities) limit your activities in any way?

Yes .... 1 ⎤ →   Q8
No .... 2 ⎦

CutDown [*] **Ask all**

Now I'd like you to think about the 2 weeks ending yesterday. During those 2 weeks, did you have to cut down on any of the things you usually do (about the house/at work or in your free time) because of (answers at LMatter) or some other illness or injury?

Yes .... 1 →   Q9
No .... 2 →   Q11

NdysCutD   **Ask if had to cut down on normal activities because of illness or injury (CutDown = 1)**

How many days was this in all during these 2 weeks, including Saturdays and Sundays?

1..14   →   Q10

Cmatter [*]   What was the matter with you?

ENTER TEXT OF AT MOST 40 CHARACTERS → Q11

DocTalk   **Ask all**

During the 2 weeks ending yesterday, apart from any visit to a hospital, did you talk to a doctor for any reason at all, either in person or by telephone?

EXCLUDE: CONSULTATIONS MADE ON BEHALF OF CHILDREN UNDER 16 AND PERSONS OUTSIDE THE HOUSEHOLD.

Yes .... 1 →   Q12
No .... 2 →   Q19

Nchats   **Ask if contact with doctor during the last 2 weeks (DocTalk = 1)**

How many times did you talk to a doctor in these 2 weeks?

1..9   →   Q13

For each consultation

WhsBhlf   On whose behalf was this consultation made?

Respondent .... 1 →   Q15
Other member of household 16 or over . 2 →   Q14

ForPerNo   **Ask if consultation was on behalf of another member of the household (WhsBhlf = 2)**

CODE WHO CONSULTATION WAS MADE FOR

(PERSON NUMBER)   →   Q15

15. NHS   **For each consultation**

Was this consultation...

Under the National Health Service .... 1 ⎤ → Q16
or paid for privately? .... 2 ⎦

16. GP   Was the doctor...

RUNNING PROMPT

A GP (ie a family doctor) .... 1 ⎤
or a specialist .... 2 →   Q17
or some other kind of doctor? .... 3 ⎦

17. DocWhere   Did you talk to the doctor...

RUNNING PROMPT

By telephone .... 1 ⎤
at your home .... 2 ⎥
in the doctor's surgery .... 3 → Q18
at a health centre .... 4 ⎥
or elsewhere? .... 5 ⎦

18. Presc   Did the doctor give (send) you a prescription?

Yes .... 1 ⎤ → Q19
No .... 2 ⎦

19. SeeNurse   **Ask all**

During the last 2 weeks ending yesterday, did you see a practice nurse at the GP surgery on your own behalf?

EXCLUDE CONSULTATIONS WITH COMMUNITY NURSES

Yes .... 1 → Q20
No .... 2 → Q21

20. Nnurse   **Ask if the respondent saw a nurse (SeeNurse = 1)**

How many times did you see a practice nurse at the GP surgery in these 2 weeks?

RECORD NUMBER OF TIMES

1..9   → Q21

21. OutPatnt   **Ask all**

During the months of (LAST 3 COMPLETE CALENDAR MONTHS) did you attend as a patient the casualty or outpatient department of a hospital (apart from straightforward ante- or post-natal visits)?

Yes .... 1 → Q22
No .... 2 → Q29

22. Ntimes1   **Ask if respondent attended outpatients (OutPatnt = 1)**

How many times did you attend in (EARLIEST MONTH IN REFERENCE ....PERIOD)?

0..97   → Q23

23. NTimes2 How many times did you attend in (SECOND MONTH IN REFERENCE PERIOD)?

0..97 → Q24

24. NTimes3 How many times did you attend in (THIRD MONTH IN REFERENCE PERIOD)?

0..97 → Q25

25. Casualty Was this visit (were any of these visits) to the Casualty department or was it (were they all) to some other part of the hospital?

At least one visit to Casualty.................. 1 → Q26
No Casualty visits .... 2 → Q27

26. NcasVis Ask if respondent visited casualty *(Casualty = 1)*

(May I just check) How many times did you go to Casualty altogether?

1..31 → Q27

27. PrVists Ask if respondent attended outpatients *(OutPatnt = 1)*

Was your outpatient visit (were any of your outpatient visits) during (REFERENCE PERIOD) made under the NHS, or was it (were any of them) paid for privately?

All under NHS .... 1 → Q29
At least one paid for privately................. 2 → Q28

28. NprVists Ask if some private visits *(PrVists = 2)*

ASK OR RECORD

(May I just check), How many of the visits were paid for privately?

1..31 → Q29

29. DayPatnt Ask all

During the last year, that is, since (DATE ONE YEAR AGO), have you been in hospital for treatment as a day patient, ie admitted to a hospital bed or day ward, but not required to remain overnight?

Yes .... 1 →See Q30
No .... 2 → Q37

30. MatDPat Ask if has been a day patient AND is a women aged between 16-49 *(DayPatnt = 1 & Sex = 2 & DVAge = 16-49), others Q34*

May I just check, was that/were any of those day patient admissions for you to have a baby?

Yes .... 1 → Q31
No .... 2 → Q34

31. NumMatDP Ask if respondent was a day patient because she was having a baby *(MatDPat = Yes)*

How many separate days have you had as a day patient for having a baby since (DATE ONE YEAR AGO)?

97 DAYS OR MORE - CODE 97

1..97 → Q

32. PrMatDP Was this day-patient stay (were any of these day-patient stays) for having a baby under the NHS, or was it (were any of them) paid for privately?

All under NHS .... 1 → Q
At least one paid for privately................. 2 →See Q

33. NprMatDP Ask if day patient stay for having a baby was pai for privately AND respondent was in hospital for more than one day *(PrMatDP = 2 & NumMatDP 1)*

ASK OR RECORD

How many of the visits were paid for privately?

1..31 → Q

34. NHSPDays Ask if the respondent was a day patient *(DayPat = 1)*

(Apart from those maternity stays) how many separa days in hospital have you had as a day patient since (DATE ONE YEAR AGO)?

97 DAYS OR MORE - CODE 97

0..97 →See Q

35. PrDptnt Ask if had one or more days in hospital *(NHSPDays > 0)*

Was this day-patient treatment (were any of these day-patient treatments) under the NHS, or was it (were any of them) paid for privately?

All under NHS .... 1 → Q
At least one paid for privately................. 2 →See Q

36. NPrDpTnt Ask if day patient stay was paid for privately AND they were in hospital for more than one day *(PrDptnt = 2 & NHSPDays > 1)*

ASK OR RECORD

How many of the visits were paid for privately?

1..31 → Q

37. InPatnt Ask all

During the last year, that is, since (DATE 1 YEAR AGO), have you been in hospital as an inpatient, overnight or longer?

Yes .... 1 →See Q
No .... 2 → Heari

38. MatInPat Ask if respondent has been an inpatient AND sh is a women aged 16-49 *(InPatnt = 1 & Sex = 2 & DVAge = 16-49)*

May I just check, was that/were any of those inpatient admissions for you to have a baby?

Yes .... 1 → Q39
No .... 2 → Q43

NmtStay   **Ask if inpatient admission was to have a baby (MatInPat = 1)**

How many separate stays in hospital as an inpatient in order to have a baby have you had since (DATE 1 YEAR AGO)?

1..6   →   Q40

MtNights   **Ask for each maternity stay**

How many nights altogether were you in hospital on your (no.) stay to have a baby?

1..97   →   Q41

MatNHSTr   Were you treated under the NHS or were you a private patient on that occasion?

NHS .... 1 → Q43
Private Patient....................................... 2 → Q42

MtPrvSty   **If private patient (MatNHSTr = 2)**

Were you treated in an NHS hospital or in a private one?

NHS hospital.. 1 ⎤
Private hospital .... 2 ⎦ → Q43

Nstays   **Ask if respondent has been an inpatient (InPatnt = 1)**

(Apart from those maternity stays) how many separate stays in hospital as an inpatient have you had since (DATE 1 YEAR AGO)?

0..6   →   Q44

Nights   **Ask for each stay**

How many nights altogether were you in hospital on your... .... (first/second/...sixth) stay?

1..97   →   Q45

NHSTreat   Were you treated under the NHS or were you a private patient on that occasion?

NHS .... 1 → Hearing
Private patient....................................... 2 → Q46

PrvStay   **Ask if a private patient (NHSTreat = 2)**

Were you treated in an NHS hospital or in a private one?

NHS hospital.. 1 ⎤
Private hospital .... 2 ⎦ → Hearing

---

HEARING DIFFICULTIES

1. HearDiff [*]   **Ask all (except proxy respondents)**

Do you ever have any difficulty with your hearing?

Yes .... 1 → Q2
No .... 2 → Child Health

2. HearAid   **If HearDiff = Yes**

(May I just check) do you ever wear a hearing aid nowadays?

Yes .... 1 → Q3
No .... 2 → Child Health

3. AidDiff   **If HearAid = Yes**

Do you ever have any difficulties with your hearing even when you're wearing an aid?

Yes .... 1 ⎤
No .... 2 ⎦ → Q4

4. NumAids   How many hearing aids have you got that you wear even if only occasionally?

1..4   →   Q5

5. AidTyp   **For each hearing aid**

Did you obtain this aid through the National Health Service or was it bought privately?

NHS .... 1 → Q7
Private.. 2 → Q6

6. Ypriv   **If AidTyp = Private**

Why did you decide to buy this aid privately?

SET [4] OF
To get a better choice............................. 1 ⎤
To get it quicker .... 2 ⎥ → Q7
Not available through NHS...................... 3 ⎥
Other.. 4 ⎦

7. AidWear   **For each hearing aid**

Do you wear this aid regularly?

Yes .... 1 ⎤ → Q8
No .... 2 ⎦

8. DontWear   **If HearDiff = Yes**

Do you have any hearing aids which still work that you no longer wear?

Yes .... 1 → Q9
No .... 2 → Child Health

9. NoNotWrn **If DontWear = Yes**

How many aids do you have that you don't wear?

1..4   →   Q10

10. NotWorn **For each aid not worn**

Did you obtain this aid through the National Health Service or was it bought privately?

NHS .... 1 ⎤
Private....................................... 2 ⎦→ Q11

11. YNtWr Why do you no longer wear this aid?

SET [5] OF
Did not help hearing.............................. 1 ⎤
Appearance... 2 ⎥→ Child
Other.. 3 ⎦   Health

CHILD HEALTH

1. AskHlth **Ask if there is a child / there are children under 16 in household (not asked of proxy respondents)**

THE NEXT SECTION IS ABOUT CHILD HEALTH. WE ONLY NEED TO COLLECT THIS INFORMATION ONCE FOR EACH CHILD IN THE HOUSEHOLD. WHO WILL ANSWER THE CHILD HEALTH SECTION FOR (CHILD'S NAME)?

INTERVIEWER ENTER PERSON NUMBER

1..14   →   Q2

2. AskNowCH INTERVIEWER: DO YOU WANT TO ASK THIS SECTION FOR (CHILD'S NAME) NOW OR LATER?

IF YOU HAVE ALREADY ASKED THIS SECTION FOR (CHILD'S NAME), DO NOT CHANGE FROM CODE 1

Yes, now/Already asked .... 1 →   Q4
Later.. 2 →   Q3

3. Cstill **If the section is to be asked later (AskNowCH = 2)**

REMINDER
THE FOLLOWING ADULTS STILL NEED TO ANSWER THE CHILD HEALTH SECTION ON BEHALF OF SOME OF THE CHILDREN

4. Genhlth [*] **For each child**

Over the last twelve months would you say (NAME's) health has on the whole been good, fairly good, or not good?

Good .... 1 ⎤
Fairly Good .... 2 ⎥→ Q5
Not Good .... 3 ⎦

5. Illness [*] Does (NAME) have any long-standing illness, disability or infirmity? By long-standing, I mean anything that has troubled them over a period of time or that is likely to affect them over a period of time?

Yes .... 1 →   C
No .... 2 →   Q

6. Lmatter [*] **Ask if child has a longstanding illness, disability or infirmity (Illness =1)**

What is the matter with (NAME)?

THIS IS TO ENSURE THAT THE RESPONDENT MENTIONS ALL LONGSTANDING ILLNESSES. YO DO NOT HAVE TO RECORD VERBATIM HERE - A SUMMARY WILL DO

ENTER TEXT OF AT MOST 40 CHARACTERS→ C

7. LMatNum HOW MANY LONGSTANDING ILLNESSES OR INFIRMITIES DOES (NAME) HAVE?

ENTER NUMBER OF LONGSTANDING COMPLAINTS MENTIONED
IF MORE THAN 6 - TAKE THE SIX THAT THE RESPONDENT CONSIDERS THE MOST IMPORTANT

1..6   →   C

8. LMatCH **For each illness mentioned at LMatNum**

WHAT IS THE MATTER WITH (NAME)?

ENTER THE (FIRST/SECOND/etc.) CONDITION/ SYMPTOM RESPONDENT MENTIONED

ENTER TEXT OF AT MOST 40 CHARACTERS→ C

9. ICDCH CODE FOR EACH COMPLAINT AT LMatCH→ Q

10. LimitAct [*] **If child has a longstanding illness, disability or infirmity (Illness =1)**

Does this illness or disability (Do any of these illnesses or disabilities) limit (NAME)'s activities in a way?

Yes .... 1 ⎤→ Q
No .... 2 ⎦

11. CutDown [*] **For each child**

Now I'd like you to think about the 2 weeks ending yesterday. During those 2 weeks, did (NAME) have cut down on any of the things he/she usually does (school or in his/her free time) because of (answer a LMatter or some other) illness or injury?

Yes .... 1 →   Q
No .... 2 →   Q

12. NdysCutD **Ask if child has had to cut down (CutDown = 1)**

How many days did (NAME) have to cut down in all during these 2 weeks, including Saturdays and Sundays?

	1..14 → Q13	

Matter [*] What was the matter with (NAME)?

ENTER TEXT OF AT MOST 80 CHARACTERS → Q14

DocTalk **For each child**

During the 2 weeks ending yesterday, apart from visits to a hospital, did (NAME) talk to a doctor for any reason at all, or did you or any other member of the household talk to a doctor on his/her behalf?

INCLUDE BEING SEEN BY A DOCTOR AT A SCHOOL CLINIC, BUT EXCLUDE VISITS TO A CHILD WELFARE CLINIC RUN BY A LOCAL AUTHORITY

INCLUDE TELEPHONE CONSULTATIONS AND CONSULTATIONS MADE ON BEHALF OF CHILDREN

Yes .... 1 → Q15
No .... 2 → Q20

Nchats **If child consulted a doctor (DocTalk = 1)**

How many times did (NAME) talk to the doctor (or you or any other member of the household consult the doctor on NAME's behalf) in those 2 weeks?

1..9 → Q16

NHS **For each consultation**

Was this consultation...

Under the National Health Service .... 1 ⎤
or paid for privately? .... 2 ⎦ → Q17

GP Was the doctor...

RUNNING PROMPT

A GP (ie a family doctor)......................... 1 ⎤
or a specialist.. 2 ⎬ → Q18
or some other kind of doctor? .... 3 ⎦

DocWhere Did you or any other member of the household (or NAME) talk to the doctor...

by telephone .... 1 ⎤
at your home .... 2 ⎥
in the doctor's surgery............................ 3 ⎬ → Q19
at a health centre .... 4 ⎥
or elsewhere? .... 5 ⎦

Presc Did the doctor give (send) (NAME) a prescription?

Yes .... 1 ⎤
No .... 2 ⎦ → Q20

Seenurse **For each child**

During the last 2 weeks ending yesterday, did (NAME)

RUNNING PROMPT
CODE ALL THAT APPLY

EXCLUDE CONSULTATIONS WITH COMMUNITY NURSES

see a practice nurse at the GP surgery,. 1 → Q21
see a health visitor at the GP surgery, ... 2 ⎤
go to child health clinic,.......................... 3 ⎬ → Q22
go to child welfare clinic, ... 4 ⎥
did not go to any of these....................... 5 ⎦

21. Nnurse **Ask if child saw a practice nurse (Seenurse = 1)**

How many times did (NAME) see a practice nurse at the GP surgery in these 2 weeks?

RECORD NUMBER OF TIMES

1..9 → Q22

22. OutPatnt **For each child**

During the months of (LAST 3 COMPLETE CALENDAR MONTHS), did (NAME) attend as a patient the casualty or outpatient department of a hospital (apart from straightforward post-natal visits)?

Yes .... 1 → Q23
No .... 2 → Q28

23. Ntimes1 **Ask if child has been an outpatient (OutPatnt = 1)**

How many times did (NAME) attend in (EARLIEST MONTH IN REFERENCE PERIOD)?

0..97 → Q24

24. NTimes2 How many times did (NAME) attend in (SECOND MONTH IN REFERENCE PERIOD)?

0..97 → Q25

25. NTimes3 How many times did (NAME) attend in (THIRD MONTH IN REFERENCE PERIOD)?

0..97 → Q26

26. Casualty Was the visit (were any of the visits) to the Casualty department or was it (were they) to some other part of the hospital?

At least one visit to Casualty.................. 1 → Q27
No Casualty visits .... 2 → Q28

27.NcasVis **Ask if child went to casualty (Casualty = 1)**

(May I just check) How many times did (NAME) go to Casualty altogether?

1..31 → Q28

28. DayPatnt **For each child**

During the last year, that is since (DATE 1 YEAR AGO) has (NAME) been in hospital for treatment as a day patient, ie admitted to a hospital bed or day ward, but not required to remain in hospital overnight?

Yes .... 1 → Q29
No .... 2 → Q30

245

29. NHSPDays Ask if child has been a day patient *(DayPatnt = 1)*

How many separate days in hospital has (NAME) had as a day patient since (DATE 1 YEAR AGO)?

1..97   Q30

30. Inpatnt   **For each child**

During the last year, that is, since (DATE 1 YEAR AGO) has (NAME) been in hospital as an inpatient, overnight or longer?

EXCLUDE: Births unless baby stayed in hospital after mother had left.

Yes .... 1 →   Q31
No .... 2 →Smoking

31. Nstays   **Ask if child has been an inpatient *(InPatnt = 1)***

How many separate stays in hospital as an inpatient has (NAME) had since (DATE 1 YEAR AGO)?

IF 6 OR MORE, CODE 6

1..6   →   Q32

32. Nights   **For each stay**

How many nights altogether was (NAME) in hospital during stay number (...)?

1..97   →Smoking

SMOKING

1. SmkIntro   **Ask this section of all adults, except proxy respondents**

The next section consists of a series of questions about SMOKING (Not asked of proxy respondents)

2. SelfCom1   **Ask all 16 and 17 year olds *(DVAge = 16-17)***

RESPONDENT IS AGED 16 OR 17 - OFFER SELF-COMPLETION FORM AND ENTER CODE

Respondent accepted self-completion... 1 →   Q3
Respondent refused self-completion ..... 2 → Drinking
Data now to be keyed by interviewer ..... 3 →   Q3

3. SmokEver   **Ask if aged 18 or over (except proxy respondents) *(DVAge ≥ 18)***

Have you ever smoked a cigarette, a cigar, or a pipe?

Yes .... 1 →   Q4
No .... 2 → Drinking

4. CigNow   **Ask if respondent has ever smoked *(SmokEver = 1)***

Do you smoke cigarettes at all nowadays?

Yes .... 1 →   Q5
No .... 2 →   Q13

5. QtyWkEnd   **Ask if respondent smokes cigarettes now *(CigN = 1)***

About how many cigarettes A DAY do you usually smoke at weekends?

IF LESS THAN 1, ENTER 0

0..97   →

6. QtyWkDay   About how many cigarettes A DAY do you usually smoke on weekdays?

IF LESS THAN 1, ENTER 0

0..97   →

7. CigType   Do you mainly smoke.....

RUNNING PROMPT

filter-tipped cigarettes .... 1 ⎤
or plain or untipped cigarettes .... 2 ⎬ →
or hand-rolled cigarettes? .... 3 ⎦ →   C

8. CigIDesc   **Ask if cigarette types include plain or filter cigarettes *(CigType = 1 or 2)***

Which brand of cigarette do you usually smoke?

GIVE 1) FULL BRAND NAME 2) SIZE, eg King, luxury, regular.
IF NO REGULAR BRAND THEN TYPE 'no reg' HE
IF RESPONDENT SMOKES TWO BRANDS EQUALLY TYPE 'two' HERE
IF RESPONDENT SMOKES SUPERKINGS (WITH NO OTHER BRAND NAME ON THE PACKET) CO AS JOHN PLAYERS SUPERKINGS

ENTER TEXT OF AT MOST 60 CHARACTERS→

9. CigCODE   Code for brand at CigIDesc   →   C

10. NoSmoke [*] Ask if respondent smokes cigarettes now *(CigN = 1)*

How easy or difficult would you find it to go without smoking for a whole day? Would you find it...

RUNNING PROMPT

Very easy .... 1 ⎤
Fairly easy.. 2 ⎬ →
Fairly difficult or.................................... 3 ⎥   C
Very difficult?....................................... 4 ⎦

11. GiveUp [*]   Would you like to give up smoking altogether?

Yes .... 1 ⎤→
No .... 2 ⎦   C

12. FirstCig   How soon after waking do you USUALLY smoke yo first cigarette of the day?

PROMPT AS NECESSARY

Less than 5 minutes 1
5-14 minutes 2
15-29 minutes 3   Q16
30 minutes but less than 1 hour 4
1 hour but less than 2 hours 5
2 hours or more 6

3. CigEver   **Ask if respondent does not smoke cigarettes now but has smoked a cigarette or cigar or pipe (SmokEver = 1 & CigNow = 2)**

Have you ever smoked cigarettes regularly?

Yes 1   Q14
No 2   Q17

4. CigUsed   **Ask if respondent has ever smoked cigarettes regularly (CigEver = 1)**

About how many cigarettes did you smoke IN A DAY when you smoked them regularly?

IF LESS THAN 1, ENTER 0.

0..97   →   Q15

5. CigStop   How long ago did you stop smoking cigarettes regularly?

PROMPT AS NECESSARY

Less than 6 months ago 1
6 months but less than a year ago 2
1 year but less than 2 years ago 3   →   Q16
2 years but less than 5 years ago 4
5 years but less than 10 years ago 5
10 years or more ago 6

6. CigAge   **Ask of all respondents who have ever smoked cigarettes (CigNow = 1 or CigEver = 1)**

How old were you when you started to smoke cigarettes regularly?

SPONTANEOUS: NEVER SMOKED CIGARETTES REGULARLY - CODE 0

0..97   →   Q17

7. CigarReg   **Ask respondents who have ever smoked (SmokEver = 1)**

Do you smoke at least one cigar of any kind per month nowadays?

Yes .... 1 →   Q18
No .... 2 →   Q19

8. CigarsWk   **Ask if respondent smokes at least one cigar per month (CigarReg = 1)**

About how many cigars do you usually smoke in a week?

IF LESS THAN 1, ENTER 0.

0..97   →See Q20

19. CigarEvr   **Ask if respondent does not smoke at least one cigar per month (CigarReg = 2)**

Have you ever regularly smoked at least one cigar of any kind per month?

Yes .... 1 →See Q20
No 2

20. PipeNow   **Ask men who have ever smoked (CigNow = 1 AND Sex = 1)**

Do you smoke a pipe at all nowadays?

Yes .... 1 → Drinking
No .... 2 →   Q21

21. PipEver   **Ask if respondent doesn't currently smoke a pipe (PipeNow = 2)**

Have you ever smoked a pipe regularly?

Yes .... 1 → Drinking
No 2

DRINKING

Ask this section of all adults except proxy respondents

1. Selfcom2   **Ask all 16 and 17 year olds (DVAge = 16-17)**

(RESPONDENT IS AGED 16 OR 17) - OFFER SELF-COMPLETION FORM AND ENTER CODE.

Interviewer asked section 1
Respondent accepted self-completion... 2 →   Q2
Data now keyed by interviewer 3

2. DrinkNow   **Ask all (except proxy respondents) (DVAge (≥18 or Selfcom2 = 1)**

I'm now going to ask you a few questions about what you drink - that is if you do drink.

Do you ever drink alcohol nowadays, including drinks you brew or make at home?

Yes .... 1 →   Q7
No .... 2 →   Q3

3. DrinkAny   **Ask if does not drink nowadays (DrinkNow = 2)**

Could I just check, does that mean you never have an alcoholic drink nowadays, or do you have an alcoholic drink very occasionally, perhaps for medicinal purposes or on special occasions like Christmas or New Year?

Very occasionally .... 1 →   Q7
Never .... 2 →   Q4

4. TeeTotal   **Ask if never drinks (DrinkAny = 2)**

Have you always been a non-drinker, or did you stop drinking for some reason?

Always a non-drinker .... 1 →   Q5
Used to drink but stopped .... 2 →   Q6

5. NonDrink [*] **Ask if respondent has always been a non-drinker (TeeTotal = 1)**

What would you say is the MAIN reason you have always been a non-drinker?

Religious reasons 1
Don't like it 2
Parent's advice/influence .... 3 → Family
Health reasons....................................... 4   Information
Can't afford it... 5
Other.. 6

6. StopDrin [*] **Ask if respondent used to drink but stopped (TeeTotal = 2)**

What would you say was the MAIN reason you stopped drinking?

Religious reasons 1
Don't like it 2
Parent's advice/influence .... 3 → Family
Health reasons....................................... 4   Information
Can't afford it... 5
Other.. 6

7. DrinkAmt [*] **Ask if respondent drinks at all nowadays (Drinknow = 1 or DrinkAny = 1)**

I'm going to read out a few descriptions about the amounts of alcohol people drink, and I'd like you to say which one fits you best. Would you say you:

hardly drink at all.................................... 1
drink a little.. 2
drink a moderate amount .... 3 →   Q8
drink quite a lot 4
or drink heavily?..................................... 5

8. Intro   INTERVIEWER - READ OUT:

I'd like to ask you whether you have drunk different types of alcoholic drink in the last 12 months. I'd like to hear about ALL types of alcoholic drinks you have had. If you are not sure whether a drink you have had goes into a category, please let me know. I do not need to know about non-alcoholic or low alcohol drinks.   →   Q9

9. Nbeer   SHOW CARD D

I'd like to ask you first about NORMAL STRENGTH beer or cider which has less than 6% alcohol.

How often have you had a drink of NORMAL STRENGTH BEER, LAGER, STOUT, CIDER or SHANDY (excluding cans and bottles of shandy) during the last 12 months?

INTERVIEWER: (NORMAL = LESS THAN 6% ALCOHOL BY VOLUME)

 IF RESPONDENT DOES NOT KNOW WHETHER BEER ETC DRUNK IS STRONG OR NORMAL, INCLUDE HERE AS NORMAL

Almost every day 1
5 or 6 days a week................................. 2
3 or 4 days a week................................. 3
once or twice a week .... 4 →   Q
once or twice a month............................ 5
once every couple of months 6
once or twice a year............................... 7
not at all in last 12 months .... 8 →   Q

10. NbeerM   **Ask if respondent drank normal strength beer (lager/stout/cider/shandy) at all this year (Nbeer 1-7)**

How much NORMAL STRENGTH BEER, LAGER, STOUT, CIDER or SHANDY (excluding cans and bottles of shandy) have you usually drunk on any o day during the last 12 months?

CODE MEASURES THAT YOU ARE GOING TO US
CODE ALL THAT APPLY
PROBE IF NECESSARY

Half pints.. 1
Small cans 2
Large cans .... 3 →   Q
Bottles.. 4

11. NbeerQ   **For each measure mentioned at NbeerM**

ASK OR RECORD:

How many (Answer AT NBeerM) of NORMAL STRENGTH BEER, LAGER, STOUT, CIDER OR SHANDY (EXCLUDING CANS AND BOTTLES OF SHANDY) have you usually drunk on any one day during the last 12 months?

1..97   →See Q

12. NbrlDesc   **Ask if respondent described measures in 'Bottle (NBeerM = 4)**

What make of NORMAL STRENGTH BEER, LAGE STOUT or CIDER do you usually drink from bottles?

IF RESPONDENT DOES NOT KNOW WHAT MAKE OR RESPONDENT DRINKS DIFFERENT MAKES OF NORMAL STRENGTH BEER, LAGER, STOUT OR CIDER, PROBE:
'What make have you drunk most frequently or most recently?'

ENTER TEXT OF AT MOST 21 CHARACTERS→ Q

13. NBrCODE   Code for brand at NBrlDesc   →   Q

14. SBeer   **Ask if respondent drinks at all nowadays (Drinknow = 1 or DrinkAny = 1)**

SHOW CARD D

Now I'd like to ask you about STRONG BEER OR CIDER which has 6% or more alcohol (eg Tennants Extra, Special Brew, Diamond White).
How often have you had a drink of STRONG BEER, LAGER, STOUT or CIDER during the last 12 month

(STRONG=6% and over Alcohol by volume)

IF RESPONDENT DOES NOT KNOW
WHETHER BEER ETC DRUNK IS STRONG OR
NORMAL, INCLUDE AS NORMAL STRENGTH AT
NBeer ABOVE

Almost every day 1
5 or 6 days a week............................... 2
3 or 4 days a week............................... 3
once or twice a week 4   Q15
once or twice a month........................... 5
once every couple of months................ 6
once or twice a year.............................. 7
not at all in last 12 months 8   Q19

SbeerM   **Ask if respondent drank strong beer (lager/stout/ cider) at all this year (SBeer = 1-7)**

How much STRONG BEER, LAGER, STOUT or
CIDER have you usually drunk on any one day during
the last 12 months?

CODE MEASURES THAT YOU ARE GOING TO USE
CODE ALL THAT APPLY
PROBE IF NECESSARY

Half pints... 1
Small cans 2   Q16
Large cans 3
Bottles.. 4

SbeerQ   **For each measure mentioned at SBeerM**

ASK OR RECORD

How many (ANSWER AT SBeerM) of STRONG
BEER, LAGER, STOUT or CIDER have you usually
drunk on any one day during the last 12 months?

1..97   See Q17

SbrlDesc   **Ask if respondent described measures in 'Bottles' (SBeerM = 4)**

What make of STRONG BEER, LAGER, STOUT or
CIDER do you usually drink from bottles?

IF RESPONDENT DOES NOT KNOW WHAT MAKE,
OR RESPONDENT DRINKS DIFFERENT MAKES
OF STRONG BEER, LAGER, STOUT OR CIDER,
PROBE:
'What make have you drunk most frequently or most
recently?'

ENTER TEXT OF AT MOST 21 CHARACTERS   Q18

SBrCODE   Code for brand at SBrlDesc   Q19

Spirits   **Ask if respondent drinks at all nowadays (Drinknow = 1 or DrinkAny = 1)**

SHOW CARD D

How often have you had a drink of SPIRITS or
LIQUEURS, such as gin,whisky, brandy, rum, vodka,
advocaat or cocktails during the last 12 months?

Almost every day 1
5 or 6 days a week............................... 2
3 or 4 days a week............................... 3   Q20
once or twice a week 4
once or twice a month........................... 5
once every couple of months................ 6
once or twice a year.............................. 7
not at all in last 12 months 8   Q21

20. SpiritsQ   **Ask if respondent drank spirits or liqueurs at all this year (Spirits = 1-7)**

How much SPIRITS or LIQUEURS (such as gin,
whisky, brandy, rum, vodka, advocaat or cocktails)
have you usually drunk on any one day during the last
12 months?

CODE THE NUMBER OF SINGLES - COUNT
DOUBLES AS TWO SINGLES.

1..97   Q21

21. Sherry   **Ask if respondent drinks at all nowadays (Drinknow = 1 or DrinkAny = 1)**

SHOW CARD D

How often have you had a drink of SHERRY or
MARTINI including port,vermouth, Cinzano and
Dubonnet, during the last 12 months?

Almost every day 1
5 or 6 days a week............................... 2
3 or 4 days a week............................... 3   Q22
once or twice a week 4
once or twice a month........................... 5
once every couple of months................ 6
once or twice a year.............................. 7
not at all in last 12 months 8   Q23

22. SherryQ   **Ask if respondent drank sherry or martini at all this year (Sherry = 1-7)**

How much SHERRY or MARTINI, including port,
vermouth, Cinzano and Dubonnet have you usually
drunk on any one day during the last 12 months?

CODE THE NUMBER OF GLASSES

1..97   Q23

23. Wine   **Ask if respondent drinks at all nowadays (Drinknow = 1 or DrinkAny = 1)**

SHOW CARD D

How often have you had a drink of WINE, including
Babycham and champagne, during the last 12
months?

Almost every day 1
5 or 6 days a week............................... 2
3 or 4 days a week............................... 3   Q24
once or twice a week 4
once or twice a month........................... 5
once every couple of months................ 6
once or twice a year.............................. 7
not at all in last 12 months 8   Q25

249

24. WineQ **Ask if respondent drank wine at all this year (Wine = 1-7)**

How much WINE, including Babycham and champagne, have you usually drunk on any one day during the last 12 months?

CODE THE NUMBER OF GLASSES.
1 BOTTLE = 6 GLASSES, 1 LITRE = 8 GLASSES

1..97 → Q25

25. Pops **Ask if respondent drinks at all nowadays (Drinknow = 1 or DrinkAny = 1)**

SHOW CARD D

How often have you had a drink of ALCOPOPS (ie alcoholic lemonade, alcoholic colas or other alcoholic fruit- or herb-flavoured drinks (eg. Hooch, Two Dogs, Alcola etc), during the last 12 months?

Almost every day 1
5 or 6 days a week................................. 2
3 or 4 days a week................................. 3
once or twice a week .... 4 → Q26
once or twice a month............................ 5
once every couple of months 6
once or twice a year 7
not at all in last 12 months .... 8 → Q27

26. PopsQ **As if respondent drank alcopops at all this year (Pops = 1-7)**

How much alcopops (ie alcoholic lemonade, alcoholic colas or other alcoholic fruit- or herb-flavoured drinks) have you usually drunk on any one day during the last 12 months?

CODE THE NUMBER OF BOTTLES

1..97 → Q27

27. DrinkOft [*] **Ask if respondent drinks at all nowadays (Drinknow = 1 or DrinkAny = 1)**

SHOW CARD D

Thinking now about all kinds of drinks, how often have you had an alcoholic drink of any kind during the last 12 months?

Almost every day 1
5 or 6 days a week................................. 2
3 or 4 days a week................................. 3
once or twice a week .... 4 → Q28
once or twice a month............................ 5
once every couple of months 6
once or twice a year 7
not at all in last 12 months 8

28. DrinkL7 You have told me what you have drunk over the last 12 months, but we know that what people drink can vary a lot from week to week, so I'd like to ask you a few questions about last week. Did you have an alcoholic drink in the seven days ending yesterday?

Yes .... 1 → Q
No .... 2 → Q

29. DrnkDay **Ask if respondent has had an alcoholic drink in the last week (DrinkL7 = 1)**

On how many days out of the last seven did you have an alcoholic drink?

1..7 →See Q

30. DrnkSame **Ask if respondent had an alcoholic drink on two more days last week (DrnkDay = 2-7)**

Did you drink more on some days than others/one o the days, or did you drink about the same on each o these/both days?

Drank more on one/some day(s)
than other(s) 1
Same each day .... 2 → Q

31. WhichDay **Ask if respondent had an alcoholic drink last we (DrinkL7 = 1)**

Which day (last week) did you last have an alcoholic drink/have the most to drink?

Sunday.. 1
Monday 2
Tuesday 3
Wednesday .... 4 → Q
Thursday 5
Friday 6
Saturday.. 7

32. DrnkType **Ask if respondent has had an alcoholic drink in the last week (DrinkL7 = 1)**

SHOW CARD E

Thinking about last (DAY AT WHICHDAY) what types of drink did you have that day?

CODE ALL THAT APPLY

Normal strength beer/lager/
cider/shandy .... 1 → Q
Strong beer/lager/cider .... 2 → Q
Spirits or liqueurs .... 3 → Q
Sherry or martini .... 4 → Q
Wine.. 5 → Q
Alcoholic lemonades/colas .... 6 → Q

33. NBrL7 **Ask if respondent drank 'normal strength beer/ lager/cider/shandy' on that day (DrnkType = 1)**

Still thinking about last (DAY AT WHICHDAY), how much NORMAL STRENGTH BEER, LAGER, STOU CIDER or SHANDY (excluding cans and bottles of shandy) did you drink that day?

CODE MEASURES THAT YOU ARE GOING TO US
CODE ALL THAT APPLY.
PROBE IF NECESSARY.

Half pints... 1
Small cans 2
Large cans .... 3 → Q
Bottles 4

. NBrL7Q   **For each measure mentioned at NBrL7**

ASK OR RECORD

How many (Answer AT NBrL7) of NORMAL
STRENGTH BEER, LAGER, STOUT, CIDER OR
SHANDY (EXCLUDING CANS AND BOTTLES OF
SHANDY) did you drink that day?

1..97   See Q35

. NB7Idesc   **Ask if respondent described measures in 'Bottles'
(NBrL7 = 4)**

ASK OR RECORD

What make of NORMAL STRENGTH BEER, LAGER,
STOUT or CIDER do you usually drink from bottles?

IF RESPONDENT DRANK DIFFERENT MAKES
CODE WHICH THEY DRANK MOST

ENTER TEXT OF AT MOST 21 CHARACTERS → Q36

. NB7CODE   Code for brand at NB7IDesc   →See Q37

. SBrL7   **Ask if respondent drank 'strong beer/lager/cider'
on that day (DrnkType = 2)**

Still thinking about last (DAY AT WHICHDAY), how
much STRONG BEER, LAGER, STOUT, CIDER did
you drink that day?

CODE MEASURES THAT YOU ARE GOING TO USE
CODE ALL THAT APPLY. PROBE IF NECESSARY

Half pints .... 1 ⎤
Small cans 2   ⎬ → Q38
Large cans .... 3 ⎟
Bottles .... 4 ⎦

. SBrL7Q   **For each measure mentioned at SBrL7**

ASK OR RECORD

How many (Answer AT SBrL7) of STRONG BEER,
LAGER, STOUT or CIDER did you drink that day?

1..97   →See Q39

. SB7Idesc   **Ask if respondent described measures in 'Bottles'
(SBrL7 = 4)**

ASK OR RECORD

What make of STRONG BEER, LAGER, STOUT or
CIDER do you usually drink from bottles?

IF RESPONDENT DRANK DIFFERENT MAKES
CODE WHICH THEY DRANK MOST

ENTER TEXT OF AT MOST 21 CHARACTERS → Q40

. SB7CODE   Code for brand at SB7IDesc§   →See Q41

41. SpirL7   **Ask if respondent drank spirits or liqueurs on that
day (DrnkType = 3)**

Still thinking about last (DAY AT WHICHDAY), how
much spirits or liqueurs (such as gin, whisky, brandy,
rum, vodka, advocaat or cocktails) did you drink on
that day?

CODE THE NUMBER OF SINGLES - COUNT
DOUBLES AS TWO SINGLES

1..97   →See Q42

42. ShryL7   **Ask if respondent drank sherry or martini on that
day (DrnkType = 4)**

Still thinking about last (DAY AT WHICHDAY), how
much sherry or martini, including port, vermouth,
Cinzano and Dubonnet did you drink on that day?

CODE THE NUMBER OF GLASSES

1..97   →See Q43

43. WineL7   **Ask if respondent drank wine on that day
(DrnkType = 5)**

Still thinking about last (DAY AT WHICHDAY), how
much wine, including Babycham and champagne, did
you drink on that day?

CODE THE NUMBER OF GLASSES
1 BOTTLE = 6 GLASSES. 1 LITRE = 8 GLASSES.

1..97   →See Q44

44. PopsL7   **Ask if respondent drank alcopops on that day
(DrnkType = 6)**

Still thinking about last (DAY AT WHICHDAY), how
much alcopops (ie alcoholic lemonade, alcoholic colas
or other alcoholic fruit- or herb-flavoured drinks) did
you drink on that day?

CODE THE NUMBER OF BOTTLES   →   Q45

45. DrAmount [*] **Ask if respondent drinks at all nowadays
(Drinknow = 1 or DrinkAny = 1)**

Compared to five years ago, would you say that on
the whole you drink more, about the same or less
nowadays?

More nowadays .... 1 ⎤
About the same .... 2 ⎬ → Family Information
Less nowadays .... 3 ⎦

FAMILY INFORMATION

1. FamIntro   **Ask this section of all aged 16-59 (except proxy
respondents)**

THE NEXT SECTION CONSISTS OF A SERIES OF
QUESTIONS ABOUT FAMILY INFORMATION
(Not asked of proxy respondents)

2. ChkFIA **To all aged 16-59, if not single or same sex cohabiting, except proxy respondents**

INTERVIEWER CODE

Respondent is married or cohabiting but their partner is NOT a household, member................. 1 → Q3
Everyone else.. 2 → Q4

3. HusbAway **Ask if married/cohabiting, but partner not a household member**

INTRODUCE AS NECESSARY

Is your husband, wife or partner absent because he/she usually works away from home, or for some other reason?

Usually works away (include Armed
Forces, Merchant Navy)......................... 1 ⎤
Marriage/partnership broken down .... 2 ⎦ → Q4

4. SelfCom3 **To all**

OFFER (COLOUR) SELF-COMPLETION FORM TO RESPONDENT AND ENTER CODE.

Interviewer asked section .... 1 ⎤
Respondent accepted self-completion... 2 ⎬ → See Q5
Data now being keyed by interviewer .... 3 ⎦
Interpreter aged under 16- section
not asked .... 4 ⎤ →View of your
Respondent refused whole section........ 5 ⎦ local area

5. WhereWed **Ask people who have been married (Marstat = 2, 3, 4 or 5)**

Thinking of your present / most recent marriage, did you get married with a religious ceremony of some kind, or at a register office or approved premises, or are / at that time were you simply living together as a couple?

Religious ceremony of some kind.......... 1 ⎤
Civil marriage in register office or
approved premises .... 2 ⎬ → Q6
Religious ceremony and register
office/ approved premises...................... 3 ⎦
Living together as a couple.................... 4 → Q7

6. NumMar **Ask if respondent has been legally married (WhereWed = 1-3)**

How many times have you been legally married?

(NUMBER INCLUDING PRESENT MARRIAGE)

1..7   → See Q7

7. CLMon **Ask all cohabiting couples, including same sex couples (exc. couples now separated) (Livewith = 1 or 3 or WhereWed = 4)**

When did you and your partner start living together as a couple?

ENTER MONTH

1..12   →   Q8

8. ClYr ENTER YEAR IN 4 DIGIT FORMAT E.G. 2000

1900..2005   → See C

9. ClMar **Ask cohabiting couples, including same sex couples, but not separated, divorced or widowed respondents (Livewith = 1 or 3 OR Wherewed = AND is NOT separated, divorced or widowed)**

Have you yourself ever been legally married?

Yes .... 1 →   Q
No .... 2 →   Q

10. ClNumMar **Ask if respondent has been legally married (CIM = 1)**

How many times have you been legally married altogether?

1..7   →   Q

11. Intro **Ask of all who are, or have been, legally married (NumMar ≥ 1 or ClNumMar ≥ 1)**

THE NEXT SCREEN CONSISTS OF A TABLE OF MARRIAGES FOR (NAME). PLEASE ENTER DETAILS OF MARRIAGES STARTING WITH THE EARLIEST AND ENDING WITH THE CURRENT OF MOST RECENT   →   Q

12. MonMar **For each marriage**

What month and year were you married?

ENTER MONTH 1..12   →   Q

13. YrMar ENTER YEAR IN 4 DIGIT FORMAT E.G. 2000

1900..2005   →   Q

14. LvTgthr Before getting married did you and your husband/wil live together as a couple?

Yes .... 1 →   Q
No .... 2 →   Q

15. MonLvTg **Ask if lived as a couple before getting married (LvTgthr = 1)**

What month and year did you start living together?

ENTER MONTH 1..12   →   Q

16. YrLvTg ENTER YEAR IN 4 DIGIT FORM E.G.2000

1900..2005   →   Q

17. Current **Ask all who are or have been legally married (NumMar ≥ 1 or ClNumMar ≥ 1)**

For last marriage entered

INTERVIEWER - IS THIS MARRIAGE CURRENT O HAS IT ENDED?

Current... 1 →   Q
Ended .... 2 →   Q

. HowEnded Ask if marriage ended *(Current = 2 or marriage number less than total marriages)*

Did your marriage end in ...

death .... 1 → Q19
divorce .... 2 ⎤→ Q21
or separation? .... 3 ⎦

. MonDie **Ask if marriage ended in death *(HowEnded = 1)***

What month and year did your husband/wife die?

ENTER MONTH 1..12   →   Q20

. YrDie ENTER YEAR IN 4 DIGIT FORMAT E.G. 2000

1900-2005   →   Q25

. MonSep **Ask if marriage ended in divorce or separation *(HowEnded = 2 or 3)***

What month and year did you stop living together?

ENTER MONTH 1..12   →   Q22

. YrSep ENTER YEAR IN 4 DIGIT FORMAT E.G. 2000

1900-2005   →See Q23

. MonDiv **Ask if marriage ended in divorce *(HowEnded =2)***

What month and year was your decree absolute granted?

ENTER MONTH 1..12   →   Q24

. YrDiv ENTER YEAR IN 4 DIGIT FORMAT E.G. 2000

1900-2005   →   Q25

. Cohab **Ask if respondent is aged 16-59 *(DVAge = 16-59)***

Have you had any previous relationships in which you lived together with someone as a couple but did not get married?

Yes .... 1 → Q26
No .... 2 → Q41

. Numcohab **Ask if respondent is aged 16-59, and has had previous cohabiting relationships *(DVAge = 16-59 & Cohab = 1)***

How many relationships have you had altogether in which you lived together with someone as a couple but did not get married?
(Please exclude your present relationship)

1..7   →   Q27

. Intro Now I would like to ask you some questions about the first three of these relationships.

RECORD DETAILS OF THE FIRST THREE RELATIONSHIPS, STARTING WITH THE FIRST

Ask each question for the first, second and third relationship   →   Q28

28. TimeCoy Thinking about the first/second/third relationship where you lived as a couple but did not get married, how long did you live together?

INTERVIEWER - ENTER NUMBER OF YEARS

0..99   →   Q29

29. Timecom INTERVIEWER - ENTER NUMBER OF MONTHS

0..11   →   Q30

30. WhencoM Can you tell me the month and year in which you started or stopped living together as a couple with your partner?

INTERVIEWER ENTER THE MONTH

1..12   →   Q31

31. WhencoY ENTER THE YEAR

1950..2005   →   Q32

32. Starten INTERVIEWER: IS THIS WHEN THE RESPONDENT AND HIS/HER PARTNER STARTED OR STOPPED LIVING TOGETHER AS A COUPLE?

ASK RESPONDENT IF YOU ARE UNSURE

Start date .... 1 ⎤→ Q33
End date .... 2 ⎦

33. Othdate If that was the date you started/stopped living together, then you stopped/started living together in ...(month) ...(year)
Does that seem about right?

Yes .... 1 → Q36
No .... 2 → Q34

34. RghtdtM **Ask if computed start/end date not correct *(Othdate1 = 2)***

What is the correct date?

INTERVIEWER ENTER THE MONTH

1..12   →   Q35

35. RghtdtY ENTER THE YEAR

1950..2005   →   Q36

36. EndCoh **Ask if respondent is aged 16-59, and has had previous cohabiting relationships *(DVAge = 16-59 & Cohab = 1)***

Ask each question for the first, second and third relationship

You said you stopped living together in ...(month) ...(year). May I just check, was this when you stopped living in the same accommodation or when the relationship ended?

Stopped living in the same
accommodation .... 1 → Q37
End of the relationship .... 2 → Q39
Both .... 3 ⎤
Partner died .... 4 ⎥ Next
Stopped living in same accommodation, ⎥ relationship
but still having a relationship .... 5 ⎦ (Q28) or Q41

37. EndrelM **Ask if date given is when they stopped living together (EndCoh = 1)**

When did the relationship end?

INTERVIEWER ENTER THE MONTH

1..12 → Q38

38. EndrelY ENTER THE YEAR

1950..2005 → Next
relationship
(Q28) or Q41

39. EndlivM **Ask if the date given is when relationship ended (EndCoh = 2)**

When did you stop living in the same accommodation?

INTERVIEWER ENTER THE MONTH

1..12 → Q40

40. EndlivY ENTER THE YEAR

1950..2005 → Next
relationship
(Q28) or Q41

CHILDREN

41. Children **Ask respondents aged 16-59 (DVAge = 16-59)**

INTERVIEWER: DOES THIS PERSON HAVE ANY CHILDREN IN THE HOUSEHOLD (INCLUDES ADULT CHILDREN AND/OR STEP OR FOSTER CHILDREN)

Yes .... 1 → See Qs
42 & 43
No .... 2 → See Q52

42. StpChldF **Ask women who have a child in the household (Sex = 2 & Children =1)**

(The next questions are about the family.) Have you any step, foster, or adopted children living with you, (including any children from your partner's previous relationship)?

Yes .... 1 → Q44
No .... 2 → Q52

43. StpChldM **Ask men who have a child in the household (Sex = 1 & Children =1)**

Have you any stepchildren of any age living with you,

(including any children from your partner's previous relationship)?

Yes .... 1 → Q4
No .... 2 → View of yo
local a

44. NumStep **Ask women with a step, foster or adopted child, a man with a stepchild living with them (StpChld = 1 or StpChldM=1)**

How many step children have you living with you altogether?

1..7 → See Q4

45. NumFost **Ask women with a step, foster or adopted child living with them (StpChldF = 1), others go to Q47**

How many foster children have you living with you altogether?

0..7 → Q4

46. NumAdop How many adopted children have you living with you altogether?

0..7 → Q

47. StepInt **Ask women with a step, foster or adopted child, a man with a stepchild living with them (StpChld = 1 or StpChldM=1)**

THE NEXT SCREEN CONSISTS OF A TABLE FOR THE STEP- CHILDREN (AND ADOPTED AND FOSTER- CHILDREN) OF (NAME) PLEASE ENTER DETAILS FOR EACH CHILD → Q4

48. ChildNo **Ask for each step/foster/adopted child**

ENTER PERSON NUMBER(S) OF THE STEP/ FOSTER/ADOPTED CHILD (INCLUDES ADULT CHILDREN)

1..20 → Q

49. ChldType ENTER CODE AS FOLLOWS

Step .... 1 ⎤
Foster.. 2 ⎥ → Q
Adopted .... 3 ⎦

50. ChLivMon DATE CHILD STARTED LIVING WITH RESPONDENT

ENTER MONTH

1..12 → Q

51. ChLivYr YEAR (IN 4 DIGIT FORMAT, E.G. 2000)

1900..2005 → See Q

52. Baby **Ask all women (Sex = 2), men go to View of your local area**

ASK OR RECORD
EXCLUDE: ANY STILLBORN
INCLUDE ANY WHO ONLY LIVED FOR A SHORT TIME

Have you ever had a baby - even one who only lived for a short time?

Yes .... 1 →   Q53
No .... 2 →See Q58

. NumBaby   **Ask women who have had a baby** *(Baby = 1)*

EXCLUDE: ANY STILLBORN

How many children have you given birth to, including any who are not living here and any who may have died since birth?

1..20   →See Q54

. BirthInt   THE NEXT SCREEN CONSISTS OF A TABLE OF CHILDREN TO WHOM (...) HAS GIVEN BIRTH PLEASE ENTER DETAILS FOR EACH CHILD

. BirthDte   **For each child**

Date of birth

PLEASE ENTER IN DATE OF BIRTH ORDER - ELDEST FIRST, YOUNGEST LAST

AS A GUIDE, THE D.O.B. OF EACH HOUSEHOLD MEMBER IS LISTED BELOW   →   Q56

. BirthSex   Sex of child

Male .... 1 ⎤
Female .... 2 ⎦→ Q57

. ChldLive   Is child living with respondent?

Yes .... 1 ⎤
No, lives elsewhere .... 2 ⎬→See Q58
No, deceased .... 3 ⎦

. Pregnant   **Ask all women aged 16-49** *(Sex = 2 & DVAge = 16-49)*, **others go to View of your local area**

(May I just check), are you pregnant now?

Yes .... 1 ⎤→
No/unsure .... 2 ⎦ Q59

MoreChld [*]   Do you think that you will have any (more) children (after the one you are expecting)? Could you choose your answers from this card.

SHOW CARD F

Yes .... 1 ⎤→
Probably yes .... 2 ⎦ Q61
Probably not... 3 ⎤→ Contraception
No .... 4 ⎦

ProbMore [*]   **Ask if respondent answered don't know above** *(MoreChld = DK)*

On the whole do you think...

You will probably have any/more children .... 1 →   Q61
Or you will probably not have any/more children? .... 2 → Contraception

---

61. TotChld [*]   **Ask if respondent is likely to have more children** *(MoreChld = 1 or 2 or ProbMore = 1)*

(Can I just check, you have ... children still alive). How many children do you think you will have born to you in all including those you have had already(who are still alive)(and the one you are expecting)?

1..14   →   Q62

62. NextAge [*]   How old do you think you will be when you have your first/next baby (after the one you are expecting)?

1..97   → Contraception

CONTRACEPTION

This section applies to women aged 16-49

1. SterilA   **Ask married women and women cohabiting with men (who are not pregnant) , others see Q3**

We've talked about how many children you think you'll have. The next questions are about ways of preventing pregnancy.

Have you or your husband/partner had a vasectomy or been sterilised - I mean have either of you ever had an operation intended to prevent you getting pregnant (again)?

Yes .... 1 →   Q2
No .... 2 → See Q6
Refused whole contraception section .... 7 → View of your local area

2. WhoStlsd   **If** *(SterilA = 1)*

Was it you who was sterilised or your husband/partner who had a vasectomy?

Respondent .... 1 ⎤
Husband or partner.............................. 2 ⎬→ Q4
Both .... 3 ⎦

3. SterilB   **Ask non-cohabiting women (who are not pregnant), others see Q9**

We've talked about how many children you think you'll have. The next questions are about ways of preventing pregnancy.

Have you ever been sterilised - I mean ever had an operation intended to prevent you getting pregnant (again)?

Yes .... 1 →   Q4
No .... 2 →   Q7

4. SterNHS   **Ask women who have been sterilised**

Was that operation carried out under the NHS?

Yes .... 1 ⎤
No .... 2 ⎬
Both sterilised - one on the NHS   ⎬→ Q5
and one not.. 3 ⎦

5. ChkFp2   **If (WhoStlsd = 1, 2 or 3) OR (SterilB = 1)**

Did this operation/either of these operations take place within the last two years?

Yes .... 1 →   Q22
No .... 2 → View of your
                                                local area

6. OtherOp1   **Ask if married/cohabiting and not pregnant or sterile**

Have you or your husband/partner had any other operation which prevents you getting pregnant (again)?

Yes respondent...................................... 1 ⎤
Yes, husband or partner .... 2 ⎬→   Q8
Yes, both.. 3 ⎦
No .... 4 →   Q15

7. OtherOp2   **If not married/cohabiting and not pregnant or sterile**

Have you had any other operation which prevents you becoming pregnant (again)?

Yes .... 1 →   Q8
No .... 2 →   Q15

8. ChkFp3   **If (OtherOp1 = 1, 2 or 3) OR (OtherOp2 = 1)**

Did this operation/either of these operations take place within the last two years?

Yes .... 1 →   Q22
No .... 2 → View of your
                                                local area

9. CcUsed   **Ask if the respondent is pregnant**

SHOW CARD [TCON1]

Here is a list of ways of preventing pregnancy - were you or your partner using any of them when you became pregnant?

Yes .... 1 →   Q10
No .... 2 →   Q20

10. CCPreg   **If CcUsed = yes**

SHOW CARD [TCON1]

Please can you look through the list to the end of the card and read out the numbers beside the methods which applied to you and your husband/partner when you got pregnant?

CODE UP TO 4 METHODS

Withdrawal .... 1 ⎤
Male sheath/condom .... 2 ⎬→See Q12
Safe period/rhythm method/Persona ..... 3 ⎬
Cap/Diaphragm...................................... 4 ⎦
Pill .... 6 →   Q11

IUD/coil/intra-uterine device................... 7 ⎤
Hormonal IUS - MIRENA .... 8 ⎟
Foams/gels/sprays/pessaries ⎟
(spermicides) .... 9 ⎬→See Q
Female condom ....10 ⎟
Going without sexual intercourse to ⎟
avoid pregnancy.....................................11 ⎟
Injections...12 ⎟
Surgically implanted hormone capsules 13 ⎟
Another method ....14 ⎦

11. PillTyp1   **Ask if CCPreg = Pill**

SHOW CARD [TCON2]

Is the pill you take one of the brands listed (MICRONOR, NORIDAY, FEMULEN, MICROVAL, NORGESTON, NEOGEST)? These are progestogen only pills (sometimes known as the mir pill) as opposed to combined pills.

Yes .... 1 ⎤
No .... 2 ⎬→See Q
Not sure .... 3 ⎦

12. CcmComb   **Ask if more than one method used**

You have mentioned that you (and your husband/ partner) usually use more than one method. Do/did you use them in combination or do/did you sometime use one and sometimes the other?

In combination .... 1 →See Q
Sometimes one, sometimes other .... 2 →   Q

13. MstFrq   **Ask if CcmComb = 2**

SHOW CARD [TCON1]

Which one do you use most often?

Withdrawal .... 1 ⎤
Male sheath/condom .... 2 ⎟
Safe period/rhythm method/Persona ..... 3 ⎟
Cap/Diaphragm...................................... 4 ⎟
Pill .... 6 ⎟
IUD/coil/intra-uterine device................... 7 ⎟
Hormonal IUS - MIRENA .... 8 ⎟
Foams/gels/sprays/pessaries ⎬→See Q
(spermicides) .... 9 ⎟
Going without sexual intercourse to ⎟
avoid pregnancy.....................................11 ⎟
Female condom ....10 ⎟
Injections...12 ⎟
Surgically implanted hormone capsules 13 ⎟
Another method ....14 ⎦

14. UsuTime   **Ask if one or more methods used**

How long has this/have these been your usual method(s)?

Under 3 months .... 1 ⎤
At least 3 months, less than 6 months ... 2 ⎬→See Q
At least 6 months, less than 1 year........ 3 ⎦
At least 1 year, less than 2 years .... 4 ⎤
At least 2 years, less than 5 years .... 5 ⎬→   Q
5 years of more .... 6 ⎦

CCMUsu   **Ask if respondent not pregnant and not sterile**

SHOW CARD [TCON3]

Here is a list of possible ways of preventing pregnancy. Which, if any, do you (and your partner) usually use at present?

CODE UP TO 4 METHODS

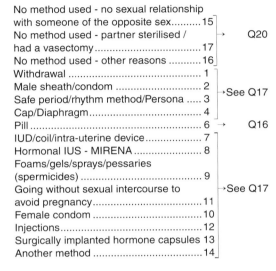

No method used - no sexual relationship
with someone of the opposite sex..........15 ⎤
No method used - partner sterilised /
had a vasectomy....................................17 →   Q20
No method used - other reasons ....16 ⎦
Withdrawal .... 1 ⎤
Male sheath/condom 2
Safe period/rhythm method/Persona ..... 3 →See Q17
Cap/Diaphragm....................................... 4 ⎦
Pill .... 6 →   Q16
IUD/coil/intra-uterine device................... 7 ⎤
Hormonal IUS - MIRENA 8
Foams/gels/sprays/pessaries
(spermicides) 9
Going without sexual intercourse to →See Q17
avoid pregnancy..................................... 11
Female condom 10
Injections... 12
Surgically implanted hormone capsules 13
Another method .... 14 ⎦

6. PillTyp2   **Ask if *CCMUsu = Pill***

SHOW CARD [TCON2]

Is the pill you take one of the brands listed
(MICRONOR, NORIDAY, FEMULEN, MICROVAL,
NORGESTON, NEOGEST)?
These are progestogen only pills (sometimes known
as the mini-pill) as opposed to combined pills.

Yes .... 1 ⎤
No .... 2 →See Q17
Not sure .... 3 ⎦

7. CcmComb1   **Ask if more than one method used**

You have mentioned that you (and your husband/
partner) usually use more than one method. Do/did
you use them in combination or do/did you sometimes
use one and sometimes the other?

In combination .... 1 →   Q19
Sometimes one, sometimes other .... 2 →   Q18

8. MstFrq1   **Ask if *CcmComb1 = 2***

SHOW CARD [TCON1]

Which one do you use most often?

Withdrawal .... 1 ⎤
Male sheath/condom 2
Safe period/rhythm method/Persona 3
Cap/Diaphragm....................................... 4 →   Q19
Pill 6
IUD/coil/intra-uterine device................... 7
Hormonal IUS - MIRENA .... 8 ⎦

Foams/gels/sprays/pessaries
(spermicides) .... 9 ⎤
Going without sexual intercourse to
avoid pregnancy..................................... 11 →   Q19
Female condom 10
Injections... 12
Surgically implanted hormone capsules 13
Another method .... 14 ⎦

19. UsuTime   **Ask if one or more methods used**

How long has this / have these been your usual
method(s)?

Under 3 months .... 1 ⎤
At least 3 months, less than 6 months... 2
At least 6 months, less than 1 year........ 3 →See Q22
At least 1 year, less than 2 years .... 4 ⎦
At least 2 years, less than 5 years .... 5 ⎤
5 years of more .... 6 ⎦ →   Q26

20. YnoCC   **Ask if no method currently used**

SHOW CARD [TCON4]

Here is a list of reasons why people do not use any
method for preventing pregnancy. Can you tell me
which reason applies/applied to you?

CODE MAIN REASON ONLY

I want to become pregnant .... 1 ⎤
Unlikely to conceive because of the
menopause 2
Unlikely to conceive because possibly
infertile 3
Don't like contraception/Find methods
unsatisfactory... 4
My partner doesn't like - or won't use - →See Q21
contraception 5
Don't know where to obtain
contraceptives / advice 6
Find access to contraceptive services
difficult.. 7
Some other reason .... 8 ⎦

21. UsedL2Yr   **Ask if no method needed or possibly pregnant**

SHOW CARD [TCON5]

Have you (or your husband/partner) ever used any of
these methods in the last 2 years?

Yes .... 1 →   Q22
No .... 2 →   Q26

22. CcBfor   **Ask if (UsedL2Yr = 1) OR (ChkFp2 = 1) OR
(ChkFp3 = 1) OR (UsuTime = 1, 2, 3, or 4)**

SHOW CARD [TCON5]

(Here is a list of ways of preventing pregnancy). Which
methods, if any, did you (or your husband/partner) use
(in the last 2 years/immediately before that)?

ENTER UP TO 4 METHODS

No method needed-no sexual
relationship 15
No method used at all .... 16 →See Q26
Pregnant 19
Withdrawal 1
Male sheath/condom .... 2 →See Q24
Safe period/rhythm method./Persona 3
Cap/diaphragm 4
Pill-not sure if mini or combined............. 6 →   Q23
IUD/Coil/intra-uterine device 7
Hormonal IUS - MIRENA 8
Foams/Gels, sprays, pessaries
(spermicides) 9
Female condom 10
Going without sexual intercourse to   →See Q24
avoid pregnancy 11
Injections 12
Surgically implanted hormone
capsules... 13
Another method 14

23. PillTyp3   **Ask if *CcBfor = Pill***

SHOW CARD [TCON2]

Is the pill you took one of the brands listed
(MICRONOR, NORIDAY, FEMULEN, MICROVAL,
NORGESTON, NEOGEST)?
These are progestogen only pills (sometimes known
as the mini-pill) as opposed to combined pills.

Yes 1
No .... 2 →See Q24
Not sure 3

24. CcmComb2   **Ask if more than one method used**

You have mentioned that you (and your husband/
partner) usually use more than one method. Do/did
you use them in combination or do/did you
sometimes use one and sometimes the other?

In combination .... 1 →See Q26
Sometimes one, sometimes other .... 2 →   Q25

25. MstFrq2   **Ask if *CcmComb2 = 2***

SHOW CARD [TCON1]

Which one do you use most often?

Withdrawal 1
Male sheath/condom 2
Safe period/rhythm method/Persona 3
Cap/Diaphragm...................................... 4
Pill 6
IUD/coil/intra-uterine device 7
Hormonal IUS - MIRENA 8
Foams/gels/sprays/pessaries   →See Q26
(spermicides) 9
Going without sexual intercourse to
avoid pregnancy.................................... 11
Female condom 10
Injections... 12
Surgically implanted hormone
capsules... 13
Another method 14

26. Empilk   **Ask if the respondent is not sterile OR they had
an operation that prevents pregnancy less than
years ago**

There are two other methods of contraception
available. Both are referred to as emergency
contraception. One is a pill based method, sometim
known as the 'morning after' pill. The other is an IUI
(intra-uterine device) method.

Before I mentioned it, had you heard of the pill meth
of emergency contraception after intercourse?

Yes 1
No .... 2 →   Q
Don't know 3

27. EmIUDk   **Ask if the respondent is not sterile OR they had
an operation that prevents pregnancy less than
years ago**

Before I mentioned it, had you heard of the IUD
method of emergency contraception after intercours

Yes .... 1 →   Q
No 2
Don't know 3   See Q

28. Emercon   **Ask if *Empilk = yes OR EmIUDk = yes***

Have you used the emergency contraception, that is
the 'morning after pill' or IUD method in the last 12
months?

Yes .... 1 →   Q
No .... 2 →See Q

29. EmerNun   **Ask if *Emercon = yes***

On how many occasions in the last 12 months have
you used emergency contraception?

ENTER NUMBER   →   Q

30. MaMeth   **Ask for each occasion used**

(And for each occasion) could you tell me the
method(s) you used?

PROMPT AS NECESSARY

Pill method, sometimes called the
'morning after pill' .... 1 →   Q
IUD coil fitted .... 2 →   Q

31. Whrpill   **Ask if *MaMeth = pill method***

SHOW CARD [TCON7]

Where did you go for this?

Your own GP or practice nurse............... 1
Another GP or practice nurse 2
Family Planning Clinic, (including
Brook Clinics) 3
Hospital Accident & Emergency   →See Q
Department 4
Directly to a chemist or pharmacy 5
A walk-in centre or minor injuries unit.... 6
Somewhere else 7

WhrIIUD Ask if *MaMeth = IUD coil fitted*

SHOW CARD [TCON8]

Where did you go for this?

Your own GP 1
Another GP 2
Family Planning Clinic, (including
Brook Clinics) .... 3 →See Q33
Hospital Accident & Emergency
Department 4
Somewhere else 5

MorePoss Ask if not pregnant and not sterile, others go to View of your local area

As far as you know, could you (and your husband/partner) have (more) children if you wanted to or would it be difficult or impossible?

Could have more children .... 1 → View of your local area
Would be difficult/impossible.................. 2 → Q34

PrDiff If *Moreposs =2*

SHOW CARD [TCON6]

Will you please look at this card and tell me what the difficulty is?

ENTER UP TO 3 REASONS

Getting pregnant .... 1 → Q36
Having a baby born alive 2
Pregnancy would endanger health .... 3 → View of your local area
Passed the menopause-change of life... 4
Other.. 5 → Q35

XprDiff If *PrDiff = 5*

RECORD OTHER DIFFICULTY

DocAdvce Ask if *PrDiff = 1 or 2*

Have you (or your husband/partner) ever consulted a doctor about the difficulty you have or would have in getting pregnant/having a baby born alive?

Yes .... 1 → View of your local area
No 2

VIEW OF YOUR LOCAL AREA
(taken from the social capital question set)

Ask selected adult

AskNow (NAME) HAS BEEN PICKED TO ANSWER THE SECTION ON VIEW OF YOUR LOCAL AREA. DO YOU WANT TO ASK THIS SECTION FOR (NAME) NOW OR LATER?

IF YOU HAVE ALREADY ASKED THIS SECTION FOR (NAME), DO NOT CHANGE FROM CODE 1

Yes, now/already asked.......................... 1 → Q2
Later.. 2

2. Areaint Now I would like to ask you some questions about your local area. (By area I mean within about a 15-20 minute walk or 5-10 minute drive from your home.) → Q3

3. Arealive **Ask selected adult**

How long have you lived in this area?

CODE YEARS
IF LESS THAN 1 CODE AS 0

0..97 → See Q4

4. Areamth **Ask if they have lived in the area for less than a year *(Arealive = 0)***

How many months have you lived in this area?

0..11 → Q5

5. Enjyliv **Ask selected adult**

Would you say this is an area you enjoy living in?

Yes 1
No .... 2 → Q6
Don't know 3

6. Locserv [*] Thinking generally about what you expect of local services, how would you rate the following:→ Q7

7. Leisyou [*] Social & leisure facilities for people like yourself

SHOW CARD TSC1

Very Good 1
Good 2
Average.. 3 → Q8
Poor 4
Very Poor 5
Don't know or have had no experience.. 6

8. Leiskids [*] Facilities for young children up to the age of 12

SHOW CARD TSC1

Very Good 1
Good 2
Average.. 3 → Q9
Poor 4
Very Poor 5
Don't know or have had no experience.. 6

9. Leisteen [*] Facilities for teenagers (those aged 13 to 17)

SHOW CARD TSC1

Very Good 1
Good 2
Average.. 3 → Q10
Poor 4
Very Poor 5
Don't know or have had no experience.. 6

10. Bins [*] Rubbish Collection

SHOW CARD TSC1

Very Good 1
Good 2
Average.. 3 → Q11
Poor 4
Very Poor 5
Don't know or have had no experience.. 6

11. Lochlth [*]   Local Health services (e.g. your GP or the local hospital)

SHOW CARD TSC1

Very Good 1
Good 2
Average.. 3 → Q12
Poor 4
Very Poor 5
Don't know or have had no experience.. 6

12. Schools [*]   Local schools, colleges and adult education

SHOW CARD TSC1

Very Good 1
Good 2
Average.. 3 → Q13
Poor 4
Very Poor 5
Don't know or have had no experience.. 6

13. Police [*]   Local police service

SHOW CARD TSC1

Very Good 1
Good 2
Average.. 3 → Q14
Poor 4
Very Poor 5
Don't know or have had no experience.. 6

14. Transport   What is your main form of transport?

Car/Motorcycle/Moped.......................... 1
Public transport (ie buses and trains) 2
Cycling .... 3 → Q15
Walking 4
Other.. 5
Never goes out...................................... 6

15. Loctrans [*]   Would you say this area has good local transport for where you want to get to?

Yes 1
No .... 2 → Q16
Don't know 3

16. Walkday [*]   How safe do you feel walking alone in this area during daytime? Do you feel ...

RUNNING PROMPT

very safe 1
fairly safe... 2
a bit unsafe 3
very unsafe .... 4 → Q17
or do you never go out alone
during daytime? 5

17. Walkdark [*]   How safe do you feel walking alone in this area after dark? Do you feel ...

RUNNING PROMPT

very safe 1
fairly safe.. 2
a bit unsafe .... 3 → Q
very unsafe 4
or do you never go out alone after dark? 5

18. Traffic [*]   Still thinking about the same area, can you tell me how much of a problem these things are.

The speed or volume of road traffic

SHOW CARD TSC2

Very big problem 1
Fairly big problem 2
Minor problem .... 3 → Q
Not at all a problem............................... 4
It happens but is not a problem.............. 5 → Q
Don't know 6

19. Parking [*]   Parking in residential streets

SHOW CARD TSC2

Very big problem 1
Fairly big problem 2
Minor problem .... 3 → Q
Not at all a problem............................... 4
It happens but is not a problem.............. 5
Don't know 6

20. Carcrime [*]   Car crime (e.g. damage, theft and joyriding)

SHOW CARD TSC2

Very big problem 1
Fairly big problem 2
Minor problem .... 3 → Q
Not at all a problem............................... 4
It happens but is not a problem.............. 5
Don't know 6

21. Rubbish [*]   Rubbish and litter lying around

SHOW CARD TSC2

Very big problem 1
Fairly big problem 2
Minor problem .... 3 → Q
Not at all a problem............................... 4
It happens but is not a problem.............. 5
Don't know 6

22. DogMess [*]   Dog mess

SHOW CARD TSC2

Very big problem 1
Fairly big problem 2
Minor problem .... 3 → Q
Not at all a problem............................... 4
It happens but is not a problem.............. 5
Don't know 6

Graffiti [*]   Graffiti or vandalism

SHOW CARD TSC2

Very big problem 1	
Fairly big problem 2	
Minor problem... 3	→ Q24
Not at all a problem.............................. 4	
It happens but is not a problem.............. 5	
Don't know 6	

NoiseNbr [*] Level of noise

SHOW CARD TSC2

Very big problem 1	
Fairly big problem 2	
Minor problem... 3	→ Q25
Not at all a problem.............................. 4	
It happens but is not a problem.............. 5	
Don't know 6	

Teenager [*] Teenagers hanging around on the streets

SHOW CARD TSC2

Very big problem 1	
Fairly big problem 2	
Minor problem... 3	→ Q26
Not at all a problem.............................. 4	
It happens but is not a problem.............. 5	
Don't know 6	

Alcdrug [*] Alcohol or drug use

SHOW CARD TSC2

Very big problem 1	
Fairly big problem 2	
Minor problem... 3	→ Q27
Not at all a problem.............................. 4	
It happens but is not a problem.............. 5	
Don't know 6	

Victim Have you personally been a victim of any of the following crimes in the past 12 months?

CODE ALL THAT APPLY

SHOW CARD TSC3

Theft or break-in to house or flat............ 1	
Theft or break-in to car parked in the area... 2	
Personal experience of theft or mugging in the area 3	→ Income
Physical attack in the area (i.e. hit or kicked in a way that hurt you)................. 4	
Racist attack in the area (either verbal or physical).. 5	
None of these 6	

INCOME

1. Intro   **Ask all adults (except proxy respondents), proxy informants go to Q49**

THE NEXT SECTION IS ABOUT BENEFITS AND OTHER SOURCES OF INCOME

2. Ben1YN   SHOW CARD G

Looking at this card, are you at present receiving any of these state benefits in your own right: that is, where you are the named recipient?

Yes 1	→ Q3
No 2	
Refused whole income section 7	→ Q49

3. Ben1Q   **Ask if receiving a state benefit (Ben1YN = 1 or 2)**

SHOW CARD G

RECORD BENEFITS RECEIVED
CODE ALL THAT APPLY (NONE OF THESE = CODE 8) ENTER AT MOST 6 CODES

Child Benefit 1	
Guardian's Allowance 2	
Invalid Care Allowance........................... 3	
Retirement pension (National Insurance), or Old Person's pension 4	
Widow's pension, Bereavement Allowance or Widowed Parents (formerly Widowed Mother's) Allowance 5	→ Q4
War Disablement Pension or War Widow's Pension (and related allowances)........................ 6	
Severe Disablement Allowance 7	
None of these 8	

4. Ben2Q   SHOW CARD H

And looking at this card, are you at present receiving any of the state benefits shown on this card - either in your own name, or on behalf of someone else in the household?

CODE ALL THAT APPLY

CARE COMPONENT of Disability Living Allowance 1	→	Q5
MOBILITY COMPONENT of Disability Living Allowance 2	→	Q6
Attendance Allowance 3	→	Q7
None of these 4	→	Q9

5. WhoReCar   **Ask if receiving CARE component of Disability Living Allowance (Ben2Q = 1)**

Whom do you receive it for?
IF CURRENT HOUSEHOLD MEMBER, ENTER PERSON NUMBER OTHERWISE ENTER 97 → See Q6

6. WhoReMob   **Ask if receiving MOBILITY component of Disability Living Allowance (Ben2Q = 2)**

Whom do you receive it for?
IF CURRENT HOUSEHOLD MEMBER, ENTER

PERSON NUMBER OTHERWISE ENTER 97→ See Q7

7. WhoReAtt   **Ask if receiving Attendance Allowance (Ben2Q = 3)**

Whom do you receive it for?
IF CURRENT HOUSEHOLD MEMBER, ENTER
PERSON NUMBER OTHERWISE ENTER 97→   Q8

8. AttAllFU   Is this paid as part of your retirement pension or do you receive a separate payment?

Together with pension .... 1 ⌉
Separate payment.................................. 2 ⌋ → Q9

9. Ben3Q   **Ask all except proxy respondents**

CODE ALL THAT APPLY
ENTER AT MOST 5 CODES

SHOW CARD I

Now looking at this card, are you at present receiving any of these benefits in your own right: that is, where you are the named recipient?

Jobseekers' Allowance............................ 1 →   Q10
Income Support .... 2 ⌉
Incapacity Benefit .... 3 │
Statutory Sick Pay.................................. 4 ├→See Q11
Industrial Injury Disablement Benefit ..... 5 │
None of these .... 6 ⌋

10. JSAType   **Ask if respondent is receiving Jobseekers' Allowance (Ben3Q = 1)**

There are two types of Jobseekers' Allowance. Is your allowance...

RUNNING PROMPT

'contributory' that is, based on your
National Insurance contributions............ 1 ⌉
or is it 'income-based' Jobseekers' │
Allowance, which is based on an ├→See Q11
assessment of your income? .... 2 ⌋

11. Ben4Q   **Ask women under 55 years (Sex = 2 & DVAge <55 years), others Q12**

SHOW CARD J
Are you currently getting either of the things shown on this card, in your own right?

Maternity Allowance............................... 1 ⌉
Statutory maternity pay from your │
employer or a former employer .... 2 ├ → Q12
Neither of these .... 3 ⌋

12. Ben4AQ   **Ask all except proxy respondents**

SHOW CARD K

Now looking at this card, are you at present receiving any of these Tax Credits, in your own right? Please include any lump sum payments received in the last six months.

Working Families' Tax Credit .... 1 ⌉
Disabled Person's Tax Credit .... 2 │
Children's Tax Credit .... 3 ├ → Q
None of these .... 4 ⌋

13. Ben5Q   **All except proxy respondents**

SHOW CARD L

In the last 6 months, have you received any of the things shown on this card, in your own right?

CODE ALL THAT APPLY

A grant from the Social Fund for funeral
expenses.. 1 ⌉
A grant from Social Fund for maternity │
expenses/Sure Start │
  Maternity Grant .... 2 │
A Social Fund loan or Community │
Care grant.. 3 │
A Back to Work bonus............................ 4 │
'Extended payment' of Housing │
Benefit/rent rebate, │
  or Council Tax Benefit .... 5 ├→See Q
Widow's payment or Bereavement │
payment- lump sum .... 6 │
Child Maintenance Bonus...................... 7 │
Lone Parent's Benefit Run-On .... 8 │
Any National Insurance or State │
Benefit not mentioned earlier................. 9 │
None of these ....10 ⌋

14. Ben1Amt   **Code for each benefit mentioned (Ben1Q, Ben2Q (except Attendance Allowance combined with pension), Ben3Q, Ben4Q, Ben4A (except for Children's Tax Credit), Ben5Q), others Q17**

How much did you get last time?

(IF COMBINED WITH ANOTHER BENEFIT AND UNABLE TO GIVE SEPARATE AMOUNT, ENTER `Don't know`)

0.00..997.00   →   Q

15. Ben1AmtDK   **If don't know or refusal at the amount of benefit received (Ben1Amt = DK or Refusal)**

INTERVIEWER: IS THIS `DON'T KNOW` BECAUSE IT'S PAID IN COMBINATION WITH ANOTHER BENEFIT, AND YOU CANNOT ESTABLISH A SEPARATE AMOUNT?

Yes (Please give full details in a Note) .. 1 ⌉→See Q
No .... 2 ⌋

16. Ben1Pd   **Ask if amount of benefit received was greater than zero (Ben1Amt > 0.00)**

How long did this cover?

one week.. 1 ⌉
two weeks .... 2 │
three weeks.. 3 │
four weeks.. 4 ├ → Q
calendar month .... 5 │
two calendar months.............................. 7 ⌋

eight times a year.................................... 8		
nine times a year..................................... 9		
ten times a year 10		
three months/13 weeks..........................13		
six months/26 weeks..............................26	→	Q17
one year/12 months/52 weeks 52		
less than one week 90		
one off lump sum 95		
none of these 97		

. OthSourc   **Ask all (except proxy respondents)**

SHOW CARD M

Please look at this card and tell me whether you are receiving any regular payment of the kinds listed on it?

Yes receiving benefits - code at next question.. 1	→	Q18
No, not receiving any 2	→	Q21

. OthSrcM   **Ask if receiving any of the benefits mentioned above (OthSourc = 1)**

SHOW CARD M

RECORD PAYMENTS RECEIVED
CODE ALL THAT APPLY
(ENTER AT MOST 4 CODES)

Occupational pensions from former employer(s) 1		
Occupational pensions from a spouse's former employer(s).................. 2		
Private pensions or annuities................ 3		
Regular redundancy payments from former employer(s)............................... 4	→	Q19
Government Training Schemes, such as YT allowance 5		

. OthNetAm   In total how much do you receive each month from (...../all these sources) AFTER tax is deducted? (ie net)

DO NOT PROBE MONTH. ACCEPT CALENDAR MONTH OR 4 WEEKLY

0.01..99999.97   →   Q20

. OthGrsAm   In total how much do you receive each month from ..... (all these sources) BEFORE tax is deducted? (ie GROSS)?

DO NOT PROBE MONTH. ACCEPT CALENDAR MONTH OR 4 WEEKLY

0.01..99999.97   →   Q21

. ReglrPay   **Ask all (except proxy respondents)**

SHOW CARD N

Now please look at this card and tell me whether you are receiving any regular payments of the kind listed on it?

Yes receiving benefits - code at next question 1	→	Q22
No, not receiving any 2	→	Q24

22. ReglrPM   **If receiving one of the benefits mentioned above (ReglrPay = 1)**

SHOW CARD N

RECORD TYPES OF PAYMENT RECEIVED

CODE ALL THAT APPLY

Educational grant.................................... 1		
Regular payments from friends or relatives outside the household 2	→	Q23
Maintenance, alimony or separation allowance 3		

23. ReglrpAm   In total how much do you receive from these each month?

0.01..99999.97   →   Q24

24. Rentpay   **Ask all (except proxy respondents)**

Are you currently receiving any rent from property or subletting?

Yes 1	→	Q25
No 2	→	See Q26

25. Rentamt   **Ask if they are receiving rent (Rentpay = 1)**

In total how much do you receive each month?

0.01..99999.97   → See Q26

The next group of questions (Q26-Q44) are only asked of those in paid work, (including those temporarily away from job or on a government scheme), but excluding unpaid family workers. ((Wrking = 1 OR JbAway = 1 OR SchemeET = 1) & (OwnBus = 2 & RelBus = 2))

The routing instructions above each question apply only to those who meet the above criteria.

26. PyPeriod   **Ask if an employee (Stat = 1), others see Q36**

THE NEXT QUESTIONS ARE ABOUT EARNINGS

How long a period does your wage/salary usually cover?

one week... 1		
two weeks 2		
three weeks.. 3		
four weeks.. 4		
calendar month 5		
two calendar months............................. 7	→	Q27
eight times a year.................................. 8		
nine times a year................................... 9		
ten times a year 10		
three months/13 weeks..........................13		
six months/26 weeks..............................26		
one year/12 months/52 weeks 52		

less than one week ....90 ⎤
one off lump sum ....95 ⎬ →   Q28
none of these ....97 ⎦

27. Takehome   Ask all, except those who are paid less than once a week, or in a one off sum, or answered none of these (PyPeriod <= 52)

How much is your usual take home pay per (period at PyPeriod) after all deductions? (Please do not include any Working Families' Tax Credit / Disabled Person's Tax Credit payment that you received)

0.00..99999.97   →   Q29

28. TakHmEst   Ask if paid less than once a week, or in a one off sum, or in none of these ways, or did not know how much money they usually took home (PyPeriod = 90, 95 or 97 or TakeHome = DK)

SHOW CARD O

Please look at this card and estimate your usual take home pay per (period at PyPeriod) after all deductions? (Please do not include any Working Families Tax Credit / Disabled Person's Tax Credit payment that you received)

0..30   →See Q29

29. GrossAm   Ask if an employee (Stat = 1)

How much are your usual gross earnings per (period at PyPeriod) before any deductions?

0.01..99999.97   →See Q30

30. GrossEst   Ask if respondent does not know how much their usual gross earnings are (GrossAm = DK)

SHOW CARD O

Please look at this card and estimate your usual gross earnings per (period at PyPeriod) before any deductions?

0..30   →   Q31

31. PaySlip   Ask if an employee (Stat = 1)

INTERVIEWER - CODE WHETHER PAYSLIP WAS CONSULTED

Pay slip consulted by respondent,
but not by interviewer............................. 1 ⎤
Pay slip consulted by interviewer .... 2 ⎬ → Q32
Pay slip not consulted .... 3 ⎦

32. PayBonus   Ask if answered PyPeriod

In your present job, have you ever received an occasional addition to pay in the last 12 months (that is since DATE 1 YEAR AGO) such as a Christmas bonus or a quarterly bonus?

EXCLUDE SHARES AND VOUCHERS

Yes .... 1 → Q33
No .... 2 →See Q40

33. HowBonus   Ask if respondent received a pay bonus (PayBonus = 1)

Was the bonus or commission paid.....

after tax was deducted (net) .... 1 →   Q
or before tax was deducted (gross) .... 2 →   Q
or some before and some after?............ 3 →   Q

34. NetBonus   If some or all tax was deducted, or they did not know if tax was deducted from pay bonus (HowBonus = 1 or 3 or DK)

What was the total amount you received in the last 1 months (that is since DATE 1 YEAR AGO) AFTER ta was deducted (ie net)?

0.01..99999.97   →See Q

35. GrsBonus   Ask if some or all tax was deducted from the pay bonus (HowBonus = 2 or 3)

What was the total amount you received in the last 1 months (that is since DATE A YEAR AGO) before ta was deducted (ie gross)?

0.01..99999.97   →See Q

36. GrsPrLTY   If self-employed less than 12 months, others see Q38

How much did you earn before tax but after deductions of any expenses and wages since becoming self-employed?

IF NOTHING OR MADE A LOSS, ENTER ZERO

0.00..999999.97   →   Q

37. PrLTYEst   SHOW CARD O

Please look at this card and estimate the amount tha you expect to earn before tax but after deductions of any expenses and wages in the first full 12 months that you will have been self-employed, that is up to the end of (month) next?

0..30   →See Q

38. GrsPrft   If self-employed more than 12 months, others Q4

How much did you earn in the last tax year, before ta but after deduction of any expenses and wages.

IF NOTHING OR MADE A LOSS, ENTER ZERO

0.00..999999.97   →See Q

39. PrftEst   If respondent does not know how much they earned last year (GrsPrft = DK)

SHOW CARD O

Please look at this card and estimate the amount tha you earned in the last tax year before tax but after th deduction of any expenses or wages?

0..30   →See Q4

. SecJob2   **Ask all (see criteria for this section above Q26)**

(Apart from your main job) do you earn any money from other jobs, from odd jobs or from work that you do from time to time?

PROMPT AS NECESSARY & INCLUDE BABYSITTING, MAIL ORDER AGENT, POOLS AGENT ETC

Yes .... 1 → Q41
No .... 2 → Q45

. SjEmplee   **Ask if respondent has other jobs (SecJob2 = 1)**

In that (those) job(s) do you work as an employee or are you self-employed?

employee .... 1 → Q42
self-employed... 2 → Q44

. SjNetAm   **Ask if doing other jobs as employee (SjEmplee = 1)**

In the last month, how much did you earn from your other/occasional job(s) after deductions for tax and National Insurance (ie net)?

0.01..99999.97   →   Q43

. SjGrsAm   In the last month, how much did you earn from your other/occasional job(s) before deductions for tax and National Insurance (ie gross)?

0.01..99999.97   →   Q45

. SjPrfGrs   **Ask if doing other jobs as self-employed (SjEmplee = 2)**

In the last 12 months (that is since DATE 1 YEAR AGO) how much have you earned from this work, before deducting income tax, and National Insurance contributions, and money drawn for your own use, but after deducting all business expenses?

IF MADE NO PROFIT ENTER 0

0.00..99999.97   →   Q45

. OthRgPay   **Ask all (except proxy respondents) NOTE: End of the section of questions which are only asked of those in paid work**

And finally, apart from anything you have already mentioned, have you received any regular payment from any of the following sources in the last 12 months (that is since DATE 1 YEAR AGO)?

ENTER AT MOST 3 CODES

EXCLUDE BENEFITS NO LONGER RECEIVED

Interest from savings, Bank or
Building Society accounts 1
Income from shares, bonds, unit trusts   → See Qs
or gilt-edged stock................................. 2   46-48
Other.. 3
None of these .... 4 → Sport & Leisure

46. Investpy   **Ask if respondent is receiving interest from savings (OthRgPay = 1)**

(Apart from interest and income from shares) how much have you received in total from interest on savings, Bank or Building Society accounts in the last 12 months?

0.01..99999.97   → See Qs
                                      47 &48

47. Sharepy   **Ask if respondent is receiving income from shares, bonds, unit trusts or gilt-edged stock (OthRgPay = 2)**

(Apart from interest and income from shares) how much have you received in total from shares, bonds, unit trusts or gilt-edged stock in the last 12 months?

0.01..99999.97   →See Q48

48. OthRgPAm   **Ask if respondent is receiving income from another source (OthRgPay = 3)**

How much have you received from other sources in the last 12 months?

0.01..99999.97   → Sport & Leisure

49. NtIncEst   **If proxy respondent or refused whole income section (proxy respondent or Ben1YN = 7)**

SHOW CARD O

I would now like to ask you about the income of (NAME).Please could you look at this card and estimate the total net income, that is after deduction of tax, National Insurance and any expenses (NAME) brings into the household in a year from all sources (benefits, employment, investments etc)?

0....30   → Sport & Leisure

SPORT AND LEISURE

Introsp   THE NEXT SECTION IS ABOUT SPORT AND LEISURE

1. Sprtyr1   **Ask all (except proxy respondents)**

On this card is a list of sports and physical activities. Please tell me if you took part in any of them in the last twelve months. Do not count any teaching, coaching or refereeing you may have done.

CODE 23 IF THE RESPONDENT HAS NOT PARTICIPATED IN ANY OF THESE SPORTS

Walking (recreational) or hiking
2 miles or more 1
Swimming or diving indoors 2
Swimming or diving outdoors 3
Cycling .... 4 → Q2
Indoor bowls... 5
Outdoor (lawn) bowls 6
Tenpin bowling 7

Keepfit, aerobics, yoga, dance exercise. 8
Martial Arts (INCLUDE
SELF DEFENCE..................................... 9
Weight training (INCLUDE BODY
BUILDING)..10
Weight lifting..11
Gymnastics 12
Snooker..13
Darts..14 → Q2
Rugby union or league.........................15
American football 16
Football indoors 17
Football outdoors 18
Gaelic sports..19
Cricket..20
Hockey 21
Netball..22
None of these 23

2. Sprtyr2   And have you taken part in any of these sports and
physical activities in the last twelve months? Again,
do not count any teaching, coaching or refereeing you
may have done.

CODE 23 IF THE RESPONDENT HAS NOT
PARTICIPATED IN ANY OF THESE SPORTS

Tennis.................................... 1
Badminton.............................. 2
Squash................................... 3
Basketball 4
Table Tennis 5
Track and field athletics......... 6
Jogging, cross country, road running..... 7
Angling, fishing 8
Yachting or dinghy sailing...... 9
Canoeing 10
Windsurfing/board sailing......11
Ice Skating ....12 → See Q3
Curling 13
Golf, pitch and putt, putting ...14
Skiing 15
Horse riding 16
Climbing, mountaineering 17
Motor sports.........................18
Shooting...............................19
Volleyball..............................20
Other....................................21
None of these 22
None at all............................23

3. MoWalks   **If *Sprtyr1 = 1***

Now thinking of the four weeks ending yesterday did
you go for a walk of 2 miles or more during these four
weeks?

Yes...................................... 1
No.. 2 } → See Q4

4. Othsprt   **If *Sprtyr2 = 21***

Please tell me what the other sport(s) or physical
activity(s) is/are

ENTER TEXT OF AT MOST 100 CHARACTERS → See Q5

5. Tuition   **Ask for each sport mentioned at SprtYr1 and
SprtYr2 (except 'walking', 'darts' and 'other')**

Over the **past twelve months** have you received
tuition from an instructor or coach to improve your
performance in (activity)?

Yes.....................   1
No.. 2 } → C
Not sure... 3

6. Comp   Thinking about (activity) have you taken part in any
organised competition in (activity) in the **last twelve
months**?

Yes... 1
No... 2 } → C

7. Sport4   Did you take part in (activity) in the past four weeks?
Again, please do not count any teaching coaching or
refereeing you may have done.

Yes... 1 } → C
No... 2 } → Q1

8. Sprtime   **If *Sport4 = 1***

On how many days in the last four weeks have you
played/gone(to) (activity).

1..28   → C

9. Spwhere   At which of these places on this card have you done,
played (activity) in the last four weeks ?

INDIVIDUAL PROMPT
CODE ALL THAT APPLY

Indoors at a facility which is **mainly
used for sport** (e.g sports centre or
gymnasium or indoor swimming pool
or commercial leisure facility................. 1
Indoors at some other location **not
mainly used for sport** (such as a   → Q1
community centre,village hall or
scout hut 2
Indoors or outdoors at home or
someone else's home .... 3 → See Q1
Outdoors on a court, course, pitch or
playing field (or outdoor swimming
pool).. 4 → Q1
Outdoors in a natural setting (such as
the countryside, rivers, lakes or
seaside) 5
Other - including roads and pathways   → See Q1
in towns and cities.................................. 6

10. SchlCol   **If *Spwhere = 1, 2 or 4***

Do any of these facilities belong to a school, college
or university?

Yes.. 1 } → See Q1
No... 2

1. SpClub   **Ask for each activity taken part in during the last 4 weeks (except 'walking', 'darts' and 'other')**

Over the **past four weeks** have you been a member of a club, particularly so that you can play/participate in (activity)?

Yes.. 1 → Q12
No.. 2 → Q13

2. ClubTyp   **If *SpClub* = 1**

What type of club was this?

ENTER AT MOST 4 CODES

Health/fitness 1
Social club (e.g employee clubs,
youth clubs... 2 → Q13
Sports club... 3
Other 4

3. Volunt   **Ask all**

Looking back over the last four weeks, have you spent any time helping to organise sport on a voluntary basis (that is, without pay except for expenses)? Please include any teaching, coaching or refereeing you may have done as a volunteer.

Yes .... 1 → Q14
No .... 2 → Q15

4. Voltime   **If *Volunt* = 1**

During the last four weeks, how many hours in total have you spent on voluntary sports work?

Less than 1 hour 1
1 hour less than 2 hours .... 2 → Q15
2 hours less than 5 hours 3
5 hours or more 4

5. Sprtnot   **Ask all**

Is there any sport or recreational activity that you do not do at the moment but would like to do?

Yes.. 1 → Q16
No... 2 → Q17

6. Sprtwch   **If *Sprtnot* = 1**

Which one activity would you like to do?

ENTER TEXT OF AT MOST 30 CHARACTERS → Q17

7. Entern   **All adults (except proxy respondents)**

Now thinking about **the four weeks ending yesterday,** could you tell me whether you have done any of these things in your leisure time or for entertainment?

8. TV   Watched TV?

Yes... 1 → Q19
No.. 2

19. Radio   Listened to the radio?

Yes.. 1 → Q20
No.. 2

20. Records   Listened to records or tapes?

Yes.. 1 → Q21
No.. 2

21. Books   Read books?

Yes.. 1 → Q22
No.. 2

22. Music   Sung, or played a musical instrument to an audience, or rehearsed for an event, or played a musical instrument for your own pleasure?

Yes.. 1 → Q23
No.. 2

23. Acting   Performed in a play or drama, or rehearsed for a performance?

Yes.. 1 → Q24
No.. 2

24. Paint   Done any painting, drawing, printmaking or sculpture?

Yes.. 1 → Q25
No.. 2

25. Dance   Done any dance (excluding fitness classes and aerobics)?

Yes.. 1 → Q26
No.. 2

26. Writing   Written any stories, plays or poetry?

Yes.. 1 → Q27
No.. 2

27. Volcult   Looking back over the last four weeks, have you spent any time helping with the running of an arts/cultural event or arts organisation on a voluntary basis (that is without pay except for expenses)?

Yes.. 1 → Q28
No.. 2 → End of Interview

28. Voltime   **If *Volcult* = 1**

During the last four weeks, how many hours have you spent on voluntary arts / cultural work in total?

Less than 1 hour........... 1
1 hour less than 2 hours........................ 2
2 hours less than 5 hours...................... 3
5 hours or more...................................... 4

END OF INTERVIEW

Appendix F Summary of main topics included in the GHS questionnaires 1971 to 2002

ACTIVITIES ON SCHOOL PREMISES   **1984**

Whether attended any event/activity
   on school premises in last 12 months
Whether activities organised by school or
   parent teacher's association
Type of activity attended (if not organised
   by school/parent teacher's association),
   number of times attended and whether
   attended a day or evening class

BURGLARIES AND THEFTS FROM PRIVATE HOUSEHOLDS

Incidence of burglaries in the 12 months before interview	**1972-73,**
Value of stolen goods and whether insured	**1979-80,** **1985-86,**
Whether incident was reported to the police	**1991, 1993, 1996**
Reasons for not reporting to the police	**1972-73,** **1979-80, 1985-86**
Incidence of attempted burglary in the 12 months before interview	**1985-86**

BUS TRAVEL   **1982**

Frequency of use of buses in the six months
   before interview
Physical and other difficulties using buses
Reasons for not using buses

CAREER OPPORTUNITIES   **1972**

Attitudes towards careers in the Armed Forces
   and the Police Force
Whether ever been in one of the Armed Forces

CAR OWNERSHIP

Number of cars or vans, if any, available to the household for private use	**1971-96, 1998, 2000-02**
Type of vehicle and whether privately/ company owned	**1998, 2000-02**
In whose name (person or firm) each car/van was registered	**1980, 1992-93**

Driving licences and private motoring   **1980**

Whether held current licence for driving a car
   or van, and for how long full licence held
Whether non-licence holders (aged 17-70) intended
   to apply for a licence (again), and reasons for
   not having done so or for not intending to do so

Frequency of use, for private motoring, of
   car/van available to the household
If household car/van not available, or not used for
   private motoring in the year before interview:
     - whether used any car/van for private motoring in
       that year
     - whether drove a car, van, lorry, or bus in the
       course of work in that year

COLOUR AND COUNTRY OF BIRTH

Colour, assessment of persons seen*	**1971-92**
Country of birth	
of adults and their parents	**1971-96, 1998, 2000-02**
of children	**1979-96, 1998, 2000-02**
Year of entry to UK	
adults	**1971-96, 1998, 2000-02**
children	**1979-96, 1998, 2000-02**
Ethnic origin	**1983-96, 1998, 2000-02**
National identity	**2001-02**

DRINKING

Rating of drinking behaviour according to quantity - frequency (QF) index based on reported alcohol consumption in the 12 months before interview	**1978, 1980,** **1982, 1984**
Rating of drinking behaviour according to average weekly alcohol consumption (AC) rating	**1986-1998** **(alternate years),** **2000-02**
Alcohol consumption on the heaviest drinking day in the 7 days before interview	**1998, 2000-02**
Personal rating of own drinking behaviour	**1978-1998** **(alternate years), 2000-02**
Whether think drinking/smoking can damage health	**1978, 1980, 1982,** **1984, 1986, 1988, 1990**

*   *Including children*

Whether non-drinkers have always been ⎤
  non-drinkers or used to drink but   **1992-1998**
  stopped, and reasons   **(alternate years),**
Whether drink more or less than the   **2000-02**
  recommended sensible amount ⎦

EDUCATION
Current education

Current education status   **1971-96, 1998, 2000-02**
Type of educational establishment currently attended
- by adults aged under 50   **1971-81, 1984-90**
- by adults aged under 70   **1991-96, 1998, 2000-02**
- by children aged 5-15   **1971-77**
Qualification/examination aimed at   **1971, 1974-76**
Expected date of completion of full-time ⎤
  education
Whether intend to do any paid work while **1971-76**
  still in full-time education, and if so when ⎦
Whether currently attending any leisure or
  recreation classes   **1973-78, 1981, 1983, 1993-96, 2000-02**

Past education

Age on leaving school   **1972-96, 1998**
Age on leaving last place of   **1971-96, 1998,**
  full-time education   **2000-02**
Type of educational establishment last attended
  full time   **1971-96, 1998**
Qualifications obtained   **1971-96, 1998, 2000-02**

Pre-school children (aged under 5)

Whether currently attending nursery/primary school,
  day nursery, playgroup, creche etc   **1971-79, 1986**
Frequency of attendance   **1979, 1986**
Whether received regular day care from person ⎤
  other than parents, and for how many hours
  per week
Whether working mothers would have to stop **1979**
  work if existing arrangements for the care of
  their children were no longer available, or
  whether they could make other care arrangements ⎦

Child care (for children aged 0-11)   **1991**

Whether uses any child care arrangements
Frequency of use and cost
Whether employer contributes towards cost,
  and if so, the amount

Child care (for children under 14)   **1998**

Whether uses any child care arrangements,
  and if so, type and frequency of use

Job training

Whether currently doing a trade apprenticeship **1971-84**

Identification of persons seriously thinking of
  taking a course of training or education for a
  particular type of job, with some details of
  the course and the source of any financial support **1973-74**

Students in institutional accommodation   **1981-87**
Estimate of numbers of full-time students at
  university or college living away from home in
  institutional accommodation, and therefore
  excluded from the GHS sample

EMPLOYMENT
Those currently working

Main job   - occupation and industry   ⎤
           - employee/self-employed   ⎦ **1971-96, 1998, 2000-02**
Subsidiary - occupation and industry   ⎤ **1971-78, 1980-84**
job   - employee/self-employed   ⎦ **1987-91**
Last job   - occupation and industry   ⎤ **1986**
           - employee/self-employed   ⎦
Whether has a second job   **1992-96, 1998**
Whether present job was obtained through
  a government scheme   **1989-92**

Youth Opportunities Programme Schemes   **1982-84**
- identification of young persons aged 16-18
  receiving training or work experience through
  the Youth Opportunities Programme or Youth
  Training Scheme

Youth Training Scheme   **1985-95**
- identification of young persons aged 16-19
  who were on the YTS and whether they were
  working with an employer or at college or
  training school

Journey time to work   **1971-76, 1978**

Usual number of hours worked per week
  (excluding overtime)   **1971-96, 1998**
Hours of paid/unpaid overtime usually
  worked per week   **1973-83, 1998**
Usual number of hours worked per week
  (including paid/unpaid overtime)   **2000-02**
Usual number of days worked per week   **1973, 1979-84**
Number of days worked in reference week   **1977-78**

Length of time with present employer/present
  spell of self-employment   **1971-96, 1998, 2000-02**

Whether self-employed during the previous
  12 months   **1986-91**

Number of changes of employer in
  12 months before interview   **1971-76, 1979-91**

Number of new employee jobs started
in 12 months before interview   **1977-78, 1983-91**

Source of hearing about present job
started in 12 months before interview   **1971-77, 1980-84**
Source of hearing about all jobs started
in 12 months before interview   **1974-77, 1980-84**

Whether paid by employer when sick   **1971-76, 1979-81**

Whether employer is in the public/private
sector   **1983, 1985, 1987**

Trade Union and Staff Association membership   **1983**

Whether people work all or part of the time at home,
reasons for doing so, whether employer makes any
financial contribution to expenses of working at
home, equipment provided by employer   **1993**

Whether does any unpaid work for members of the
family and if so, for whom, number of hours a
week, type of work and where   **1993-95**

Whether has ever been a company director   **1987**

Type of National Insurance contribution paid by:
- married and widowed women
    aged 16 or over   **1972-79**
    aged 16-59   **1980**
- married, widowed, and separated women
    aged 16-59   **1981-82**
    aged 20-59   **1983**

Level of satisfaction with present job as a whole   **1971-83**
Level of satisfaction with specific aspects of
present job   **1974-83**
Whether thinking of leaving present employer, and
if so why   **1971-76**

Whether signed on at an Unemployment
Benefit Office in the reference week, either
to claim benefit or to receive National
Insurance credits   **1984-90, 1994-96**

Absence from work in the reference week
- reasons for absence   **1971-72, 1974-84**
- length of period of absence   **1971-72, 1974-80, 1984**
- number of working days off last week   **1981-84**
- whether absent because of illness or
  accident, and length of absence   **1973**
- whether in receipt of National Insurance
  sickness benefit (and supplementary
  allowance) for the absence   **1971-76**

Sickness absence in the four weeks before
interview   **1981-84**
Sickness absence in the 3 months before interview   **1992**

Whether registered as unemployed in the reference
week (if had worked less than full week)   **1977-82**
Unemployment experience in 12 months
before interview   **1975-77, 1983-84**

Economic activity status 12 months before
interview and, if economically inactive then,
reasons for (re-)entering the labour force   **1979-81**

Economic activity status 12 months before
interview, including whether a full-time student
and working   **1982-91**

Whether in employment prior to present job,   **1986**
and if so, whether that job was full/part time
and reasons for leaving

Whether on any government schemes   **1985-96**

Usual job of father
- of all persons aged 16 or over   **1971-76**
- of persons aged 16-49 in full-time or part-
  time education   **1977-78**
- of all persons aged 16-49   **1979-89**
- of all persons aged 16-59   **1989-91**

Those currently unemployed
Most recent job - occupation and industry   **1971-96,**
                - employee/self-employed   **1998, 2000-02**

Whether most recent job was obtained through
a government scheme   **1989-92**
Whether has ever had a paid job   **1986-96, 1998, 2000-02**
Whether has ever worked for an employer
as part of a government scheme   **1989-91**
Whether registered as unemployed in the
reference week   **1971-83**
Methods of seeking work in the reference week

Whether signed on at an Unemployment
Benefit Office in the reference week, either
to claim benefit or to receive National
Insurance credits   **1984-90, 1994-96**

Whether looking for full or part-time work   **1983**

Whether taking part in either the Youth Training
Scheme or the Youth Opportunities Programme
last week   **1984**

Whether last job was organised through the Youth
Opportunities Programme (persons aged 16-19)   **1982**

For those who in the reference week were looking
for work
- would they have been able to start within
2 weeks if a job had been available   **1991-96, 1998, 2000-02**

For those who in the reference week were waiting
to take up a new job already obtained:
- would they have started that job in the
reference week if it had been available then,
or would they have chosen to wait   **1977-82**
- when was the new job obtained and when did
they expect to start it   **1979**

Whether paid unemployment benefit (and
supplementary allowance) for reference week   **1971-74**

When last worked and reasons for
stopping work   **1971-73, 1974-79, 1986**
Reasons for leaving last job   **1981-82, 1986**
Whether last job was full/part time   **1986**
Length of current spell of unemployment   **1974-96, 1998**
Unemployment experience in 12 months
before interview   **1975-77, 1983-84**

Economic activity status 12 months before
interview and, if economically inactive then,
reasons for (re-)entering the labour force   **1979-81**

Economic activity status 12 months before
interview, including whether a full-time
student and working   **1982-91**

Number of new employee jobs started in
12 months before interview   **1977, 1982-91**
Source of hearing about all jobs started in
12 months before interview   **1982-84**

Whether on any government schemes   **1985-96, 1998, 2000-02**

Whether does any unpaid work for members of the
family and if so: **1993-96,**
number of hours a week and where   **1998, 2000-02**
for whom and type of work   **1993-96**

Whether has ever been a company director   **1987**

Type of National Insurance contribution paid in the
preceding two completed tax years by:
- married, widowed, and separated women
aged 20-59, who were not working
in the week before interview   **1982-83**

Usual job of father
- of all persons aged 16 or over   **1971-76**
- of persons aged 16-49 in full-time or part-
time education   **1977-78**
- of all persons aged 16-49   **1979-88**
- of all persons aged 16-59   **1989-92**

The economically inactive
Major activity in the reference week
Last job - occupation and industry   **1971-96, 1998, 2000-02**
     - employee/self-employed

Usual job (of retired persons)
- occupation and industry   **1973-76, 1979-88**
- employee/self-employed

When finished last job   **1971-73, 1977-78, 1986**
Reasons for stopping work   **1971-73, 1978-82, 1986**

Whether registered as unemployed in the
reference week   **1972-83**
Whether signed on at an Unemployment
Benefit Office in the reference week, either
to claim benefit or to receive National
Insurance credits   **1984-90, 1994-96**
Whether paid unemployment benefit (and
supplementary allowance) for reference week   **1972-74**

Whether would like a regular paid job, whether
looking for work, and if a job had been
available would they have been able to start
within 2 weeks   **1991-96, 1998, 2000-02**
Length of time currently out of employment   **1993-96, 1998, 2000-02**

Main reason for not looking for work   **1986-87**
Whether would like regular paid job   **1986-87**
Whether has ever had a paid job   **1986-96, 1998, 2000-02**
Whether has had a paid job in last 12 months   **1987-91**
Whether has ever worked for an employer as
part of a government scheme   **1989-91**
Whether has had a paid job in previous 3 years   **1986**
Whether last job was full/part time   **1986**

Unemployment experience in 12 months   **1975-77,**
before interview   **1983-84**

Economic activity status 12 months before
interview (persons aged 16-69)   **1980-81**

Economic activity status 12 months before
interview including whether a full-time
student and working   **1982-91**

Number of new employee jobs started
in 12 months before interview　　　　　**1977, 1984-91**
Source of hearing about all jobs started
in 12 months before interview　　　　　**1977**

Whether on any government schemes　**1985-96, 1998, 2000-02**

Whether does any unpaid work for members of
the family and if so:
number of hours a week and where　**1993-96, 1998, 2000-02**
for whom and type of work　　　　　**1993-96**

Whether has ever been a company director　**1987**

Type of National Insurance contribution paid in
the preceding two completed tax years by:
- married, widowed, and separated women
aged 20-59, who were not working in the
week before interview　　　　　　　**1982**

Future work intentions, including whether would
seek work earlier if satisfactory arrangements
could be made for looking after children　**1971-76**

Usual job of father
- of all persons aged 16 or over　　　**1971-76**
- of persons aged 16-49 in full-time or part-
time education　　　　　　　　　　**1977-78**
- of all persons aged 16-49　　　　　**1979-88**
- of all persons aged 16-59　　　　　**1989-92**

FAMILY INFORMATION/FERTILITY
Marriage, cohabitation and childbirth
Marital history　　　**1979-96, 1998, 2000-02**
Date of present marriage　　　　　**1971-78**
Whether first marriage　　　　　　**1974-78**
Expected family size:
at time of present marriage
at time of interview
Whether woman thinks she has completed
her family　　　　　　　　　　　**1971-78**
Age when most recent baby was born
Age when expects to have last baby
Date of birth and sex of each child born in
present marriage
Date of birth and sex of all liveborn children
and whether they live with mother　**1979-96, 1998, 2000-02**
Where children under 16, not living with
mother, are currently living　　　　**1979**
Where children under 19, not living with
mother, are currently living　　　　**1982**

Date of birth of step, foster, and adopted children
living in the household, and how long
they have lived there　**1979-87, 1989-96, 1998, 2000-02**
Whether women think they will have any (more)
children, how many in all, and age at which
they think will have their first/next baby **1979-96, 1998, 2000-02**
Current cohabitation　　　**1979-96, 1998, 2000-02**
Cohabitation before current or most recent
marriage　　　　　　　　　**1979, 1981-88**
Cohabitation before all marriages　**1989-96, 1998, 2000-02**
Number of cohabiting relationships that
did not lead to marriage　　　　**1998, 2000-02**

Contraception and sterilisation
Whether woman/partner has been　　　**1983-84,**
sterilised for contraceptive reason　**1986-87,**
Whether woman/partner has had　**1989,1991,1993,**
other sterilising operation　　　**1995, 1998, 2002**

Details of any reversal of sterilisation　**1983-84,**
operations　　　　　　　　　　**1986-87**
Current use of contraception/reason for not
using contraception　**1983, 1986, 1989, 1991, 1993,**
1995, 1998, 2002
Previous usual method of
contraception　**1989, 1991, 1993, 1995, 1998, 2002**
Use of contraception in the previous 12 months　**1989**
Use of contraception in previous 2 years　**1991, 1993,**
1995, 1998, 2002
Use of emergency contraception in previous
2 years　　　　**1993, 1995, 1998, 2002**
Whether woman/partner would have
difficulties in having (more) children　**1983-84**
Reasons for difficulties and whether　**1986-87, 1989**
consulted a doctor about difficulties　**1991,1993,**
in getting pregnant　　　　**1995, 1998, 2002**

FORESTS
Whether ever visits forests or woodland areas,
facilities visitors would like to see there　**1987**

HEALTH
Chronic sickness (longstanding illness or disability)
Prevalence of longstanding illness or
disability*　　**1971-76, 1979-96, 1998, 2000-02**

Causes of the illness or disability*　　**1971-75**
When the illness or disability started*　　**1971**

Type of illness or disability　**1988-89, 1994-96, 1998, 2000-02**

* *Including children*

Prevalence of limiting longstanding
    illness or disability*   **1972-76, 1979-96, 1998, 2000-02**

When it started to limit activities and whether
    housebound or bedfast because of it*   **1972-76**

**Acute sickness (restricted activity in a two-
week reference period)**
Prevalence and duration of restricted
    activity*   **1971-76, 1979-96, 1998, 2000-02**

Causes of restricted activity*   **1971-75**
Number of days in bed and number of days of
    (certificated) absence from work/school*   **1971-76**
Help from people outside household with
    housework or shopping   **1971-74**

**Health in general in the 12 months before
interview**   **1977-96, 1998, 2000-02**

Chronic health problems   **1977-78**
Prevalence of chronic health problems
Constant effects of chronic health problems
    (eg taking things easy, using prescribed/non-
    prescribed medication, watching diet, taking
    account of weather)

Contact with health services in 12 months before
    interview because of chronic health problems

Effect of chronic health problems in the 14 days
    before interview (eg resting more than usual,
    using prescribed/non-prescribed medication,
    changing eating or drinking habits, cutting
    down on activities, consulting GP, seeking
    advice from other persons)

**Short-term health problems (in the 14 days
before interview)**   **1977-78**
Prevalence of short-term health problems
Effects of short-term health problems in the 14
    days before interview

GP consultations
Consultations in the two weeks before interview:
    number of consultations*
    NHS or private*
    type of doctor*   **1971-96, 1998, 2000-02**
    site of consultation*

    cause of consultation*   **1971-75**

whether consulted because something was the
    matter, or for some other reason*   **1981**

whether consultation about reported long-   **1983-84,**
    standing illness or restricted activity*   **1986-87**

whether was given a prescription*   **1981-96, 1998, 2000-02**
whether was referred to hospital*   **1981-85, 1988-90**
whether was given National Insurance   **1981-85**
    medical certificate
whether saw a practice nurse and, if so, the   **2000-02**
    number of times*

Access to GPs: **1977**
    whether own doctor worked alone or with other
    doctors
    whether could usually see doctor of own
    choice at surgery
    most recent consultation at surgery:
    - when it took place
    - NHS or private
    - by appointment or not
    - how far ahead appointment made
    - time spent waiting at surgery
    - attitudes towards waiting time for
        appointment, waiting time at surgery, and
        length of consultation

Outpatient (OP) attendances
Attendances at hospital OP departments in a three-
    month reference period:
    number of attendances*   **1971-96, 1998, 2000-02**
    NHS or private   **1973-76, 1982-83, 1985-87, 1995-96,
                                   1998, 2000-02**
    nature of complaint causing attendance*   **1974-76**
    whether claimed for under private
    medical insurance   **1982-83, 1987, 1995**
    number of casualty visits*   **1995-96, 1998, 2000-02**

Appointments with OP departments: **1973-76**
    whether had (or was waiting for) an
    appointment* how long ago since told
    appointment would be made*

Day patient visits
Number of separate days in hospital as a
    day patient in the last year*   **1992-96, 1998, 2000-02**
    whether NHS or private   **1995-96, 1998, 2000-02**

Inpatient spells
Spells in hospital as an inpatient in a three-
    month reference period:
    number and length of spells*   **1971-76**
    NHS or private patient*   **1973-75**

Including children

Stays in hospital as an inpatient in a 12-month
　　reference period:
　　　　number of stays*　　　　　　　　　**1982-96, 1998, 2000-02**
　　　　number of nights on each stay*　　**1992-96, 1998, 2000-02**
　　　　NHS or private patient　**1982-83, 1985-87, 1995-96, 1998,**
　　　　　　　　　　　　　　　　　　　　　　　　　　　　　2000-02

　　　　whether private patients were treated
　　　　in an NHS/private hospital　　　　　**1998, 2000-02**
　　　　whether claimed for under private
　　　　medical insurance　　　　　　　　　　**1982-83, 1987**

Whether on waiting list for admission to
　　hospital and length of time on list*　　　**1973-76**

Mobility aids　　　　　　　　　　**1993, 1996, 2001**
Whether has any difficulty getting about without
　　assistance, and if so, what help is needed,
　　whether the problem is temporary or
　　permanent, the number and types of walking
　　aids, and who supplied them

Accidents　　　　　　　　　　　　　　　　**1987-89**
Accidents in the three-month reference period
　　that resulted in seeing a GP or going to a hospital:
　　　　whether saw GP or went to hospital or did both
　　　　　　and in the last case, which first*
　　　　type of accident and where occurred*
　　　　whether occurred during sport*
　　　　whether occurred during working hours*
　　　　time off work as a result of accident
　　　　whether went to hospital A & E Department
　　　　　　(Casualty) or other part of hospital*
　　　　whether stayed in hospital overnight as a
　　　　　　result of accident, and if so how many nights*

Accidents at home　　　　　　　　　　**1981, 1984**
Accidents at home, in a three-month reference period,
　　that resulted in seeing a GP or going to hospital:
　　　　whether saw GP or went to hospital or did both
　　　　　　and, in the last case, which first*
　　　　whether went to hospital A & E Department
　　　　　　(Casualty) or other part of hospital*

Health and personal social services
Use of various services:
- by adults and children　　　　　　　　**1971-76**
- by persons aged 60 or over　　　　　　　**1979**
- by persons aged 65 or over　**1980-85, 1991, 1994, 1998, 2001**

Elderly persons
Whether any relatives living nearby:
- persons aged 60 or over　　　　　　　**1979-80**
- persons aged 65 or over　　　　　　　　**1994**

Persons aged 65 or over:
- whether need help in getting about　　**1980, 1985,**
　　inside the house and outside, and　　**1991, 1994,**
　　with a range of personal and　　**1996, 1998, 2001**
　　household tasks
- if help is needed, who usually helps ⎤
- frequency of social contacts with　　　　**1980, 1985**
　　relatives and friends　　**1991, 1994, 1998, 2001**
- use of public transport　　　　　　　⎦
- whether needs a regular daily carer　⎤　**1998, 2001**
- whether lives in sheltered accommodation ⎦
- when illness started to limit activities　　**2001**
- health compared to a year ago　　　　　**2001**

Informal carers
Whether looks after a sick, handicapped or　**1985, 1990,**
　　elderly person in same or other household, nature　**1995,2000**
　　of care provided and time spent, whether help
　　receivedfrom other people or statutory services
Reasons for not receiving help from statutory
　　services　　　　　　　　　　　　　　　　　**1995**
Whether dependent receives respite care　　**1995, 2000**
Whether carer's health has been affected　　**2000**

Informal carers aged 8-17　　　　　　**1996**
- whether looks after a sick, handicapped or
　　elderly person in the same household, nature of
　　care provided and time spent, whether help
　　received from other people or statutory services

Sight and hearing
Difficulty with sight and whether wears glasses
　　or contact lenses:
- persons aged 16 or over　　**1977-79, 1981-82, 1987, 1994**
- persons aged 65 or over **1980, 1985, 1987, 1991, 1994, 1998, 2001**

Whether wears glasses or contact lenses*　⎤
Whether obtained new glasses in previous　│　**1987,**
　　12 months and number of pairs*　　⎬　**1990-1994**
Whether had a sight test in previous
　　12 months*　　　　　　　　　　⎦
Whether sight test was NHS or private　　　**1990-94**
Whether sight test was paid for by informant
　　or employer, provided free by optician, or
　　covered by insurance　　　　　　　　　　**1991-94**
Whether obtained any ready made reading
　　glasses in the previous 12 months　　　　**1992-94**

Types of contact lens worn, and whether　⎤
　　obtained through NHS or privately　　│
Reasons for trying contact lenses　　　⎬　**1982**
Reasons stopped wearing contact lenses　│
Care of contact lenses　　　　　　　　⎦

Including children

275

Difficulty with hearing and whether wears an aid:
- persons aged 16 or over **1977-79, 1981, 1992, 1995, 1998, 2002**
- persons aged 65 or over   **1980, 1985, 1991, 1994, 1998, 2001**

Types of hearing aid worn, and whether
obtained through NHS or privately   **1979**

Reasons for not wearing an aid   **1979, 1992, 1995, 1998, 2002**

Whether hearing aid was obtained through NHS
or bought privately, and if bought privately,
the reason(s)   **1992, 1995, 1998, 2002**

Tinnitus (sensation of noise in the ears or head)

Prevalence of tinnitus, frequency and duration of
symptoms, whether ever consulted a doctor
about it   **1981**

Dental health

Whether has any natural teeth   **1983, 1985, 1987,
1989, 1991, 1993, 1995**

To those aged under 18, how long since
last visit to the dentist, and whether registered
with a dentist*   **1993, 1995**

How long since last visit to the dentist*
Treatment received*   **1983**

Whether goes to the dentist for check-ups,   **1983, 1985,
or only when having trouble with teeth*   1987, 1989,
1991, 1993, 1995**

Medicine-taking   **4th qtr 1972, 1973**

Medicines taken in the seven days before
interview:
- categories of medicine
- patterns of consumption of analgesics

Private medical insurance   **1982-83, 1986-87, 1995**

Whether covered by private medical insurance
and, if so:
- whether policy holder or dependant on
someone else's policy*
- whether subscription paid by employer

Whether covered by private medical
insurance in the last 12 months   **1987**

Whether company director's private medical
insurance subscription is paid for by the
company of which he is a director   **1987, 1995**

HOUSEHOLD COMPOSITION

Age*, sex*, marital status of household
members   **1971-96, 1998, 2000-02**

Relationship to head of household*

Family unit(s)

Housewife   **1971-80**

Including children

HOUSING (see also MIGRATION)

Present accommodation: amenities

Length of residence at present address*
Age of building
Type of accommodation   **1971-96,
Number of rooms and number of bedrooms   1998, 2000-02**
Whether have separate kitchen

Bath/WC: sole use, shared, none   **1971-90**
WC: inside or outside the accommodation

Installation/replacement of bath or WC
Cost of improvements made to the   **1971-76**
accommodation

Floor level of main accommodation   **1973-96,
Whether there is a lift   1998, 2000-02**

Tenure

Whether present home is owned or rented   **1971-96, 1998, 2000-02**

Whether in co-ownership housing association
scheme   **1981-95**

Change of tenure on divorce or remarriage   **1991-93**
Change of tenure on marriage or cohabitation   **1998**

Housing history of local authority tenants and
owner occupiers who had become owners in
the previous five years   **1985-86**

Whether ever rented from a local authority,
and if so, whether bought that accommodation,
source of finance, whether have since moved
and distance moved   **1991-93**

Owner occupiers:
- in whose name the property is owned   **1978-96, 1998, 2000-02**
- whether property is owned outright or being
bought with a mortgage or loan   **1971-96, 1998, 2000-02**
- how outright owners originally acquired
their home   **1978-80, 1982-83, 1985-86**
- source of mortgage or loan   **1978-80, 1982-86, 1992-93**
- whether currently using present home as
security for a (second) mortgage or loan
of any kind, and if so, details   **1980-82, 1992-93**
- whether owner occupiers with a mortgage
have taken out a remortgage on their
present home, and if so, details   **1985-87, 1992-93**
- whether recent owner occupiers had previously
rented this accommodation and, if so,
from whom and for how long   **1981-82, 1985-86**
- whether had rented present accommodation
before deciding to buy   **1992-93**
- whether previous accommodation was owned
and if so, details of the sale   **1992-93**

Renters:
- in whose name the property is rented   **1985-96, 1998, 2000-02**
- from whom the accommodation is rented **1971-96, 1998, 2000-02**
- whether landlord lives in the same
  building   **1971-72, 1975-76, 1979-96, 1998, 2000-02**
- whether have considered buying present
  home and, if not why not   **1980-89**
- tenure preference   **1985-88**
- whether previously owned/buying
  accommodation and reasons for leaving   **1995-96**

Local authority renters:
- whether expect to move soon, and if so
  whether expect to rent or buy
- whether expect to buy present home   **1990-91**
- landlord preference
- awareness of Tenants' Choice Scheme

Housing costs

Gross value		**1971-86**
Net rateable value	Scotland	**1971-86**
Yearly rate poundage	only	**1972-86**

Type of mortgage   **1972-77, 1979, 1981, 1984-86**
Current mortgage payments   **1972-77, 1979, 1981, 1984**
Purchase price of present home, amount of
  mortgage or loan and date mortgage
  started   **1985-86, 1992-93**
Current rent
Amount of any rent rebate/allowance   **1972-77,**
  and/or rate rebate received   **1979, 1981**
Whether in receipt of housing benefit   **1985-95, 1998, 2000-02**
Whether rent paid by DSS or local authority   **1998, 2000-02**

Method of obtaining mortgage tax relief   **1984**
Council Tax band for households containing   **2001**
person(s) aged 65 or more

Central heating and fuel use

Whether have central heating   **1971-96, 1998, 2000-02**
Type of fuel used for central heating   **1978-92**
Type of fuel mainly used for central heating **1993-96, 1998, 2000-02**
Type of fuel mainly used for room heating   **1978-81,**
  in winter   **1983, 1985**

Consumer durables

Possession of various consumer
  durables   **1972-76, 1978-96, 1998, 2000-02**
Possession of a telephone   **1972-76, 1979-96, 1998, 2000-02**

Possession of a mobile telephone: **1992, 2000-02**
- number available for use
- in whose name each is owned or rented   **1992**
- whether fitted in a car or van
Access to the Internet   **2000-02**

Deep frying   **1986**
Whether does any deep frying, frequency and
  methods used

HOUSING SATISFACTION

Overall level of satisfaction with present
  accommodation   **1978, 1988, 1990**
Reasons for dissatisfaction
Satisfaction with specified aspects of   **1978**
  accommodation
Troublesome features
Housing preferences   **1978, 1987, 1988**
Satisfaction with landlord   **1990**

INCOME

Income over 12 months before interview
Gross earnings as employee, from self-
  employment
Income from state benefits, investments,   **1971-78**
  and other sources
Number of weeks for which income
  received from each source

Whether currently receiving income from each
  source   **1974-78**

Current income
Current earnings (gross, take-home, usual) as
  employee, from self-employment, and from
  second or occasional jobs
   1979-96,
Current income from state benefits,   **1998, 2000-02**
  occupational pensions (own or husband's),
  rents, savings and investments, and any
  other regular sources

Current income from maintenance, alimony
  or separation allowance   **1981-96, 1998, 2000-02**

Financial help received from former husband
  towards household bills   **1982-83**

INHERITANCE — **1995**

Number, type, value and dates of
   inheritances received
Details of property inheritance

LEISURE

Holidays away from home in the four	**1973, 1977,**
weeks before interview:	**1980, 1983, 1986**
length of holiday	
countries visited (in UK)	
Leisure activities in the four weeks before	**1973, 1977,**
interview:	**1980, 1983,**
types of activity	**1986**
number of days on which engaged in each activity	
whether activity done while away on holiday	

Sports activities in the four weeks and year	**1987, 1990,**
before interview:	**1993, 1996**
- number of days on which engaged in each sport	**2002**
- where activities took place	**1996, 2002**
- whether member of a sports club	
- whether took part in organised competition	
- whether received tuition	
- whether did any voluntary sports work in the	**2002**
4 weeks before interview, and amount of	
time spent	

Arts and entertainments, museums, galleries,	**1987**
historic buildings:	
- whether visited in the 4 weeks before interview	
- number of days on which visited	

Social activities and hobbies in the	**1973, 1977, 1980,**
four weeks before interview	**1983, 1986, 1987,**
	1990, 1993, 1996, 2002

Whether did any voluntary arts/cultural work	
in the 4 weeks before interview, and	
amount of time spent	**2002**

LIBRARIES — **1987**

Whether visited a public library in the
  4 weeks before interview:
  - number of visits
  - library services used

LONG-DISTANCE TRAVEL — **1971-72**

Number of long-distance journeys made in
  the 14 days before interview

* Including children

278

Starting and finishing points of journeys
Type of transport used for longest part of journeys
Main purpose of journeys
Number of people travelled with

MIGRATION

Past movement

Length of residence at previous address*	**1971-77**
Previous accommodation:	
- tenure	**1971-73, 1978-80**
- household composition	**1971**
- number of rooms	
- bath/WC: sole use, shared, none	
- WC: inside or outside accommodation	
Reasons for moving from previous address	**1971-77**
Number of moves in last five years*	
1971-77, 1979-96, 1998, 2000-02	

Potential movement

Identification of households containing	
persons who are currently thinking	**1971-78,**
of moving*	**1980-81,**
Whether will be moving as whole	**1983**
household or splitting up*	
Reasons for moving	**1971-76,1978, 1980-81**
Proposed future tenure	**1980-81, 1983**
Actions taken to find somewhere to	**1971-76, 1980-81**
live	
Whether had experienced difficulties	
- in finding somewhere else to live	**1980-81**
- in raising a mortgage/loan or in finding	
a deposit	

Frustrated potential movement

Identification of households containing	
persons who, though not currently	
thinking of moving, had seriously	
thought of doing so in the	**1974-76, 1980,**
two years before interview*	**1983**
Whether would have moved as whole	
household or would have split up*	
Proposed tenure	**1974-76, 1980**
Reasons for deciding not to move	**1974-76, 1980, 1983**
Whether decision not to move was	
connected with rise in house prices	**1974-76, 1980**
Whether reasons for thinking about moving	
were work-related	**1983**
Whether had experienced difficulties in raising	
a mortgage/loan or in finding a deposit	**1980**

PENSIONS

Whether covered by employer's pension scheme	**1971-76, 1979, 1982-83, 1985, 1987-96, 1998, 2000-02**
Whether the scheme is contributory, reasons for not belonging to the scheme	**1971-76, 1979, 1982-83, 1985, 1987**
Whether ever belonged to present employer's pension scheme	**1985, 1987**
Length of time in present employer's pension scheme	
Whether transferred any previous pension rights to present employer's pension scheme	**1983,**
	1985,
Whether in receipt of a pension from a previous employer, and if so, at what age they first drew it	**1987**
Whether ever belonged to a previous employer's pension scheme	
Length of time in last employer's pension scheme and in last job	**1985**
Whether retained any pension rights from any previous employer	**1971-76 1979, 1982-83, 1985, 1987**
Whether pays Additional Voluntary Contributions into employer's pension scheme	**1987**
Whether has a stakeholder pension	**2001-02**
Whether currently belongs to a personal pension scheme and whether employer contributes	**1991-96, 1998, 2000-02**
Whether has ever contributed towards a personal pension	**1987-96, 1998, 2000-02**
Date the personal pension was taken out	**1989-90**
Whether belonged to an employer's pension scheme during the 6 months prior to taking out a personal pension	**1989-90**
Whether makes any other income tax deductible pension contributions	**1993-96, 1998, 2000-02**
- whether free standing additional voluntary contributions	**2000-02**
Whether receiving an occupational pension, and if so, how many	
Age first drew occupational pension and whether this was earlier or later than the usual age	**1990**
Reasons for drawing the pension early or late, and whether the amount of pension was affected	

SHARE OWNERSHIP

Whether owns any shares	**1987-88**
Whether shares are owned solely or jointly with spouse	**1987**
Whether shares owned are in employer's company	**1987-88**
Whether has a Personal Equity Plan	**1988, 1992-96, 1998**

SMOKING

Cigarette smoking

Prevalence of cigarette smoking	**1972-76, 1978-1998 (alternate years), 2000-02**
Current cigarette smokers:	
number of cigarettes smoked per day	**1972-76, 1978-1998 (alternate years),**
type of cigarette smoked mainly	**2000-02**
usual brand of cigarette smoked	**1984-1998 (alternate years), 2000-02**
age when started to smoke cigarettes regularly	**1988-1998 (alternate years), 2000-02**
whether would find it difficult to not smoke for a day	
whether would like to give up smoking altogether	**1992-1998 (alternate years), 2000-02**
when is the first cigarette of the day smoked	

Regular cigarette smokers:

- age when started smoking cigarettes regularly	**1972-73**

Occasional cigarette smokers:

- whether ever smoked cigarettes regularly	
- age when started to smoke cigarettes regularly	
- number smoked per day when smoking regularly	**1972-73**
- how long ago stopped smoking cigarettes regularly	

Current non-smokers:

whether ever smoked cigarettes regularly	**1972-76, 1978-1998 (alternate years), 2000-02**
age when started to smoke cigarettes regularly	**1972-73,**
number smoked per day when smoking regularly	**1980-1998 (alternate years),**
how long ago stopped smoking cigarettes regularly	**2000-02**

Cigar smoking

Prevalence of cigar smoking	**1972-76, 1978-1998 (alternate years), 2000-02**
Current cigar smokers:	
number of cigars smoked per week	**1988-1998 (alternate years), 2000-02**
number of cigars smoked per month	**1972-73**
type of cigar smoked	
age when started to smoke cigars regularly	**1972**

279

Current non-smokers:

whether ever smoked cigars regularly   **1972-76, 1978-1998 (alternate years), 2000-02**

age when started to smoke cigars regularly

how long ago stopped smoking cigars regularly   **1972**

Pipe smoking

Prevalence of pipe smoking among males   **1972, 1978, 1986-1998 (alternate years), 2000-02**

Current pipe smokers:

amount of tobacco smoked per week   **1972-75**

age when started to smoke a pipe regularly   **1972**

Current non-smokers:

whether ever smoked a pipe regularly   **1972-76, 1978, 1986-1998 (alternate years), 2000-02**

age when started to smoke a pipe regularly

how long ago stopped smoking a pipe regularly   **1972**

SOCIAL CAPITAL

Opinion of local services, amenities, organisations,

safety in the area, local problems   **2000-02**

TRAINING

Whether received any job training in the previous

4 weeks, and if so:

- the type of training

- hours spent in last 4 weeks   **1987-89**

- whether paid by employer while training

- whether compulsory   **1987**

- reasons for doing training

VOLUNTARY WORK

Whether did any voluntary work in the

12 months before interview and, if so:

- what kind of work, whether also done

in the last 4 weeks, and amount of

time spent   **1981, 1987, 1992**

- whether done regularly or from time to time   **1981**

- on how many days   **1987, 1992**

- number of hours spent   **1992**

- whether any organisation was involved   **1981**

- which organisations were involved   **1987, 1992**

- whether the organisation was a trade union

or political party   **1987**

- who mainly benefited from the work   **1981**

Whether did any voluntary sports/arts/cultural work

in the 4 weeks before interview, and amount of time

spent   **2002**

Appendix G List of tables

Previous volumes in the GHS series

General Household Survey: Introductory report
Origin and development of the survey - Population - Housing - Employment - Education - Health

HMSO 1973

General Household Survey 1972
Population - Household theft - Housing - Employment - Education - Health - Medicine-taking - Smoking - Sampling error

HMSO 1975

General Household Survey 1973
Population - Housing - Employment - Leisure - Education - Health - Medicine-taking - Smoking

HMSO 1976

General Household Survey 1974
Population - Housing and migration - Employment - Education - Health - Smoking

HMSO 1977

General Household Survey 1975
Population - Housing and migration - Employment - Education - Health - Smoking

HMSO 1978

General Household Survey 1976
Trends 1971 to 1976 - Population - Housing and migration - Employment - Education - Health - Smoking - Sampling error

HMSO 1978

General Household Survey 1977
Population - Housing and migration - Employment - Education - Health - Leisure

HMSO 1979

General Household Survey 1978
Population - Housing and migration - Housing satisfaction - Employment - Education - Health - Smoking, drinking, and health

HMSO 1980

General Household Survey 1979
Population - Housing - Burglaries and thefts from private households - Employment - Education - Health - Family information - Income

HMSO 1981

General Household Survey 1980
Population - Housing and household mobility - Burglaries and thefts from private households - Employment - Education - Health - Smoking - Drinking - Elderly people in private households

HMSO 1982

General Household Survey 1981
Population - Housing - Employment - Education - Health - The prevalence of tinnitus - Voluntary work

HMSO 1983

General Household Survey 1982
Population - Marriage and fertility - Housing - Employment - Education - Health - Smoking - Drinking - Cigarette smoking, drinking and health - Bus travel - Non-government users of the GHS

HMSO 1984

General Household Survey 1983
Population - Marriage and fertility - Contraception, sterilisation and infertility - Housing - Employment - Education - Health - Leisure

HMSO 1985

General Household Survey 1984
Population - Marital history, fertility and sterilisation - Housing - Employment - Education - Health - GP consultations in relation to need for health care - Cigarette smoking: 1972 to 1984 - Drinking

HMSO 1986

General Household Survey 1985
Population - Marital status and cohabitation - Housing - Employment - Education - Health

HMSO 1987

General Household Survey 1985　HMSO 1988
Supplement A: Informal carers
by Hazel Green

General Household Survey 1986　HMSO 1989
Population - Marriage and
fertility - Contraception, sterilisation
and infertility - Housing -
Burglary - Employment -
Education - Health - Smoking -
Elderly people in private households 1985 -
Leisure

General Household Survey 1986　HMSO 1989
Supplement A: Drinking
by Hazel Green

General Household Survey　HMSO 1990
Report on sampling error
Based on 1985 and 1986 data
by Elizabeth Breeze

General Household Survey 1987　HMSO 1989
People, households and families -
Housing - Health - Sterilisation and
infertility - Entertainments, libraries,
forests - Occupational pension scheme
coverage - Share ownership - Employment -
Education - Family information and fertility

General Household Survey 1987　HMSO 1990
Supplement A: Voluntary work
by Jil Matheson

General Household Survey 1987　HMSO 1991
Supplement B: Participation in sport
by Jil Matheson

General Household Survey 1988　HMSO 1990
by Kate Foster, Amanda Wilmot and
Joy Dobbs
People, households and families -
Family information and fertility - Health -
Smoking - Drinking - Education -
Share-ownership - Employment -
Occupational and personal pensions -
Housing

General Household Survey 1989　HMSO 1991
by Elizabeth Breeze, Gill Trevor and
Amanda Wilmot
People, households and families -

Employment and pension schemes -
Health - Accidents - Marriages and
cohabitation - Fertility and contraception -
Housing

General Household Survey 1990　HMSO 1992
by Malcolm Smyth and Fiona Browne
People, households and families -
Housing - Occupational pension scheme
coverage and receipt of occupational
pensions - Health - Smoking - Drinking -
Sport, physical activities and entertainment

General Household Survey:　OPCS 1992
Carers in 1990
OPCS Monitor SS 92/2

General Household Survey 1991　HMSO 1993
by Ann Bridgwood and David Savage
People, households and families -
Housing - Burglaries in private households -
Employment - Occupational and personal
pension scheme coverage - Childcare -
Health - Contraception - Education -
Family information

General Household Survey 1991　HMSO 1994
Supplement A: People aged 65 and over
by Eileen Goddard and David Savage

General Household Survey 1992　HMSO 1994
by Margaret Thomas, Eileen Goddard,
Mary Hickman and Paul Hunter
People, families and households -
Health - Smoking - Drinking - Occupational
and personal pension scheme coverage -
Employment - Education - Family
Information - Housing

General Household Survey 1992　HMSO 1994
Supplement A: Voluntary work
by Eileen Goddard

General Household Survey 1993　HMSO 1995
by Kate Foster, Beverley Jackson,
Margaret Thomas, Paul Hunter, Nikki Bennett
People, families and households -
Housing - Burglaries in private households -
Employment - Health - Contraception -
Sport and leisure activities - Family
Information - Education - Pensions

Living in Britain   HMSO 1995
Preliminary results from the 1994
General Household Survey

Living in Britain   HMSO 1996
Results from the 1994
General Household Survey
by Nikki Bennett, Lindsey Jarvis,
Olwen Rowlands, Nicola Singleton,
Lucy Haselden
  Households, families and people- Health -
  Smoking - Drinking - Elderly people in
  private households - Employment -
  Pensions - Family information -
  Education - Housing

1994 General Household Survey: ONS 1998
follow-up survey of the health of
people aged 65 and over
by Eileen Goddard

Living in Britain   The Stationery
Preliminary results from the 1995   Office 1996
General Household Survey

Living in Britain   The Stationery
Results from the 1995   Office 1997
General Household Survey
by Olwen Rowlands, Nicola Singleton,
Joanne Maher, Vanessa Higgins
  Households, families and people - Housing
  and consumer durables - Employment -
  Pensions - Education - Health - Private
  medical insurance - Dental health -
  Hearing - Contraception - Family information

General Household Survey 1995   The Stationery
Supplement A: Informal carers   Office 1998
by Olwen Rowlands

Living in Britain   The Stationery
Preliminary results from the 1996   Office 1997
General Household Survey

Living in Britain   The Stationery
Results from the 1996   Office 1998
General Household Survey
by Margaret Thomas, Alison Walker,
Amanda Wilmot, Nikki Bennett
  Households, families and people - Housing
  and consumer durables - Burglaries in private
  households - Employment - Pensions - Education -
  Health - Mobility and mobility aids - Smoking -
  Drinking - Marriage and cohabitation - Sports
  and leisure activities

First release of results from the   ONS 1999
1998 General Household Survey

Living in Britain   The Stationery
Results from the   Office 2000
1998 General Household Survey
by Ann Bridgwood, Robert Lilly,
Margaret Thomas, Jo Bacon,
Wendy Sykes, Stephen Morris
  Households, families and people - Housing
  and consumer durables - Marriage and
  cohabitation - Pensions - Health - Smoking -
  Drinking - Contraception - Day care - Cross-topic
  analysis

People aged 65 and over - results from the   ONS 2000
1998 General Household Survey
by Ann Bridgwood

Living in Britain   The Stationery
Results from the 2000   Office 2001
General Household Survey
by Alison Walker, Joanne Maher, Melissa Coulthard,
Eileen Goddard, Margaret Thomas
  Changes over time - Households, families
  and people - Housing and consumer durables -
  Marriage and cohabitation - Pensions - Health -
  Smoking - Drinking

People's perceptions of their   The Stationery
neighbourhood and community   Office 2002
involvement - results from the social
capital module of the General Household
Survey 2000
by Melissa Coulthard, Alison Walker, Antony Morgan

Carers 2000   The Stationery
by Joanne Maher, Hazel Green   Office 2002

Disadvantaged households: results   ONS
from the 2000 General Household   2002
Survey: Supplement A
by Wendy Sykes, Alison Walker

Living in Britain   The Stationery
Results from the 2001   Office 2003
General Houshold Survey
by Alison Walker, Maureen O'Brien, Joe Traynor,
Kate Fox, Eileen Goddard, Kate Foster
  Changes over time - Households, families
  and people - Housing and consumer durables -
  Marriage and cohabitation - Pensions - Health -
  Smoking - Drinking - Mobility aids

People aged 65 and over   The Stationery
Results from the 2001 General   Office 2003
Household Survey
by Joe Traynor, Alison Walker